(continued from previous page)

"Your books have saved my GPA, and quite possibly my sanity. My course grade is now an 'A', and I couldn't be happier."

Student, Winchester, IN

"These books are the best review books on the market. They are fantastic!"

Student, New Orleans, LA

"Your book was responsible for my success on the exam. . . I will look for REA the next time I need help."

Student, Chesterfield, MO

"I think it is the greatest study guide I have ever used!"

Student, Anchorage, AK

"I encourage others to buy REA because of their superiority. Please continue to produce the best quality books on the market."

Student, San Jose, CA

"Just a short note to say thanks for the great support your book gave me in helping me pass the test . . . I'm on my way to a B.S. degree because of you !"

Student, Orlando, FL

Super Review™

All You Need to Know!

ORGANIC CHEMISTRY II

By the Staff of
Research & Education Association
Dr. M. Fogiel, Director

Research & Education Association
61 Ethel Road West
Piscataway, New Jersey 08854

SUPER REVIEW™
OF ORGANIC CHEMISTRY II

Printed in the United States of America

Library of Congress Catalog Card Number 00-131300

International Standard Book Number 0-87891-283-5

SUPER REVIEW is a trademark of
Research & Education Association, Piscataway, New Jersey 08854

WHAT THIS Super Review WILL DO FOR YOU

This **Super Review** provides all that you need to know to do your homework effectively and succeed on exams and quizzes.

The book focuses on the core aspects of the subject, and helps you to grasp the important elements quickly and easily.

Outstanding **Super Review** features:

- Topics are covered in logical sequence

- Topics are reviewed in a concise and comprehensive manner

- The material is presented in student-friendly language that makes it easy to follow and understand

- Individual topics can be easily located

- Provides excellent preparation for midterms, finals and in-between quizzes

- In every chapter, reviews of individual topics are accompanied by Questions **Q** and Answers **A** that show how to work out specific problems

- At the end of most chapters, quizzes with answers are included to enable you to practice and test yourself to pinpoint your strengths and weaknesses

- Written by professionals and test experts who function as your very own tutors

Dr. Max Fogiel
Program Director

CONTENTS

vi

9 SPECTROSCOPY

CHAPTER 1

Arenes

1.1 Structure and Nomenclature

Arenes are compounds that contain both aromatic and aliphatic units.

The simplest of the alkyl benzenes, methyl benzene, has the common name toluene. Compounds that have longer side chains are named by adding the word "benzene" to the name of the alkyl group.

Toluene   Isobutylbenzene   m-Ethylisopropylbenzene

The simplest of the dialkyl benzenes, the dimethyl benzenes, have the common name *xylenes*. Dialkyl benzenes that contain one methyl group are named as derivatives of toluene.

o-Xylene   m-Xylene   p-Xylene   p-Ethyltoluene

A compound that contains a complex side chain is named as a phenyl alkane (C_6H_5 = phenyl). Compounds that contain more than one benzene ring are named as derivatives of alkanes.

$$\underset{\text{2-Methyl-3-phe-}\atop\text{nylpentane}}{\underset{|}{\overset{\overset{\text{CH}_3}{|}}{\text{CH}_3\text{CHCHCH}_2\text{CH}_3}}}$$

1,2-Diphenylethane

Styrene is the name given to the simplest alkenyl benzene. Others are named as substituted alkenes. Alkynyl benzenes are named as substituted alkynes.

—CH≡CH₂
Styrene
(Vinylbenzene)
(Phenylethylene)

—CH₂ CH=CH₂
Allyl benzene
(3–Phenylpropene)

—C≡CH
Phenylacetylene

Problem Solving Examples:

  Draw the structure of:

(a) m-xylene

(b) mesitylene

(c) o-ethyltoluene

(d) p-di-tert-butylbenzene

(e) cyclohexylbenzene

(f) 3-phenylpentane

(g) isopropylbenzene (cumene)

(h) trans-stilbene

(i) 1,4-diphenyl-1,3-butadiene

(j) p-dibenzylbenzene

(k) m-bromostyrene

(l) diphenylacetylene

To draw these structures, it will be necessary to consider the nomenclature of arenes (compounds that contain both aromatic and aliphatic units).

The simplest of the alkylbenzenes (compounds made up of aromatic and alkane units), methylbenzene, is given the name of toluene. Compounds with longer side chains are named by prefixing the name of the alkyl group to the word "-benzene." For example, n-propylbenzene:

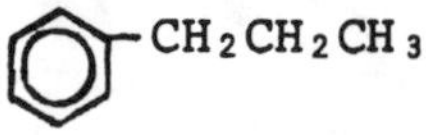

The name xylene is the special name given to the simplest of the dialkylbenzenes, the dimethylbenzenes. There is o-xylene, m-xylene, and p-xylene. Dialkylbenzenes containing one methyl group are named as derivatives of toluene, while others are named by prefixing the names of both alkyl groups to the word "-benzene." A compound containing a very complicated side chain might be named as a phenylalkane (C_6H_5 = phenyl).

Examples:

| p-ethyltoluene | m-ethylisopropyl-benzene | 2-methyl-3-phenyl-pentane |

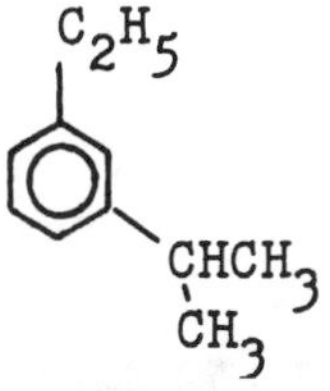

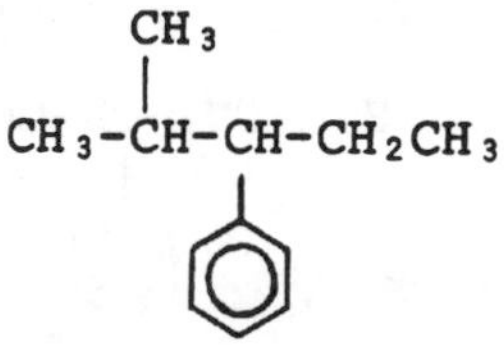

Styrene is the special name given to the simplest alkenylbenzene. Others are generally named as substituted alkenes (sometimes as substituted benzenes). Alkynylbenzenes are named as substituted alkynes. Examples:

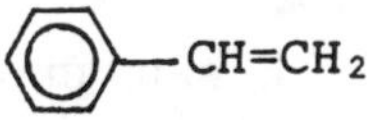

| Styrene | Phenylacetylene |

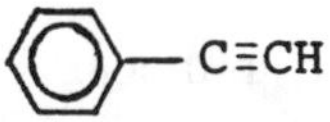

The following structures can now be written:

(a) m-xylene (b) mesitylene 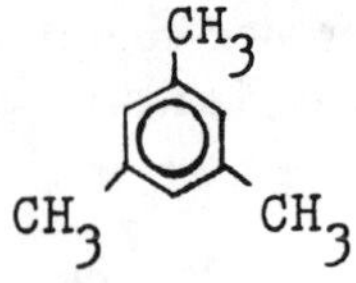 (c) o-ethyltoluene

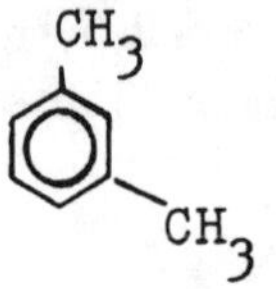

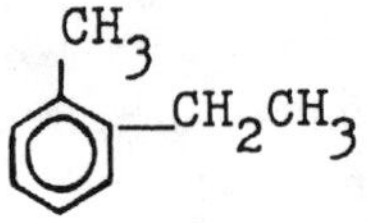

(d) p-di-tert-butylbenzene 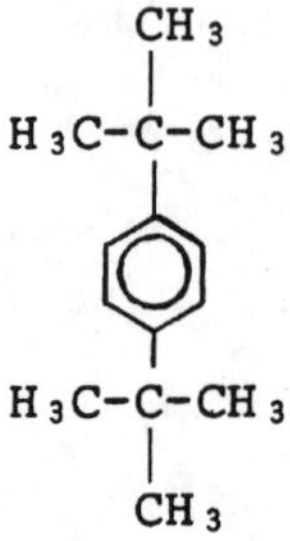 (e) cyclohexylbenzene

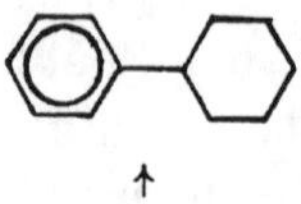

↑
equatorial bond

(f) 3-phenylpentane (g) isopropylbenzene

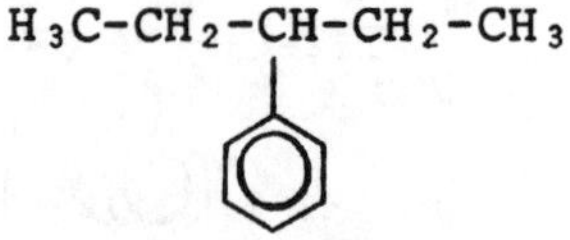

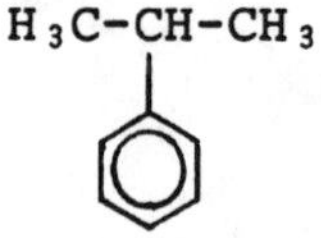

$H_3C-CH_2-CH-CH_2-CH_3$

$H_3C-CH-CH_3$

(h) trans-stilbene 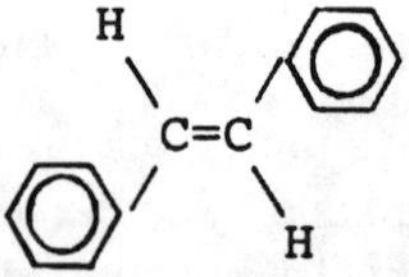 (i) 1,4-diphenyl-1,3-butadiene

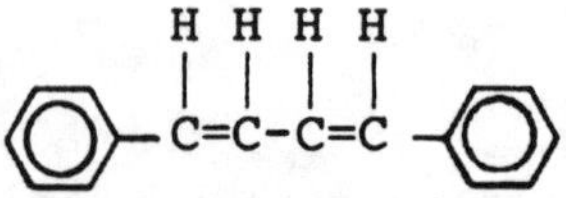

(all isomers possible
–Z,Z– or E,E– or E,Z–)

(j) p-dibenzylbenzene (k) m-bromostyrene

(l) diphenylacetylene

Name each of the following compounds by an accepted system.

(a) $(C_6H_5)_2CHCl$ (b) $C_6H_5CHCl_2$

(c) CH$_3$ (d) CH$_2$CH=CH$_2$ ... NO$_2$ (e) CH=CHCH$_2$OH ... Cl ... Cl

C_6H_5

This problem requires a knowledge of arene nomenclature. Benzene (C_6H_6) is the simplest aromatic hydrocarbon. Benzene compounds can be named as a derivative of benzene or as a phenyl-substituted compound.

(a) In IUPAC (International Union of Pure and Applied Chemistry) nomenclature, a compound having two benzene rings separated by a carbon is named as a methane derivative. Since the compound $(C_6H_5)_2CHCl$ has a chloro group on the middle carbon, it is named as a chloro-substituted compound. There are two phenyl groups (the benzene rings) attached to the methane carbon. Two phenyl substituents can be written as a diphenyl-substituted compound. In IUPAC nomenclature, the names of substituents appear in alphabetical order. Hence, the IUPAC name for the compound $(C_6H_5)_2CHCl$ is chlorodiphenyl-methane. There is no numbering system used for naming methane derivatives because the parent chain is only one carbon long. The compound can alternatively be named as a benzhydrol derivative. Benzhydrol has the formula $(C_6H_5)_2CHOH$. The $(C_6H_5)_2CH-$ portion is called the benzhydryl group. A chloro group in place of the hydroxyl group (–OH) in benzhydrol can be named as a chloride compound. The compound $(C_6H_5)_2CHCl$ can be called benzhydryl chloride.

(b) Benzaldehyde has the formula C_6H_5CHO. The $C_6H_5-\overset{|}{C}H-$ group is called the benzal group. The compound $C_6H_5CHCl_2$ is named as a chloride. The presence of the benzal group gives it the name benzalchloride. In IUPAC nomenclature, the compound $C_6H_5CHCl_2$ would be named as a benzene derivative and not as a methane derivative. The $-CHCl_2$ group is a chloro substituted methyl group. The presence of two chloro groups gives it the name "dichloromethyl." Hence, the IUPAC name for the compound $C_6H_5CHCl_2$ is dichloromethylbenzene.

(c) The compound $C_6H_5 - C_6H_5$ is called biphenyl. The compound whose name we desire $\left(CH_3 \underline{\hspace{0.3cm}}\bigcirc\underline{\hspace{0.3cm}} C_6H_5 \right)$ can be named as a derivative of biphenyl. The compound is a methyl substituted biphenyl compound. The numbering system of biphenyl is:

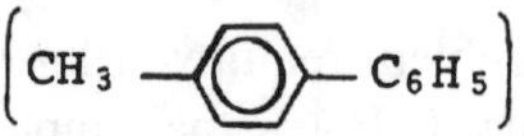

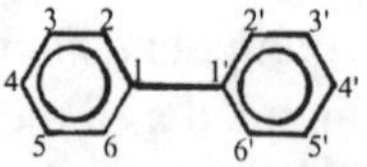

The n,n´ nomenclature of biphenyl compounds is used only when there is a substituent on each ring. If only one ring bears substituents, the n nomenclature is used and not the n´. In this problem, the methyl group is the only substituent in the biphenyl compound

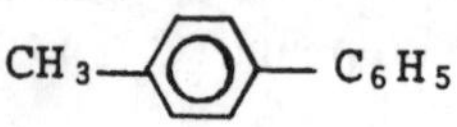

$$CH_3-\bigcirc- C_6H_5$$

Hence, the methyl is considered to be in the 4 position and not the 4′ position. Therefore, the compound is named as 4-methylbiphenyl.

(d) The compound $\underset{\displaystyle NO_2}{\bigcirc}$ – CH –CH=CH₂ is named as a benzene derivative. There are two substituents on the ring: allyl (–CH₂ – CH = CH₂) and nitro (NO₂). These groups are located meta with respect to each other.

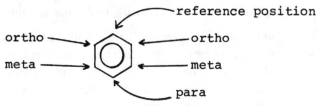

Orientations on a disubstituted benzene ring.

The names of the two groups on the compound appear in alphabetical order. Their relative orientation on the ring is meta (abbreviated as m-) and precedes the name of the compound.

Hence the compound $\underset{\displaystyle NO_2}{\bigcirc}\!\!-\!\!\underset{\displaystyle CH=CH_2}{\overset{\displaystyle CH_2}{|}}$ is named m-allylnitrobenzene.

(e) The compound $Cl-\overset{\diagup Cl}{\bigcirc}\diagdown_{CH=CH-CH_2OH}$ can be named as a derivative of allyl alcohol (CH₂ = CH – CH₂OH). The numbering system of allyl alcohol is $\overset{3}{CH_2} = \overset{2}{CH} - \overset{1}{CH_2OH}$. In the compound we want to name, there is a substituted phenyl substituent at carbon 3 of allyl alcohol. The benzene carbon bonded to carbon 3 of allyl alcohol will be carbon # 1 of the aromatic ring. Hence the substituent

$$Cl-\overset{\displaystyle Cl}{\bigcirc}-$$

is called 2,4-dichlorophenyl. The compound

Cl —⟨O⟩— CH=CH–CH₂OH is therefore named as 3-(2,4-

dichlorophenyl)-allyl alcohol.

1.2 Physical Properties of Arenes

Alkyl benzenes are insoluble in water, but they are soluble in nonpolar solvents like ether; they are generally less dense than water, and their boiling points rise with increasing molecular weight.

Melting points not only depend on the molecular weight, but also on the molecular shape. There exist relationships between melting points and structures of aromatic compounds; for example, in isomeric disubstituted benzenes, the para isomer generally melts at a considerably higher temperature than do the other two. Since dissolution, like melting, involves overcoming the intermolecular forces of the crystal, the para isomer that forms the most stable crystals, is the least soluble in a given solvent.

The higher melting point and lower solubility of a para isomer is an example of the effect of molecular symmetry on intracrystalline forces. The more symmetrical a compound is, the better it fits into a crystal lattice; hence, the melting point is increased, and the solubility is lowered.

1.3 Preparation of Alkylbenzenes

A)   Attachment of alkyl groups: Friedel-Crafts alkylation

$$\bigcirc + RX \underline{\text{Lewis acid}} \bigcirc^{R} + HX \quad R \text{ may rearrange.}$$

Lewis acid: $AlCl_3$, BF_3, HF, etc.
Ar–X cannot be used in place of R–X.

Mechanism of Friedel-Crafts alkylation.

a) $RCl + AlCl_3 \rightleftharpoons AlCl_4^- + R_+$ Carbonium ions from alkyl halides

b) $R_+ + C_6H_6 \rightleftharpoons C_6H_5 \overset{+}{<}{\overset{R}{\underset{H}{}}}$

c) $C_6H_5\overset{+}{<}{\overset{R}{\underset{H}{}}} + AlCl_4^- \rightleftharpoons C_6H_5R + HCl + AlCl_3$

A carbonium ion may: (1) eliminate a hydrogen ion to form an alkene; (2) rearrange to form a more stable carbonium ion; (3) combine with a basic molecule or a negative ion; (4) form a larger carbonium ion by adding to an alkene; (5) abstract a hydride ion from an alkene; and (6) alkylate an aromatic ring.

The use of the Friedel-Crafts alkylation is limited by: (1) the risk of polysubstitution; (2) a possibility of rearrangement of the alkyl group; (3) the impossibility of replacement of the alkyl halides by the aryl halides; (4) aromatic rings that are less reactive than halobenzenes do not undergo Friedel-Crafts alkylation; and (5) aromatic rings that contain $-NH_2$, $-NHR$, or $-NR_2$ groups do not undergo Friedel-Crafts alkylation.

Conversion of Side Chain

a) ⬡—C(=O)-R $\xrightarrow[\text{or N}_2\text{H}_4, \text{ base, heat}]{\text{Zn(Hg), HCl, heat}}$ ⬡—CH$_2$R   Clemmensen or Wolff-Kishner reduction

A Ketone

b) ⬡—CH=CHR $\xrightarrow{\text{H}_2/\text{Ni}}$ ⬡—CH$_2$CH$_2$R

B) Dehydrocyclization Reaction

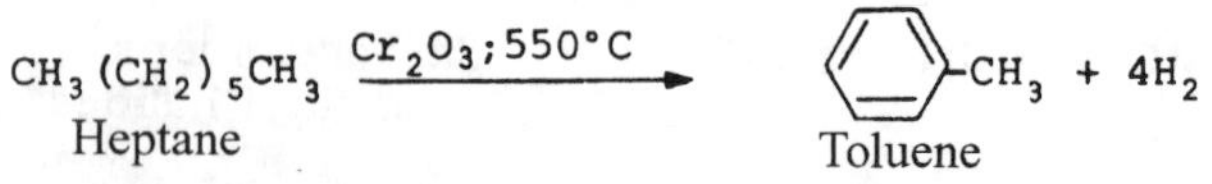

$$CH_3(CH_2)_5CH_3 \xrightarrow{Cr_2O_3;550°C}$$

Heptane

⬡—CH_3 + $4H_2$

Toluene

C) Ullmann Reaction

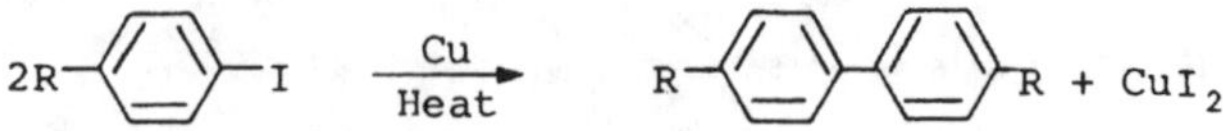

$$2R\text{—}⬡\text{—}I \xrightarrow[\text{Heat}]{Cu} R\text{—}⬡\text{—}⬡\text{—}R + CuI_2$$

D) Electrophilic Aromatic Substitution

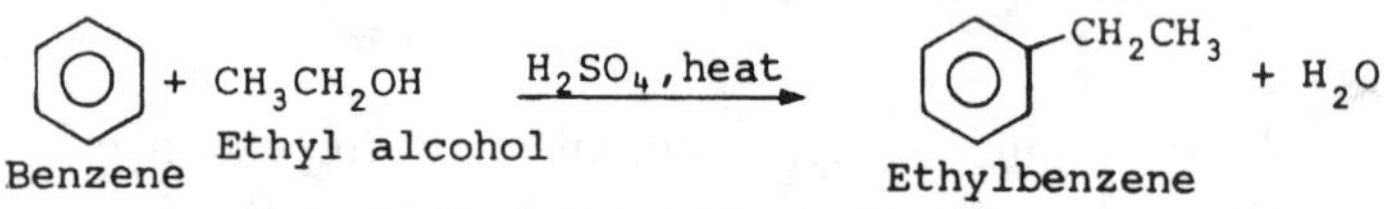

⬡ + CH_3CH_2OH $\xrightarrow{H_2SO_4,\text{heat}}$ ⬡—CH_2CH_3 + H_2O

Benzene   Ethyl alcohol   Ethylbenzene

E) Hydrogenation

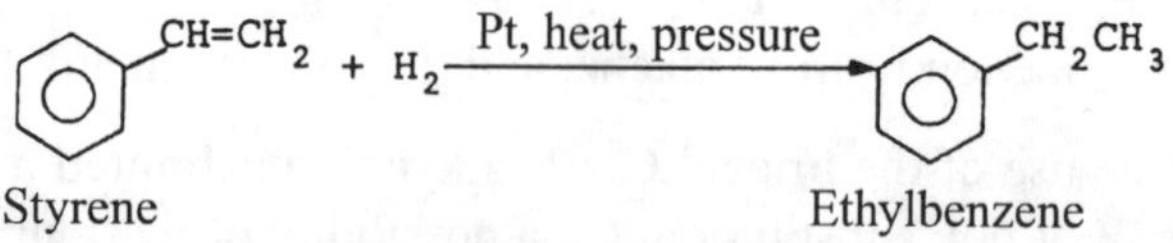

⬡—$CH=CH_2$ + H_2 $\xrightarrow{\text{Pt, heat, pressure}}$ ⬡—CH_2CH_3

Styrene   Ethylbenzene

Problem Solving Examples:

Q How might you prepare ethylbenzene from: (a) benzene and ethyl alcohol; (b) acetophenone, $C_6H_5COCH_3$; (c) styrene, $C_6H_5CH = CH_2$; (d) a-phenylethyl alcohol, $C_6H_5CHOHCH$; and (e) b-phenylethyl chloride, $C_6H_5CH_2CH_2Cl$?

A

(a) ⬡ + CH_3CH_2OH $\xrightarrow{?}$ ⬡—CH_2CH_3

Benzene   Ethyl alcohol   Ethylbenzene

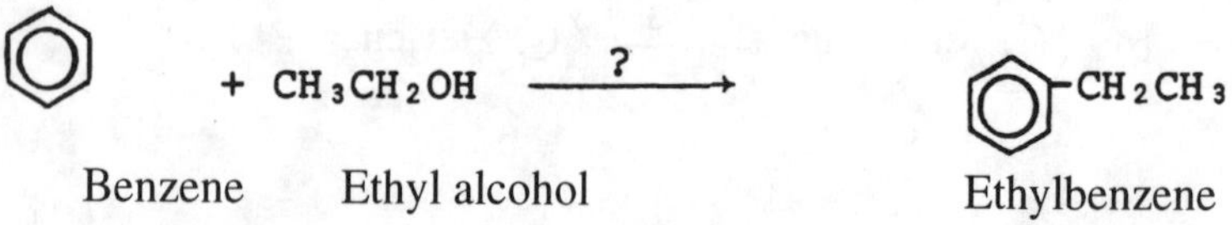

In this problem we want to add an alkyl group to an aromatic ring to obtain an alkylbenzene. Attachment of an alkyl group may be accomplished by Friedel-Crafts alkylation. The general reaction may be written as:

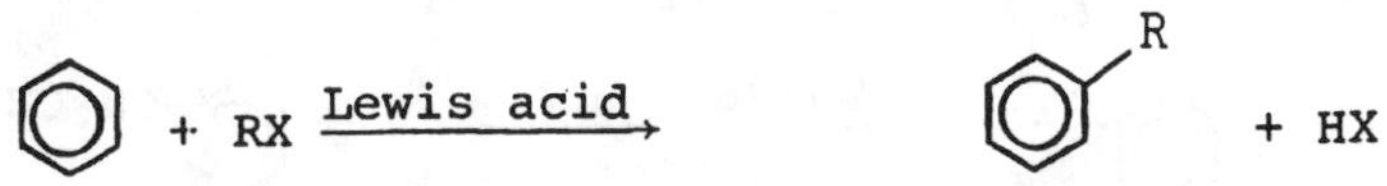

The Lewis acid may be $AlCl_3$, BF_3, HF, etc. RX represents an alkyl halide, alcohol, or alkene, but not an aryl halide. The Lewis acid functions by generating a carbonium ion (R+) from RX, which subsequently attacks the electron rich aromatic (benzene) ring to give the alkylbenzene,

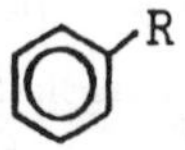

Hence, to synthesize ethylbenzene from ethyl alcohol and benzene, add an acid to the solution (for example, H_2SO_4 and heat). This results in formation of the ethyl cation ($CH_3CH_2^+$) that now attacks the benzene ring to give the desired product. Hence,

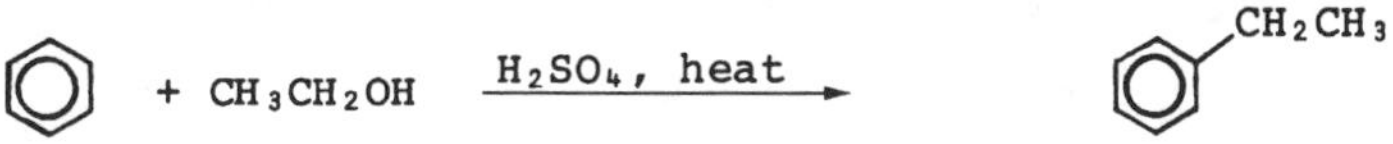

(b)

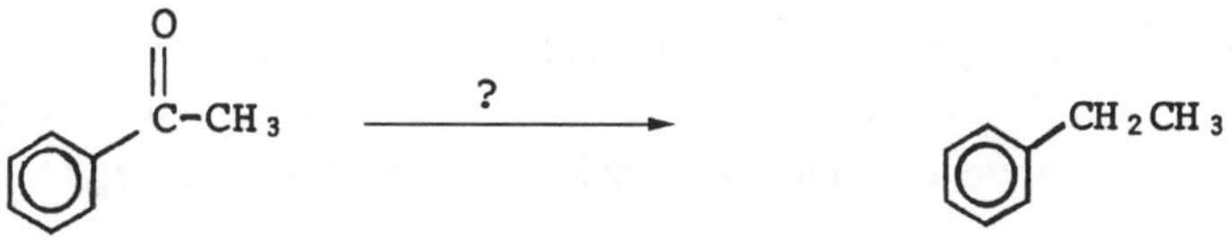

Acetophenone

The task in this case is to reduce the carbonyl group $\left(\diagdown\diagup C=O \right)$ to the methylene ($-CH_2-$) group. Such a side chain conversion may be accomplished by Clemmensen or Wolff-Kishner reduction. In the former, we have amalgamated zinc and concentrated hydrochloric acid, whereas the latter consists of hydrazine (NH_2NH_2) and a strong base like KOH or potassium tert-butoxide. The reaction may be generalized as:

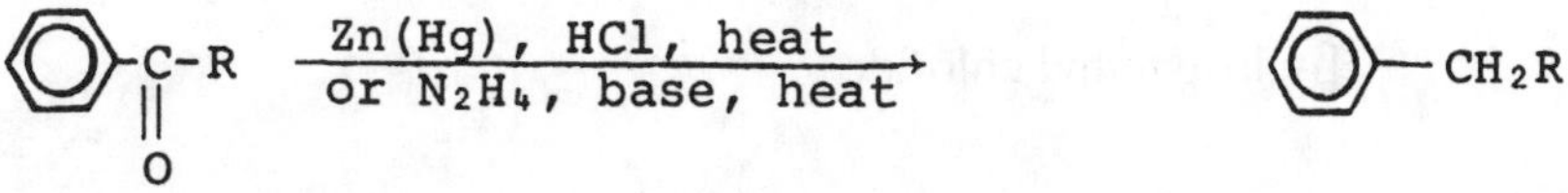

Therefore, we could synthesize ethylbenzene from acetophenone as follows:

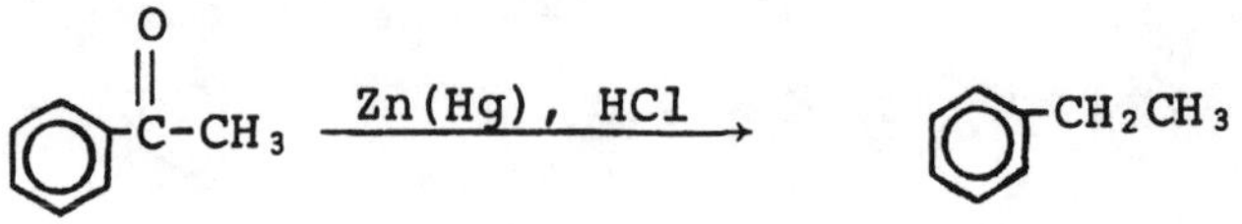

(c)

Styrene

This conversion can readily be made by the addition of hydrogen (H_2) across the carbon-carbon double bond of styrene. This is done by catalytic hydrogenation as shown:

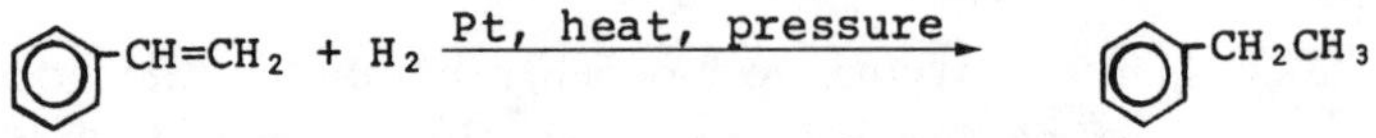

(d)

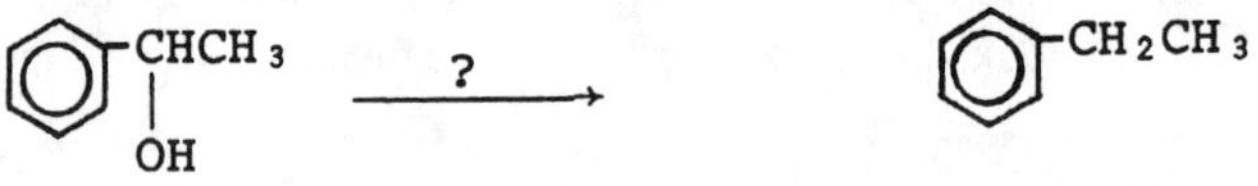

a-phenylethyl alcohol

If the alcohol is dehydrated with the use of acid such as H_2SO_4, styrene would be produced. From styrene, we go to ethylbenzene by catalytic hydrogenation as mentioned above.

Overall:

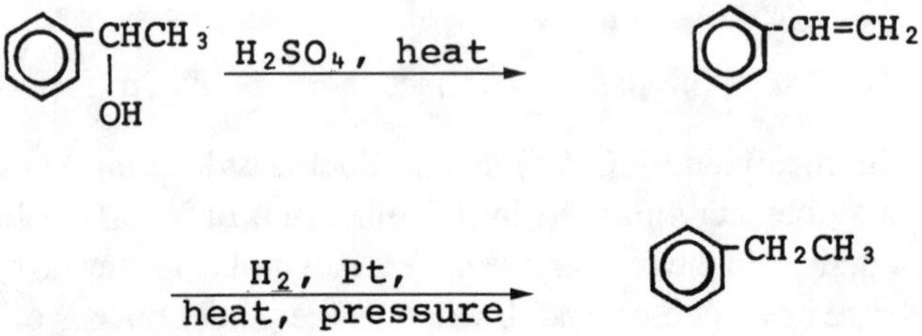

(e)

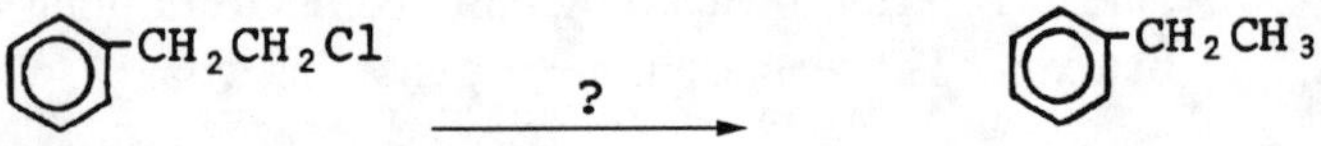

β-phenylethyl chloride

This conversion can be carried out by using the reagent KOH (alc) so that a dehydrohalogenation reaction occurs to yield styrene.

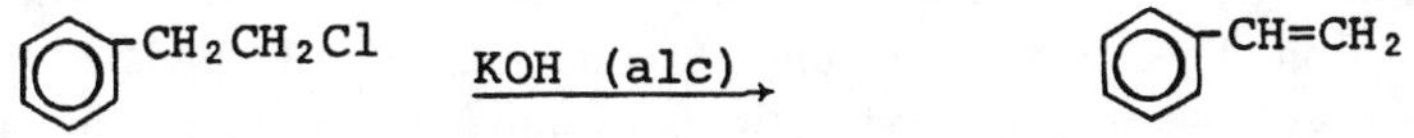

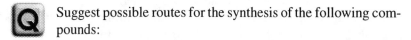

Styrene can now undergo catalytic hydrogenation to produce ethylbenzene.

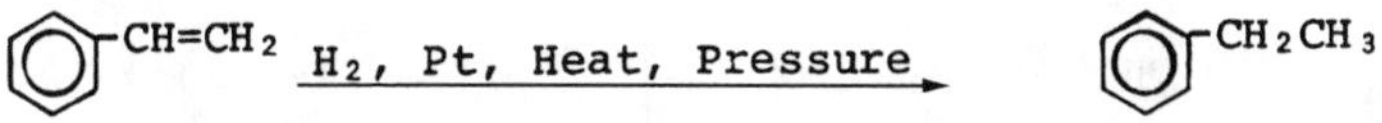

Q Suggest possible routes for the synthesis of the following compounds:

(a)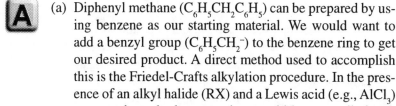

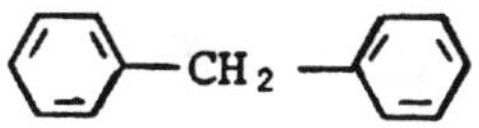

(b)

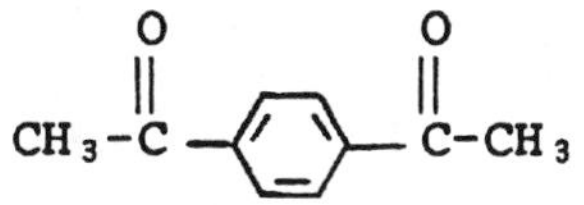

(c)

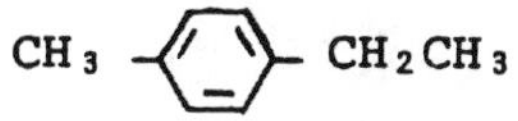

A (a) Diphenyl methane ($C_6H_5CH_2C_6H_5$) can be prepared by using benzene as our starting material. We would want to add a benzyl group ($C_6H_5CH_2^-$) to the benzene ring to get our desired product. A direct method used to accomplish this is the Friedel-Crafts alkylation procedure. In the presence of an alkyl halide (RX) and a Lewis acid (e.g., $AlCl_3$) as a catalyst, the benzene ring would become alkylated. Hence, the desired product could be synthesized by reacting benzene with benzyl chloride in the presence of $AlCl_3$:

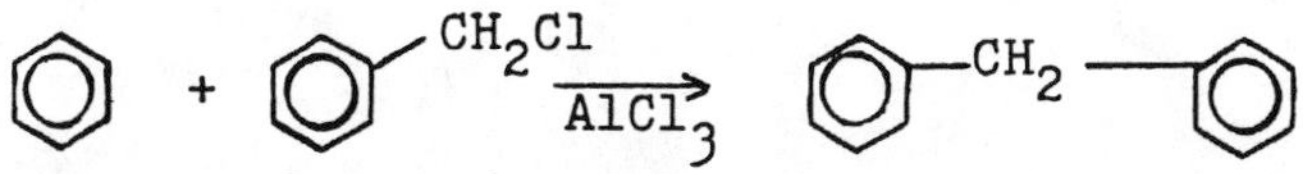

There is a problem associated with the synthesis. When the benzene ring becomes alkylated, the alkyl group activates the ring. This makes the ring more susceptible to further alkylation; that is, the presence of an activating group on the ring will enhance its susceptibility to alkylation. Therefore, the reaction produces more polyalkylated than monoalkylated product. One method of synthesis that will not encounter this problem is the Friedel-Crafts acylation. When a benzene ring is treated with an acyl halide $\left(\begin{array}{c} O \\ \parallel \\ R-C-X \end{array}\right)$ in the presence of a Lewis acid (e.g., $AlCl_3$) catalyst, an acyl group $\left(R-\overset{|}{C}=O \right)$ will be added onto the ring. Since an acyl group deactivates an aromatic ring (due to the electron-withdrawing carbonyl), diacylation is more difficult than monoacylation. Hence, the monoacylated product is the major product in the reaction. When benzene is reacted with benzoyl chloride in the presence of $AlCl_3$, benzophenone is the major product:

Benzoyl chloride   Benzophenone

To convert benzophenone to our desired product, we must reduce the carbonyl $\left(\begin{array}{c} O \\ \parallel \\ -C- \end{array}\right)$ to a methylene ($-CH_2-$). This can be done by using the Clemmensen reduction. If there is a carbonyl group attached to a benzene ring, it will be converted into a methylene group when treated with zinc amalgam and hydrochloric acid. When benzophenone undergoes a Clemmensen reduction, diphenylmethane is produced:

(b) p-diacetylbenzene $\left(CH_3\overset{\displaystyle O}{\overset{\|}{C}}-\text{⟨O⟩}-\overset{\displaystyle O}{\overset{\|}{C}}-CH_3 \right)$ can be synthesized

by starting with p-diethylbenzene $\left(CH_3CH_2 -\text{⟨O⟩}- CH_2CH_3 \right)$.
Free radical chlorination will chlorinate only the methylene groups. This occurs because the reaction proceeds through the most stable benzylic radical:

$$Cl_2 \xrightarrow{h\upsilon} Cl^\bullet + Cl^-$$

Mechanism of free radical chlorination.

$$Cl\bullet \; + \; CH_3CH_2 -\text{⟨O⟩}- CH_2CH_3 \;\rightarrow\; HCl \; + \; CH_3\overset{\bullet}{C}H -\text{⟨O⟩}- CH_2CH_3$$

$$Cl_2 \; + \; CH_3\overset{\bullet}{C}H-\text{⟨O⟩}- CH_2CH_3 \;\rightarrow\; Cl\bullet \; + \; \underset{\displaystyle Cl}{CH_3CH} -\text{⟨O⟩}- CH_2CH_3$$

The overall reaction is:

$$CH_3CH_2 -\text{⟨O⟩}- CH_2CH_3 \xrightarrow[h\nu]{Cl_2} \underset{\displaystyle Cl}{\overset{\displaystyle Cl}{CH_3C}} -\text{⟨O⟩}- \underset{\displaystyle Cl}{\overset{\displaystyle Cl}{C-CH_3}}$$

If the chlorinated product is treated with water, a substitution reaction will occur to produce a gem diol.

$$\left(HO-\overset{\displaystyle |}{\underset{\displaystyle |}{C}}-OH \right)$$

This gem diol will undergo water loss to form the carbonyl:

$$\underset{\underset{Cl}{|}}{\overset{\overset{Cl}{|}}{CH_3C}} - \!\!\!\!\bigcirc\!\!\!\!- \underset{\underset{Cl}{|}}{\overset{\overset{Cl}{|}}{CCH_3}} \xrightarrow{H_2O} \underset{\underset{Cl}{|}}{\overset{\overset{OH}{|}}{CH_3-C}} - \!\!\!\!\bigcirc\!\!\!\!- \underset{\underset{Cl}{|}}{\overset{\overset{Cl}{|}}{C-CH_3}} + HCl$$

$$\downarrow H_2O$$

$$HCl + \underset{\underset{OH}{|}}{\overset{\overset{OH}{|}}{CH_3-C}} - \!\!\!\!\bigcirc\!\!\!\!- \underset{\underset{Cl}{|}}{\overset{\overset{OH}{|}}{C-CH_3}} \xleftarrow{H_2O} \underset{\underset{Cl}{|}}{\overset{\overset{OH}{|}}{CH_3C}} - \!\!\!\!\bigcirc\!\!\!\!- \underset{\underset{Cl}{|}}{\overset{\overset{OH}{|}}{C-CH_3}} + HCl$$

$$\downarrow H_2O$$

$$HCl + \underset{\underset{OH}{|}}{\overset{\overset{OH}{|}}{CH_3-C}} - \!\!\!\!\bigcirc\!\!\!\!- \underset{\underset{OH}{|}}{\overset{\overset{OH}{|}}{C-CH_3}} \xrightarrow{-2 H_2O} \underset{}{\overset{\overset{O}{\|}}{CH_3-C}} - \!\!\!\!\bigcirc\!\!\!\!- \overset{\overset{O}{\|}}{C-CH_3}$$

p-diacetylbenzene

The overall reaction is:

$$\underset{\underset{Cl}{|}}{\overset{\overset{Cl}{|}}{CH_3- C}} - \!\!\!\!\bigcirc\!\!\!\!- \underset{\underset{Cl}{|}}{\overset{\overset{Cl}{|}}{C}} -CH_3 \xrightarrow{H_2O} \underset{}{\overset{\overset{O}{\|}}{CH_3-C}} - \!\!\!\!\bigcirc\!\!\!\!- \overset{\overset{O}{\|}}{C-CH_3}$$

(c) p-ethyltoluene $\left(CH_3 - \!\!\!\!\bigcirc\!\!\!\!- CH_2CH_3 \right)$ can be synthesized from toluene. An ethyl group can be added onto the ring by using a Friedel-Crafts alkylation. There is the problem of polyalkylation in this case. Hence, we can use the method described in part (a) Friedel-Crafts acylation followed by a Clemmensen reduction. If toluene is reacted with acetyl chloride (CH_3COCl) in the presence of $AlCl_3$, the acetyl group will add onto the ring. Toluene will be acetylated at its ortho or para position due to the ortho, para directing methyl group. The reaction is:

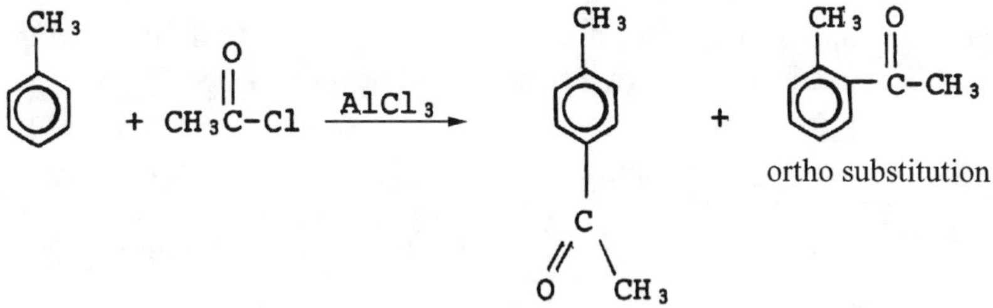

$$CH_3\text{-benzene} + CH_3\overset{O}{\underset{\|}{C}}\text{-Cl} \xrightarrow{AlCl_3}$$

CH₃-benzene-C\
para substitution

$+$

CH₃-benzene $\overset{O}{\underset{\|}{C}}$-CH₃\
ortho substitution

The product mixture can be separated and the para isomer can undergo Clemmensen reduction to give the desired product:

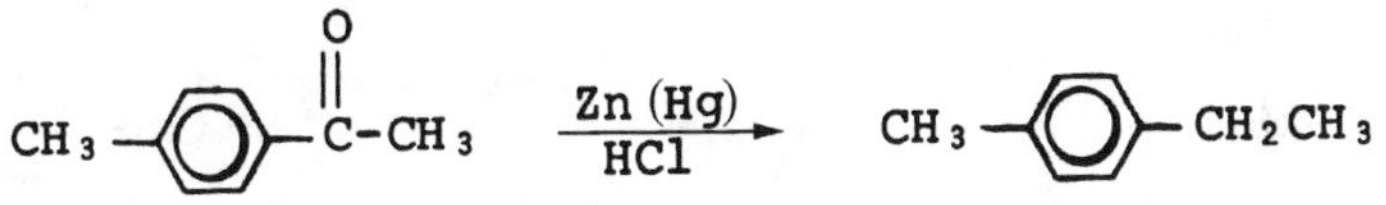

$$CH_3\text{-benzene-}\overset{O}{\underset{\|}{C}}\text{-CH}_3 \xrightarrow[\text{HCl}]{\text{Zn (Hg)}} CH_3\text{-benzene-}CH_2CH_3$$

p-ethyltoluene

Q How can n-butyl benzene be prepared from n-proply phenyl ketone?

A n-propyl phenyl ketone can be converted to n-butyl benzene by a Clemmensen reduction using amalgamated zinc and hydrochloric acid or by a Wolff-Kishner reduction using hydrazine and base.

1. Clemmensen reduction

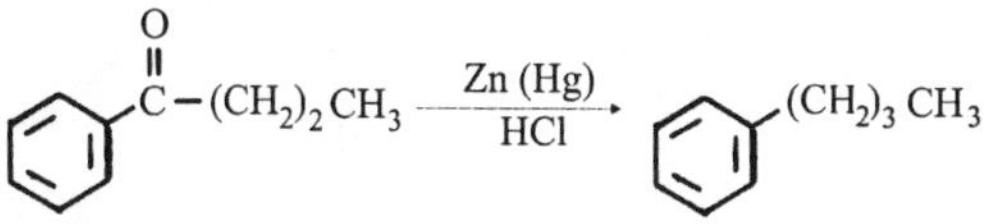

$$\text{benzene-}\overset{O}{\underset{\|}{C}}\text{-(CH}_2)_2CH_3 \xrightarrow[\text{HCl}]{\text{Zn (Hg)}} \text{benzene-(CH}_2)_3CH_3$$

n-butyl benzene

2. Wolff-Kishner reduction

$$\text{benzene-}\overset{O}{\underset{\|}{C}}\text{-(CH}_2)_2CH_3 \xrightarrow[-200°]{\text{NH}_2\text{ NH}_2,\text{OH}} \text{benzene-(CH}_2)_3CH_3$$

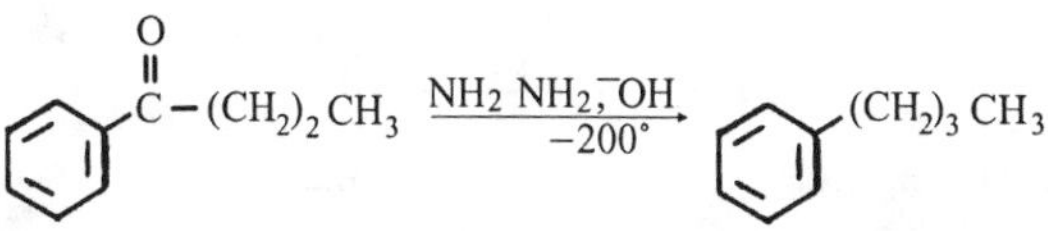

The Clemmensen reduction and the Wolff-Kishner reduction are a good method of preparing straight-chain alkyl benzenes. The starting acyl benzene is prepared by Friedel-Crafts acylation.

Q Why would the Ullmann reaction not be the method of choice in preparing CH_3CH_2—⟨benzene⟩—⟨benzene⟩—CH_3 from CH_3CH_2—⟨benzene⟩—I and CH_3—⟨benzene⟩—I?

A The Ullmann reaction is a useful coupling reaction between identical aryl halides that results in the production of biaryl compounds.

A coupling reaction between CH_3—⟨benzene⟩—I and CH_3CH_2—⟨benzene⟩—I would result in the desired product as well as the coupling products of CH_3CH_2—⟨benzene⟩—I coupling with itself and CH_3—⟨benzene⟩—I coupling with itself.

CH_3CH_2—⟨benzene⟩—I + CH_3—⟨benzene⟩—I + $2Cu$ ⟶

CH_3CH_2—⟨benzene⟩—⟨benzene⟩—CH_3 + CH_3CH_2—⟨benzene⟩—⟨benzene⟩—CH_2CH_3

desired product   and

CH_3—⟨benzene⟩—⟨benzene⟩—CH_3

1.4   Reactions of Alkylbenzenes

A) Hydrogenation

⟨benzene⟩CH_2CH_3 + $3H_2$ $\xrightarrow{\text{Ni,Pt or Pd}}$ ⟨cyclohexane⟩CH_2CH_3

Ethylbenzene   Ethylcyclohexane

B) Oxidation

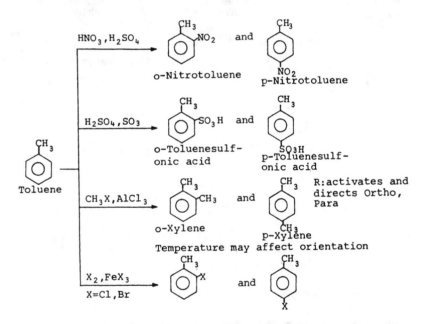

The oxidation reaction is used for two purposes: (a) synthesis of carboxylic acids and (b) identification of alkylbenzenes.

C) Substitution in the ring. Electrophilic aromatic substitution.

D) Substitution in the side chain. Free-radical halogenation.

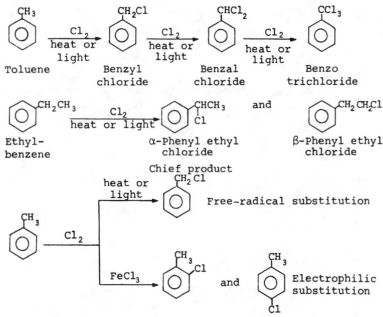

Hydrogenation atoms attached to a carbon joined directly to an aromatic ring are called benzylic hydrogens.

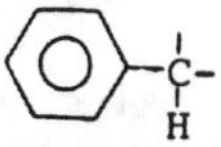

Ease of abstraction of hydrogen atoms:

allylic benzylic > 3° > 2° >1° CH_4, vinylic.

Problem Solving Example:

Q   When 2,4,6-trinitroanisole is treated with methoxide in methanol, a red anion having the composition $(C_8H_8O_8N_3)-$ is produced. Such anions are called Meisenheimer complexes after the chemist who first suggested the correct structure. What structure do you think he suggested? One of Meisenheimer's experiments compared the product of reaction of 2,4,6-trinitroanisole with ethoxide ion with the product of 2,4,6-trinitrophenyl ethyl ether with methoxide ion. What do you think he found?

A Benzene compounds that possess a potential leaving group and also contain substituents that can stabilize a negative charge by resonance will undergo a nucleophilic aromatic substitution reaction upon treatment with a nucleophilic reagent. The reaction proceeds through an isolable intermediate benzene anion called a Meisenheimer complex.

Example: The nucleophile attacks the carbon in the aromatic ring that bears the leaving group and destroys the aromaticity of the compound. The negative charge is resonance stabilized by the groups on the ring. The greatest stabilization occurs when the stabilizing groups are ortho and para to the leaving group. When 2,4,6-trinitroanisole is treated with methoxide in methanol, the nucleophilic methoxide anion attacks the carbon bearing the methoxyl group. The resulting negative charge will be resonance stabilized by the three nitro groups:

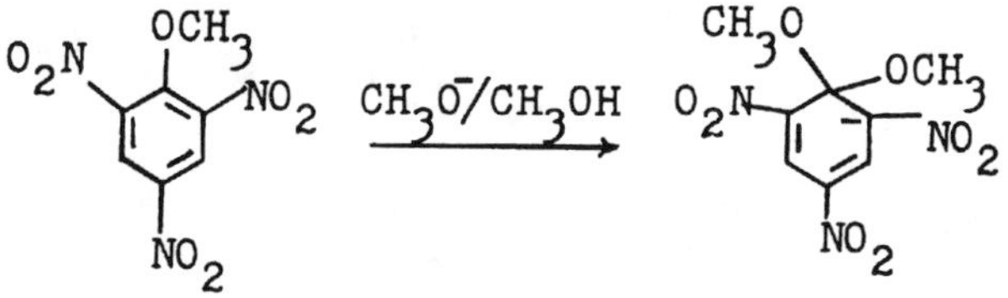

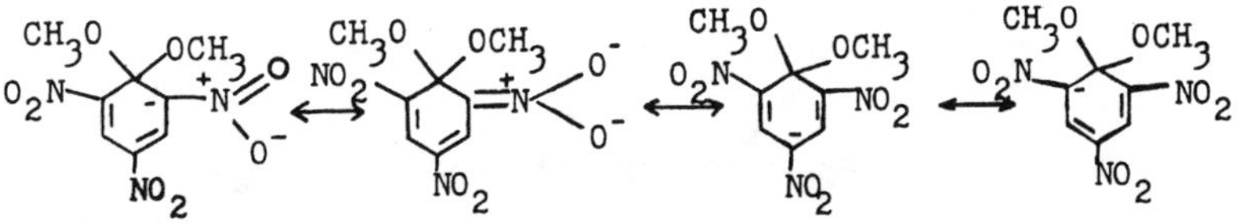

Above: Resonance stabilization of intermediate benzene anion; shown only for one nitro group. The overall structure for the resonance stabilized species is:

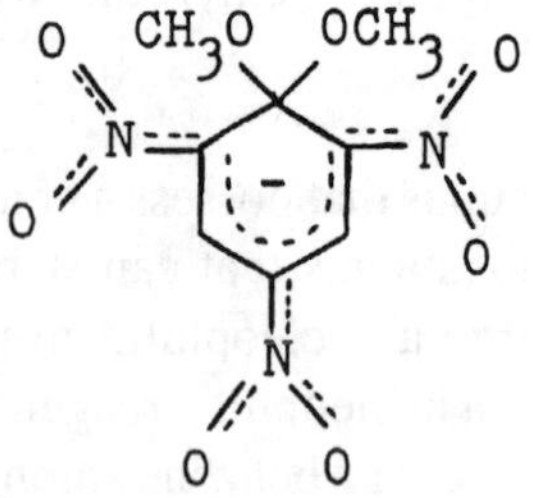

Above: Meisenheimer Complex

When 2,4,6-trinitroanisole is treated with ethoxide ion, it will produce the same Meisenheimer complex that 2,4,6-trinitrophenyl ethyl ether produces when reacted with methoxide ion. This Meisenheimer complex can be represented as follows:

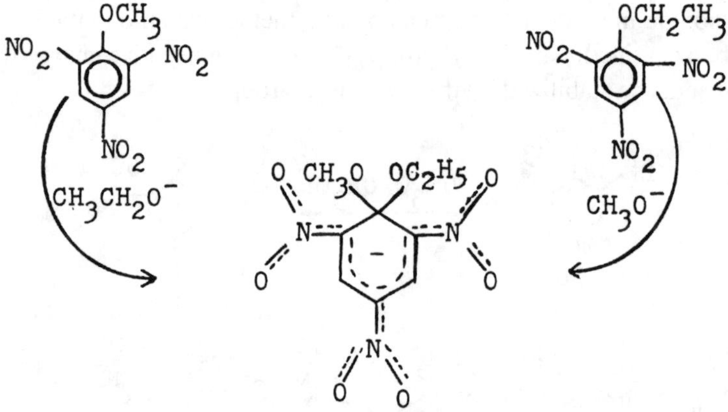

Meisenheimer Complex

1.5   Resonance Stabilization of Benzyl Radicals

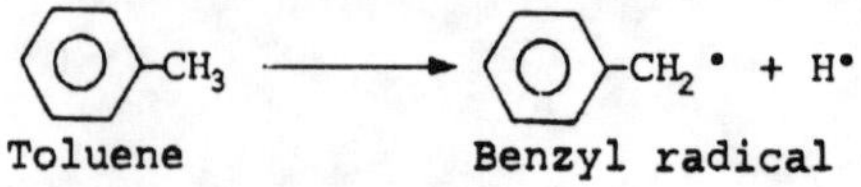

Toluene   Benzyl radical

Ease of formation of free radicals:

allyl benzyl > 3° > 2° > 1° CH_3, vinyl

The more stable the radical, the more rapidly it is formed; the less energy the radical contains, the more stable it is.

Stability of free radicals:

allyl benzyl > 3° > 2° > 1° CH_3, vinyl

Resonance stabilizes and lowers the energy content of a benzyl radical. Through resonance the benzyl radical becomes more stable than the hydrocarbon it is formed from.

Benzyl and allyl-free radicals are extremely reactive and unstable particles.

Problem Solving Example:

 It is believed that the odd electron in the benzyl radical is not localized on the side chain but is delocalized about the ring. Draw resonance structures of the benzyl radical to support this view.

Resonance structures of the benzyl radical can be drawn as follows:

$$CH_2\cdot \quad CH_2\cdot \quad CH_2 \quad CH_2 \quad CH_2$$

I   II   III   IV   V

The benzyl radical is a hybrid of the Kekulé structures I and II. Structures III, IV, and V show that the odd electron is delocalized about the ring. This accounts for the high stabilization of the benzyl radical.

1.6 Preparation and Reactions of Alkenylbenzenes

Preparation of Alkenylbenzenes

Alkenylbenzenes are aromatic hydrocarbons with a side chain containing a double bond.

A) Dehydrogenation

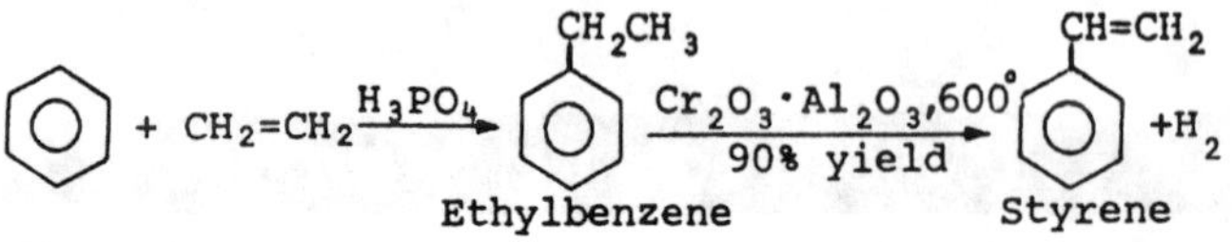

B) Dehydrohalogenation and dehydration

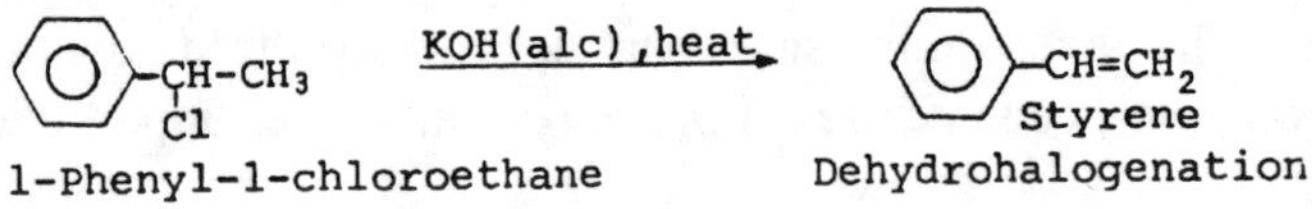

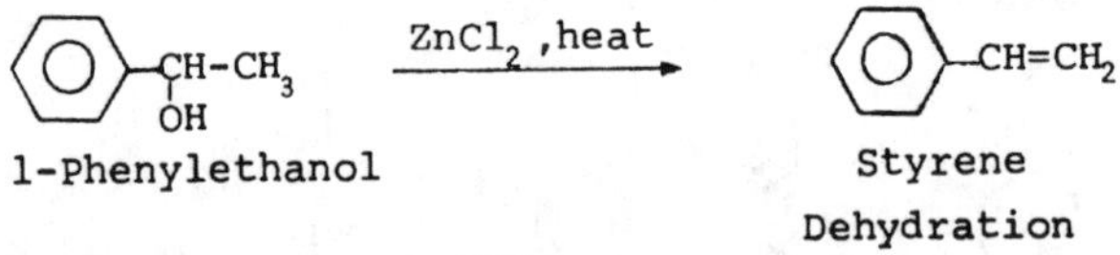

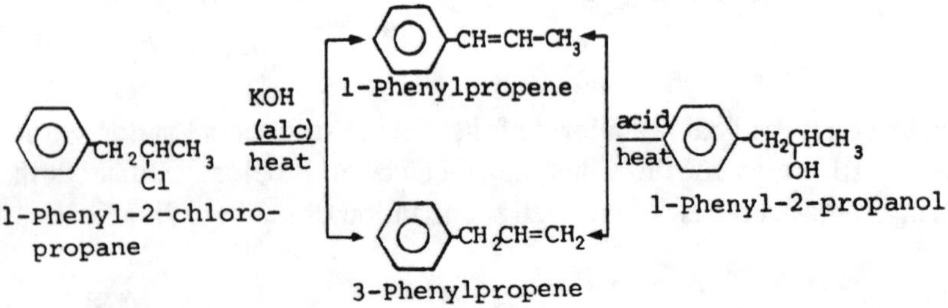

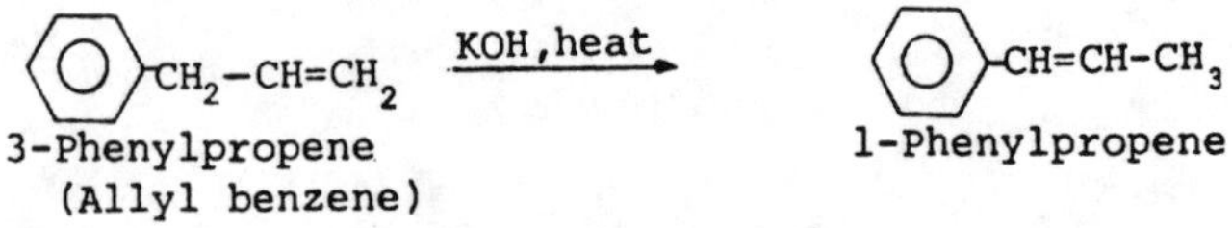

A double bond is conjugated with the ring when it is separated from the benzene ring by one single bond. Conjugation gives unusual stability to a molecule. This stability has an effect on the orientation and the ease of elimination.

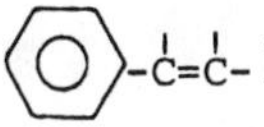

Double bond conjugated with ring.

Reactions of Alkenylbenzenes

The double bond shows higher reactivity than the resonance-stabilized benzene ring.

A) Substitution in the Ring

a) Catalytic hydrogenation

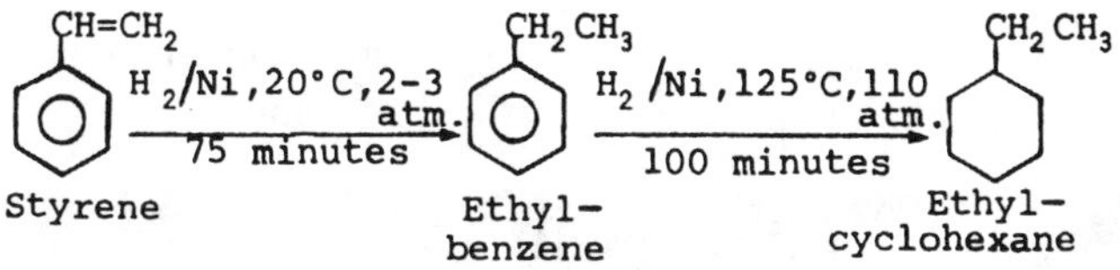

CH=CH2
H2/Ni, 20°C, 2-3 atm.
75 minutes
CH2 CH3
H2 /Ni, 125°C, 110 atm.
100 minutes
CH2 CH3
Styrene
Ethyl-benzene
Ethyl-cyclohexane

b) Ring halogenation

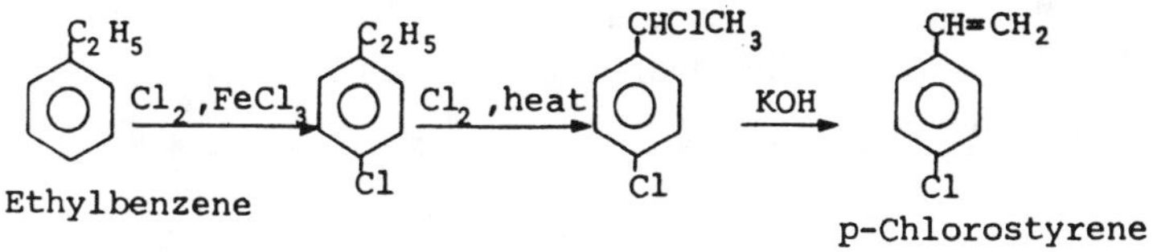

C2 H5
Cl2 ,FeCl3
C2H5
Cl2 ,heat
CHClCH3
KOH
CH=CH2
Cl
Cl
Cl
Ethylbenzene
p-Chlorostyrene

B) Addition to conjugated alkenylbenzenes: Orientation. Stability of benzylcation. In either electrophilic or free-radical additions, the first step takes place in the way that yields the more stable particle.

No peroxides

$$C_6H_5CH = CHCH_3 \xrightarrow{HBr} C_6H_5\underset{+}{C}HCH_2CH_3 \xrightarrow{Br}$$

a benzylcation

$$C_6H_5\underset{\displaystyle Br}{\underset{|}{C}}HCH_2CH_3$$

Peroxides present

$$C_6H_5CH = CHCH_3 \xrightarrow{Br} C_6H_5\overset{\bullet}{C}HCHCH_3 \xrightarrow{HBr}$$
$$\underset{Br}{|}$$

a benzyl free-radical

$$C_6H_5CH_2CHCH_3$$
$$\underset{Br}{|}$$

Addition to conjugated alkenylbenzenes: reactivity.

Stability of   benzyl allyl
carbonium ions: $3° > 2° > 1°\ CH_3^+$

Conjugated alkenylbenzenes are more stable than simple alkenes.

Conjugated alkenylbenzenes are much more reactive than simple alkenes toward both ionic and free-radical addition.

Problem Solving Examples:

Q What arene derivatives are expected from the oxidation of the following with hot, alkaline potassium permanganate followed by acidification?

(a)

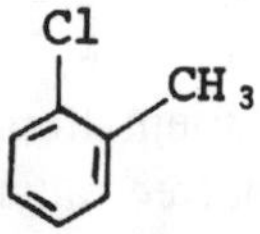

(e)

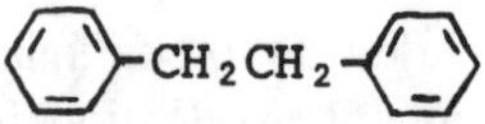

(b)

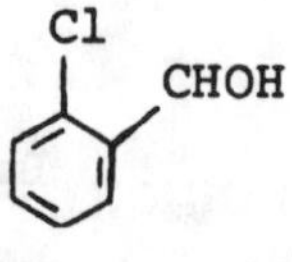

(f)

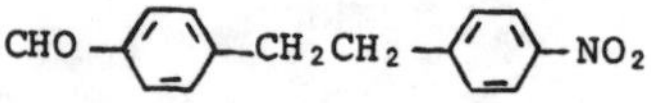

(c)

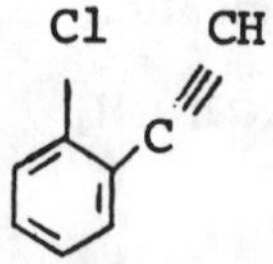

(g)

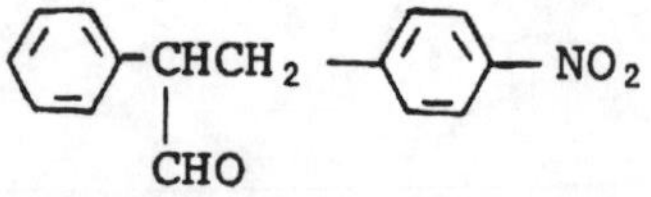

(d)

(h)

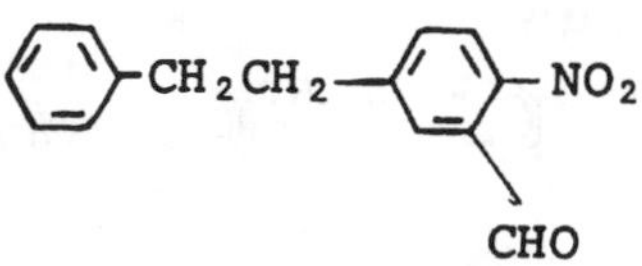

A It is known that alkanes and benzene are quite inert to oxidation with potassium permanganate in aqueous alkali. However, alkyl and aryl groups attached to benzene rings can be oxidized to carboxyl ($-COOH$) groups with hot, alkaline permanganate. For example,

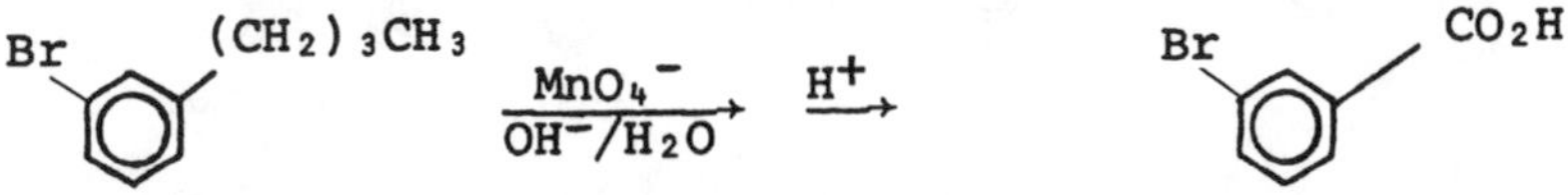

With this in mind, we can predict the products of the following reactions:

(a)

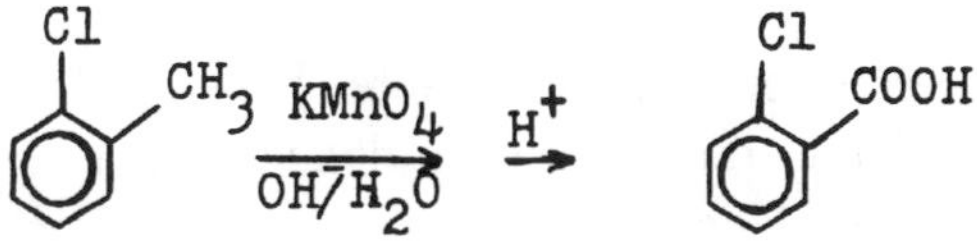

(b)

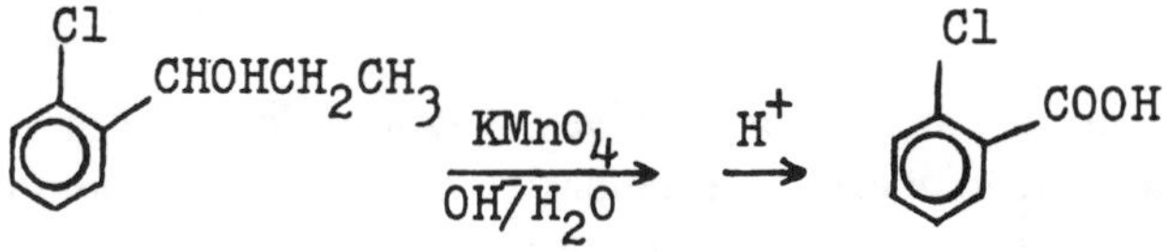

(c)

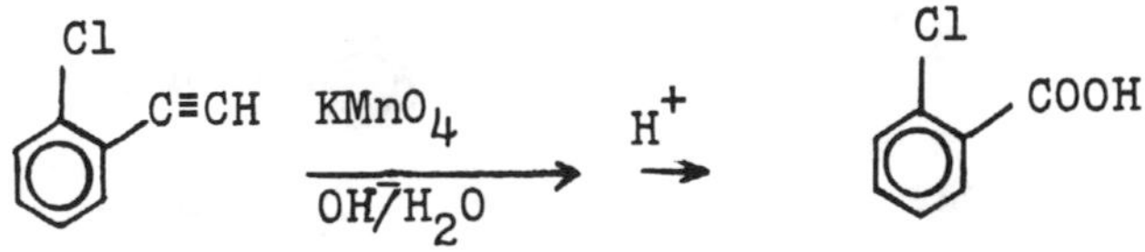

(d)

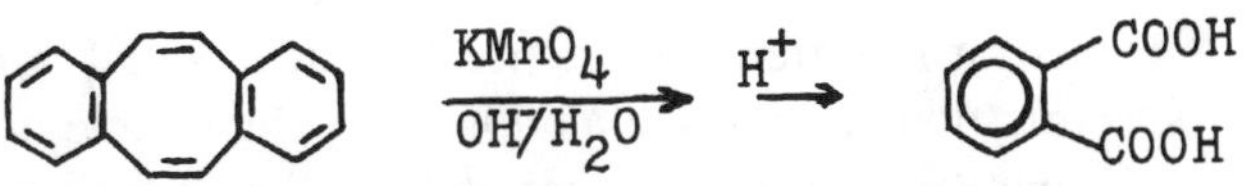

(e)

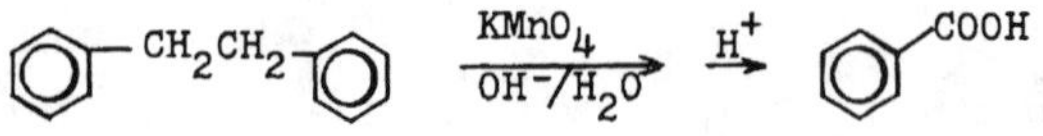

(f)

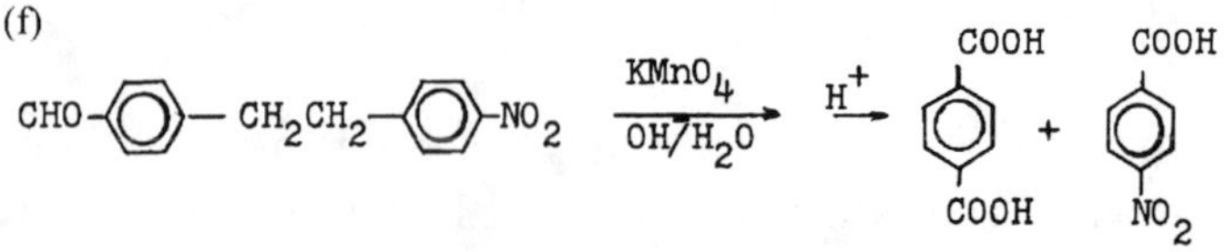

(g)

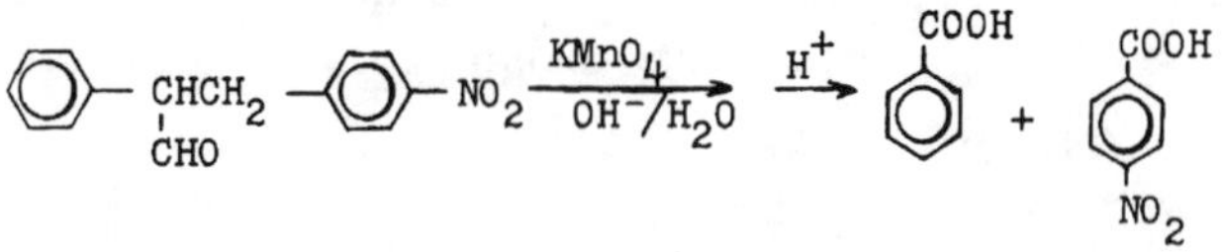

(h)

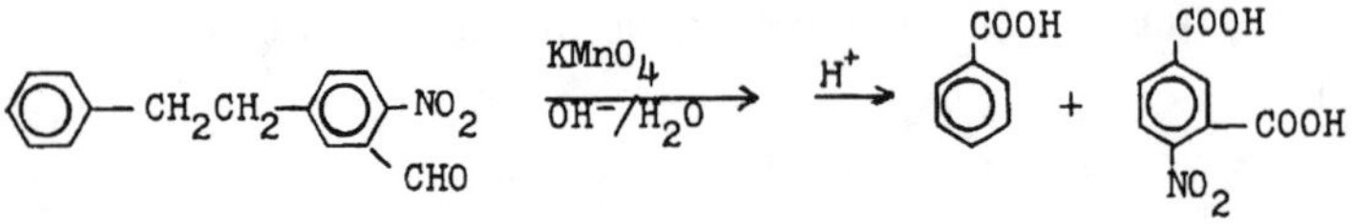

Q Describe simple chemical tests (if any) that would distinguish between: (a) styrene and ethylbenzene; (b) styrene and phenylacetylene; (c) allylbenzene and 1-nonene; and (d) allylbenzene and allyl alcohol ($CH_2 = CH - CH_2OH$). Tell exactly what you would do and see.

A This problem can be solved by writing down the chemical structure to note the particular functional group the compound possesses and then devising characterization tests to distinguish between the functional groups and, as such, the compounds.

(a)

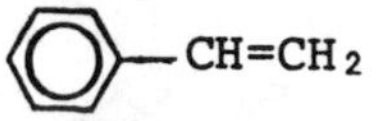

Styrene 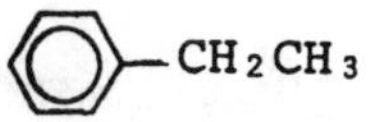   Ethylbenzene

Both of these arenes (compounds containing both an aromatic and aliphatic unit) possess double bonds. However, only styrene has the double bond (for unsaturation) in its side group where it is extremely reactive to reagents that normally add across double bonds as in alkenes. (The double bonds of the aromatic ring are resonance stabilized so that they are not reactive to these reagents.) Hence, these two compounds can be distinguished by the ability to decolorize both a solution of bromine in carbon tetrachloride and a cold, dilute, neutral permanganate solution. Styrene will decolorize since the aliphatic portion ($-CH = CH_2$) behaves as an alkene, whereas ethylbenzene will not (the side chain, $-CH_2CH_3$, behaves as an alkane so that it is inert to these reagents). The following will be seen:

styrene   red   colorless

Ethylbenzene + Br_2/CCl_4 → no reaction

    red   red

          purple   brown ppt.   colorless

Ethylbenzene + MnO_4^- → no reaction

          purple   purple

(b) Styrene   Phenylacetylene

Phenylacetylene can be distinguished from styrene by the former's ability to form heavy metal acetylides. The acidic acetylenes such as phenylacetylene react with certain heavy metal ions, chiefly Ag^+ and Cu^+, to form insoluble acetylides. The reaction is indicated by the formation of a precipitate. Styrene will not form this precipitate if placed in a solution such as $Ag(NH_3)_2OH$ or $Cu(NH_3)_2OH$. Hence, detection of a precipitate in either of these solutions is the simple chemical test desired.

(c) 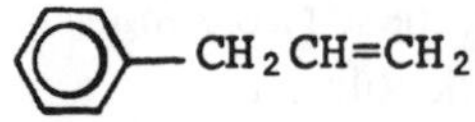 $\bigcirc$—$CH_2CH=CH_2$ $H_2C=CH(CH_2)_6CH_3$

   Allylbenzene   1-nonene

These two compounds can be distinguished by the physical property (melting point) of the products of oxidation. Upon oxidation with hot $KMnO_4$ or $K_2Cr_2O_7$, allylbenzene will yield $\bigcirc$—COOH, mp 122°C while the oxidation of 1-nonene will yield $C_7H_{15}COOH$, mp 16°C.

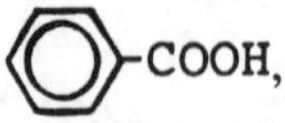

(d) 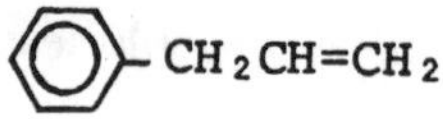 $\bigcirc$—$CH_2CH=CH_2$ $CH_2=CH—CH_2OH$

   Allylbenzene   Allyl alcohol

Since allyl alcohol is a primary alcohol, it will give a positive chromic anhydride test, whereas allylbenzene will not. Primary and secondary alcohols are oxidized by chromic anhydride, CrO_3, in aqueous sulfuric acid. The clear orange solution turns blue-green and becomes opaque within two seconds in the presence of these alcohols. Hence, allyl alcohol and allylbenzene may be distinguished by the former's ability to give the color change in chromic anhydride.

1.7 Electrophilic Aromatic Substitution of Substituted Benzene

A) Ortho, Para Directing Substituents

Strongly Activating

   $-NH_2$ ($-NHR$, $-NR_2$)

Moderately Activating

   $-OCH_3$ ($-OCH_2CH_3$, etc.)

$-NHCOCH_3$

Weakly Activating

$-C_6H_5$
$-CH_3 \, (-CH_2CH_3, \text{ etc.})$

B) Ortho, Para Directing Deactivating

$-F, -Cl, -Br, -I$

C) Meta Directing Deactivating

$-NO_2$
$-N^+(CH_3)_3$
$-CN$
$-CO_2H \, (-CO_2R)$
$-SO_3H$
$-CHO, COR$

By knowing the directing influence of these substituents, we can predict the outcome of most aromatic substitution reactions. For example, to predict the product of nitration on toluene, we see from above that the methyl group is an ortho, para director and is weakly activating. It can therefore be predicted that the nitration of toluene will occur slightly more rapidly than the nitration of benzene. The expected products of this reaction would be ortho-nitro toluene and para-nitro toluene.

The following examples illustrate the orientation of electrophilic substitution reactions on substituted benzenes.

1.

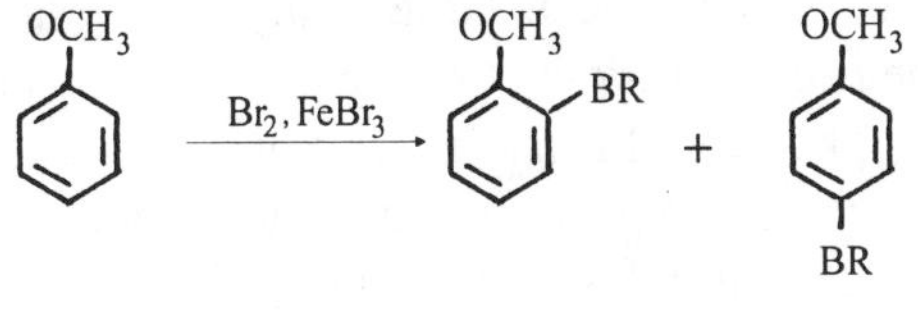

2.

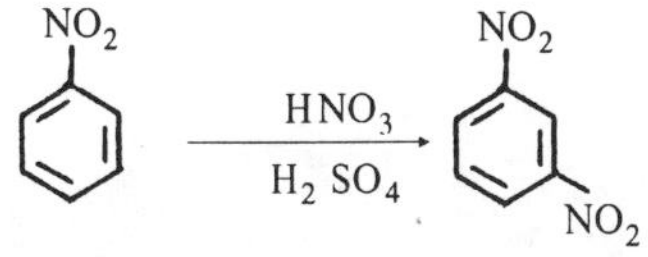

3.

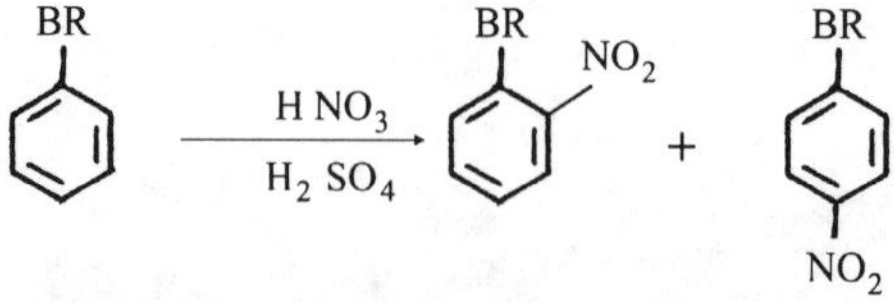

$$\text{BR} \xrightarrow[\text{H}_2\text{SO}_4]{\text{H NO}_3} \text{BR-NO}_2 \;+\; \text{BR-NO}_2$$

Problem Solving Examples:

Q Give the structure and principal organic products expected from the following nitration reactions. In each case tell whether the nitration will occur faster or slower than with benzene itself.

(a) tertiary butyl benzene

(b) nitro benzene

(c) acetophenone ($C_6H_5 - COCH_3$)

(d) bromo benzene

A (a)

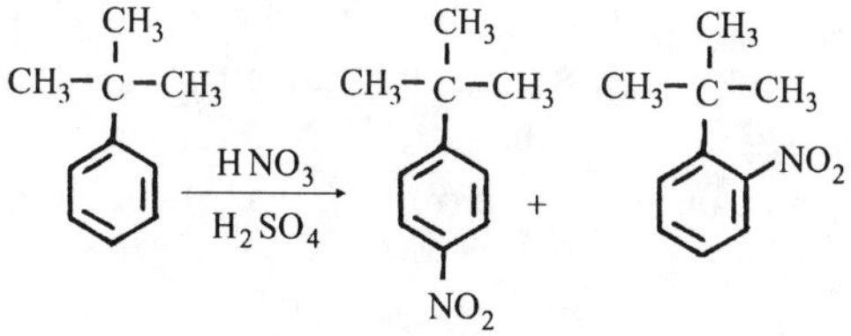

Principal product

The nitration reaction of t-butyl benzene will occur faster than the nitration of benzene because the t-butyl group is slightly activating. The principal product expected would be p-nitro t-butyl benzene. The t-butyl group is large and would sterically hinder attack at the ortho position. Substitution at the para position would be the more favored product.

(b)

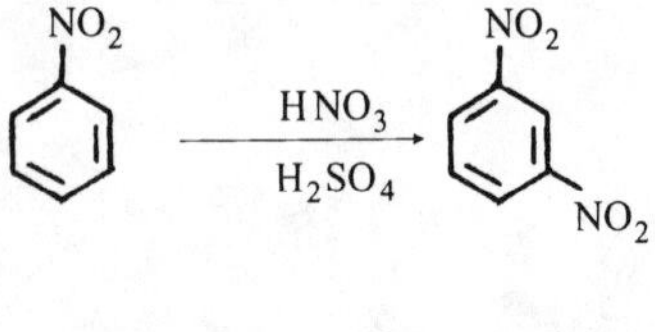

The nitration of nitro benzene would occur slower than that of benzene because the nitro group tends to deactivate the rate of reaction. The nitro group is a meta director. The principal product is therefore 1,3-dinitrobenzene.

(c) $\underset{H_2SO_4}{\overset{HNO_3}{\longrightarrow}}$

The –$COCH_3$ group is a meta directing, deactivating group. The reaction will proceed more slowly than with benzene. The expected product would be meta-nitro acetophenone.

(d) $\underset{H_2SO_4}{\overset{HNO_3}{\longrightarrow}}$ +

The bromo group is ortho, para directing and deactivating. The reaction would proceed more slowly than the nitration of benzene. The expected products are p-nitro bromo benzene and o-nitro bromo benzene.

Q Arrange the following in order of reactivity toward ring bromination, listing the most reactive compound at the top and the least reactive at the bottom.

benzene, toluene, nitro benzene, anisole, aniline

A The order of reactivity is as follows:

1. aniline

2. anisole

3. toluene

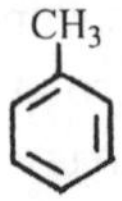

CH$_3$

4. benzene

5. nitro benzene

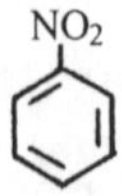

NO$_2$

Aniline contains the –NH$_2$ group, which is strongly activating. The methoxyl group of anisole is moderately activating. The methyl group of toluene is weakly activating. The nitro group of nitro benzene deactivates the ring toward substitution.

Q In the following compounds, indicate which ring you would expect to be attacked in bromination. Draw the structures of the principal products.

(a) 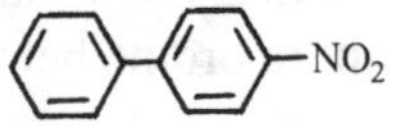 —NO$_2$

(b) 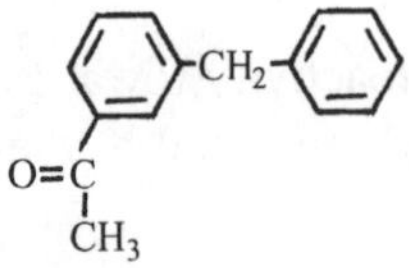 –CH$_2$–

O=C
|
CH$_3$

A Bromination would be expected to occur in the ring not deactivated by the nitro or acyl groups.

(a) 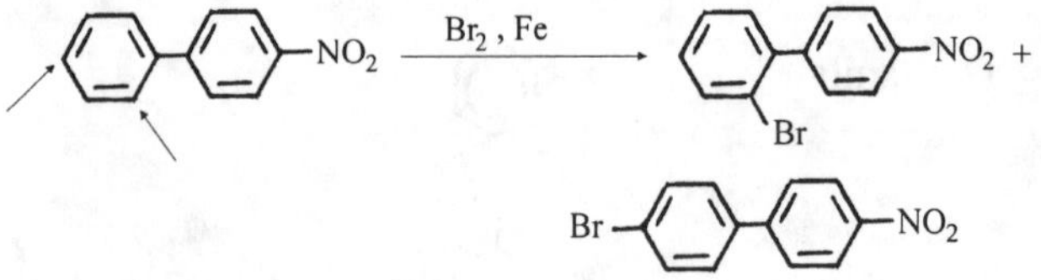   Br$_2$, Fe
        —NO$_2$ ⟶   —NO$_2$ +
                                                              Br

        Br—   —NO$_2$

(b)

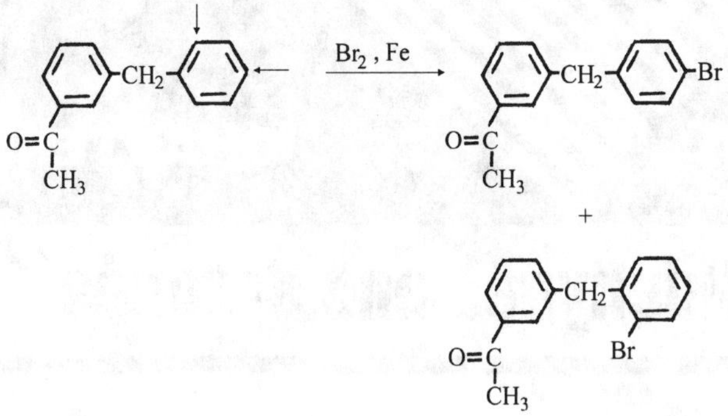

$$\xleftarrow{} \quad \xrightarrow{\text{Br}_2 \text{ , Fe}} \quad$$

+

CHAPTER 2

Aldehydes and Ketones

Carboxylic acids, aldehydes, and ketones have a carboxylic group $\diagdown C{=}O$ in common. The general formula for aldehydes is RCHO, and that for ketones is RCOR'.

2.1 Nomenclature (IUPAC System)

A) Aldehydes: The longest continuous chain containing the carbonyl group is considered the parent structure and the "-e" ending of the corresponding alkane is replaced by "-al."

B) Ketones: The "-e" ending of the corresponding alkane is replaced by "-one."

Example:

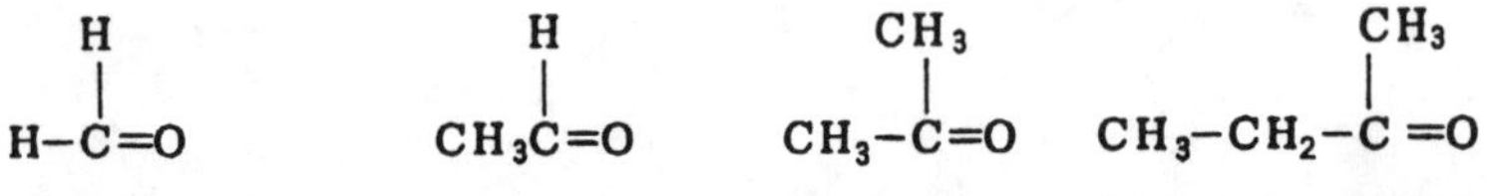

$$\overset{\displaystyle H}{\underset{\displaystyle}{H-\overset{|}{C}{=}O}} \qquad \overset{\displaystyle H}{CH_3\overset{|}{C}{=}O} \qquad \overset{\displaystyle CH_3}{CH_3-\overset{|}{C}{=}O} \qquad \overset{\displaystyle CH_3}{CH_3-CH_2-\overset{|}{C}{=}O}$$

methanal   ethanal   propanone   butanone (methyl
(formaldehyde)   (acetaldehyde)   (acetone)   ethyl ketone)

Problem Solving Examples:

 What is a ketone?

 The main functional group of a ketone is the carbonyl group

$$\left[\diagdown C{=}O \diagup \right].$$ In a ketone there are two alkyl or aryl groups attached

to the carbonyl carbon. They can either be the same or different groups. The traditional representation of a ketone is

$$R - \underset{\underset{O}{\|}}{C} - R'$$

where R and R' represent alkyl or aryl groups. In general ketones are named by replacing the final "e" of the hydrocarbon name with the suffix "one." For example,

$$H_3C - \underset{\underset{O}{\|}}{C} - CH_3$$

is called propanone (its trivial name is acetone) and

$$H_3C-\underset{\underset{O}{\|}}{C}-CH_2CH_2CH_3$$

is called 2-pentanone.

In case there are two or more functional groups present, the suffix takes the form of "dione," "trione," etc., and the "e" ending remains. For example,

$$H_3C-\underset{\underset{O}{\|}}{C}-CH_2-CH_2-\underset{\underset{O}{\|}}{C}-CH_3$$

is called 2,5 hexanedione.

What is an aldehyde?

The functional group that differentiates an aldehyde from other organic molecules is a carbonyl atom (C = O) joined to another carbon atom and to a hydrogen atom. In general, the formula for an aldehyde is

$$
\text{R-C}\!\!\begin{array}{c}\text{O}\\ \parallel\\ \diagdown\text{H}\end{array}
$$

where R symbolizes an alkyl or aryl group. Aldehydes are named by adding either "al" or "aldehyde" to the end of the name of the parent compound. Substituent groups attached to the alkyl group are taken into account by prefixes before the name of the parent compound.

An example of an aldehyde is acetaldehyde, which is composed of two carbons. It can be drawn as

$$
\text{CH}_3\text{-C}\!\!\begin{array}{c}\text{O}\\ \parallel\\ \diagdown\text{H}\end{array}
$$

acetaldehyde

2.2 Physical Properties

A) Formaldehyde and acetaldehyde are colorless liquids.

B) The boiling points of aldehydes are much lower than those of corresponding alcohols, and their solubility in water decreases with an increase in carbon content.

C) Aldehydes are easily oxidized.

D) Acetone is a colorless liquid and although less dense than water, it is completely miscible with it.

E) Ketones and aldehydes cannot form intermolecular hydrogen bonding, and thus they have lower boiling points than alcohols and carboxylic acids of comparable molecular weight.

With an increase in the molecular size of the carbonyl compound, the influence of a nonpolar alkyl group predominates, and solubility decreases.

2.3 Preparation of Aldehydes

By removal of water from a primary alcohol group linked to the desired radical through:

A) Oxidation

$$R - CH_2 - OH + \text{mild oxidation} \rightarrow R - CHO + H_2O$$

Example

$$3CH_3 - CH_2 - OH + K_2Cr_2O_7/4H_2SO_4, aq \rightarrow 3CH_2-CHO$$

$$\qquad \text{ethanol} \qquad\qquad\qquad \text{ethanal}$$

$$+ Cr_2(SO_4)_3 + K_2SO_4 + 7H_2O$$

B) Catalytic Dehydrogenation

$$R - CH_2 - OH + Cu, 300°C \rightarrow R - CHO + H_2$$

By heating calcium salts of fatty acids containing the desired radical with calcium formate:

$$(R - CO - O)_2Ca + (H - CO-O)_2Ca, \text{fuse} \rightarrow 2R - CHO + 2CaCO_3$$

Example $\quad (CH_3 - CO - O)_2 Ca +$

$$(H - CO - O)_2 Ca \rightarrow 2CH_3 - CHO + 2CaCO_3$$

$$\text{ethanal}$$

By passing the vapors of the acid mixed with formic acid vapors over manganous oxide at 300°C:

$$R - CO - OH + H - CO - OH/(MnO, 300°C) \rightarrow R - CHO + CO_2 + H_2O$$

By the hydrolysis of the corresponding dihaloalkanes:

$$R - CHCl_2 + H_2O/(PbO, boil) \rightarrow R - CHO + 2HCl$$

By reaction of Grignard reagent with ethyl formate (ethyl orthoformate) or HCN in ether, followed by hydrolysis:

a) $R-MgX + H-CO-O-CH_2-CH_3 \rightarrow R-CHO + CH_3-CH_2-O-MgX$

b) $R-MgX + HCN \xrightarrow{\text{ether}} R-C(=N-Mg-X)H$

$$R - C(= N - Mg - X)H + H_2O/2HX \rightarrow RCHO + MgX_2 + NH4X$$

By cleavage; glycols react with lead tetracetate in anhydrous benzene solution:

$$R-CH-CH-R' + Pb(CH_3COO)_4 \xrightarrow[\text{anhydrous}]{C_6H_6} R-CHO + R'-CHO$$
$$\underset{OH \quad OH}{|\quad |}$$
$$+ Pb(CH_3COO)_2$$
$$+ 2CH_3COOH$$

Problem Solving Examples:

Q Propose a reasonable mechanism for the synthesis of benzaldehyde diethyl acetal from benzene.

A Benzene can be halogenated in the presence of a catalytic amount of iron (Fe). The halobenzene can form a Grignard reagent, which can be carbonated to give a carboxylic acid. Reduction of a carboxylic acid can give an aldehyde; an acetal is an aldehyde derivative. The complete synthesis is broken down into various steps and is shown below:

1) Chlorobenzene synthesis:

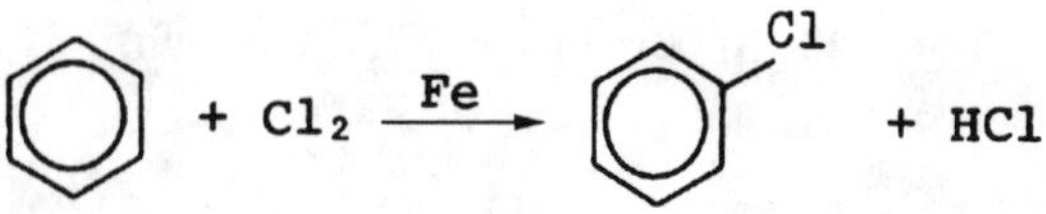

$$\bigcirc + Cl_2 \xrightarrow{Fe} \bigcirc^{Cl} + HCl$$

2) Grignard formation:

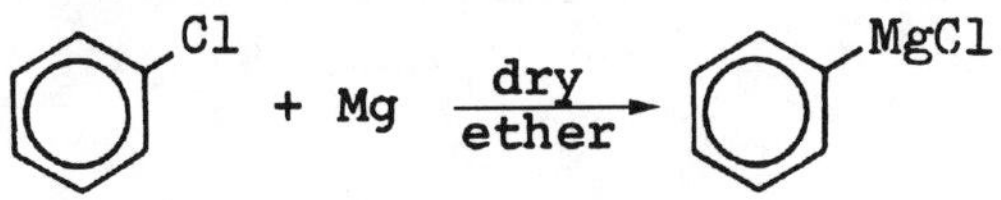

3) Carbonation:

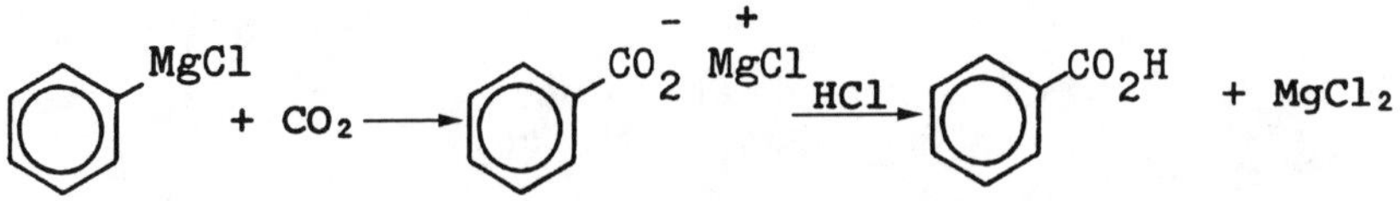

4) Rosenmund reduction of carboxylic acid:

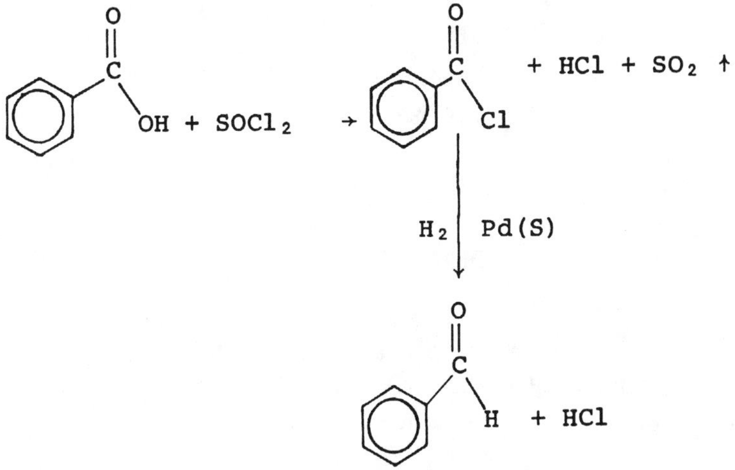

Benzaldehyde

5) Acetal synthesis:

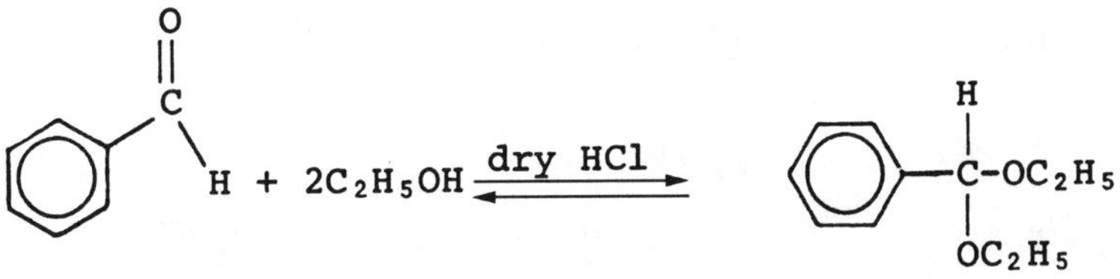

Benzaldehyde diethyl
acetal

Q One of the compounds that was initially formed in the primitive atmosphere was formaldehyde ($H_2C = O$). It is one of the precursor molecules that came to make up living organisms. It can be formed photolytically as shown in the following equation:

$$CH_4 + H_2O \xrightarrow{h\upsilon} H_2C = O + 4H \cdot$$

Propose a mechanism for this reaction.

A When ultraviolet light ($h\upsilon$) is the energy source for a reaction, free radicals are usually involved.

When methane (CH_4) is exposed to ultraviolet light, the following fragmentation will occur.

(i) $CH_4 \xrightarrow{h\nu} \cdot CH_3 + H$
methyl   hydrogen
radical   radical

Since there is water in this system, it will also be exposed to the ultraviolet rays, causing it to break up into radicals.

(ii) $2H_2O \xrightarrow{h\nu} 2HO \cdot + 2H \cdot$

In designing a mechanism, one attempts to build molecules that look increasingly like the desired products. The system now contains the HO•, H•, and CH_3• radicals. If HO• and CH_3• combine, a product resembling formaldehyde is made.

(iii) $\cdot CH_3 + HO \cdot \rightarrow HO - CH_3$
methanol

To produce formaldehyde two hydrogen atoms must be removed from this product; one from the oxygen and one from the carbon.

(iv) $HO - CH_3 \xrightarrow{h\upsilon} HO - \overset{\bullet}{C}H_2 + H \cdot$

Remember that ultraviolet light will cause the formation of radicals. In the next reaction the other hydrogen will be removed in a different way. An •OH radical will add to the product formed in reaction (iv) and a transition state will form which will lead to the production of one molecule of formaldehyde and one molecule of water. The brackets indicate the intermediate transition state.

(v)

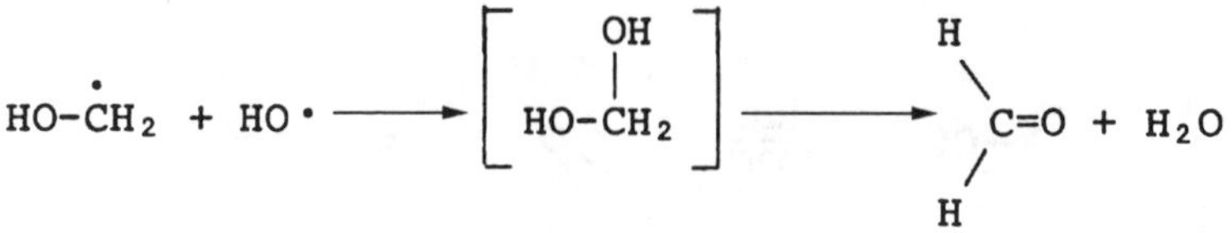

$$HO-\overset{\bullet}{C}H_2 + HO^{\bullet} \longrightarrow \left[\begin{array}{c} OH \\ | \\ HO-CH_2 \end{array} \right] \longrightarrow \overset{H}{\underset{H}{\diagdown}}C=O + H_2O$$

intermediate
transition state

If reactions (i) through (v) are added together, one regains the given overall reaction.

2.4   Preparation of Ketones

Removal of two hydrogen atoms from a secondary alcohol group linked to the desired radicals through:

A) Oxidation

$$RCH_2OH + \text{mild oxidation} \rightarrow R-\overset{\overset{\displaystyle O}{\|}}{C}-R + H_2O$$

Example

$$3(CH_3- CHOH- CH_3) + Cr_2O_7 + 8H^+ \rightarrow CH_3-\overset{\overset{\displaystyle O}{\|}}{C}-CH_3$$
$$+ 2CR^{++} + 7H_2O$$

B) Catalytic Dehydrogenation

$$RCH + (Cu, 300°C) \rightarrow RCOR + H_2$$

By reaction of Grignard reagent with esters (other than those of formic acid or acyl halides) or alkyl nitriles (or amides) in ether, followed by hydrolysis:

a)   R Mg X + R CO—O Et, ether → R—CO R + Et OMg X

b) $R \, Mg \, X + R \, CN \rightarrow R–C(= N – Mg – X)R$

$R–C \, (= N–Mg–X)R + H_2O/2HX \rightarrow R–CO–R + MgX_2 + NH_4X$

By the hydrolysis of the corresponding dihaloaldanes:

$$R_2CCl_2 + H_2O/(PbO, \, boil) \rightarrow R_2C = O + 2HCl$$

Ketones from carboxylic acids and their derivatives:

$$R–COOH \xrightarrow[300°C]{MnO} \underset{R}{\overset{R}{>}} C = O + CO_2 + H_2O$$

Nucleophilic substitution of the halides by the alkyl or aryl group of organocadmium compounds:

$$\underset{O}{2R–\overset{\|}{C}–X} + \underset{R'}{R'–\overset{|}{Cd}} \rightarrow \underset{O}{2R–\overset{\|}{C}–R'} + CdX_2$$

R = Alkyl, aryl

R' must be aryl or 1° alkyl

Friedel-Crafts acylation of aromatic compounds:

$$ArH + R–C\overset{\displaystyle O}{\underset{\displaystyle Cl}{<}} \xrightarrow[\text{or other Lewis acid}]{AlCl_3} \underset{O}{Ar–\overset{\|}{C}–R} + HCl$$

Cleavage of glycols having the hydroxyl group attached to tertiary carbon atoms:

$$\underset{OH \, OH}{\overset{R \quad R'}{R–\overset{|}{\underset{|}{C}}–\overset{|}{\underset{|}{C}}–R'}} + Pb(CH_3COO)_4 \xrightarrow[\text{anhydrous}]{C_6H_6} \underset{O}{R–\overset{R}{\underset{\|}{C}}} + \underset{R'}{\overset{R'}{C}}\overset{}{\underset{}{<}}$$

$$+ Pb(CH_3COO)_2 + 2CH_3COOH$$

Problem Solving Examples:

Q Outline all steps in a possible laboratory synthesis of each of the following from alcohols of four carbons or fewer, using any needed inorganic reagents:

(a) 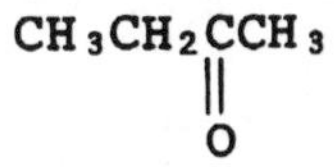 $CH_3CH_2CCH_3$
             $\overset{\|}{O}$

(b) CH_3-C-CH_3
             $\overset{\|}{O}$

Methyl ethyl ketone   **Acetone**

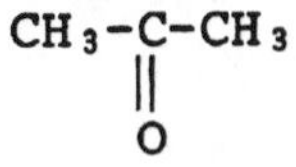

A Ketones can be prepared by oxidation of secondary alcohols with a variety of reagents, such as CrO_3 (chromium trioxide) or $K_2Cr_2O_7$ (potassium dichromate) with acid. The compound that is formed by oxidation of an alcohol depends upon the number of hydrogens attached to the carbon bearing the –OH group, that is, upon whether the alcohol is primary, secondary, or tertiary. For example, in the first problem, to synthesize the product, methyl ethyl ketone, a secondary alcohol is used that has the same number of carbon atoms as the product. In this case sec-butyl alcohol is used as the precursor.

(a) 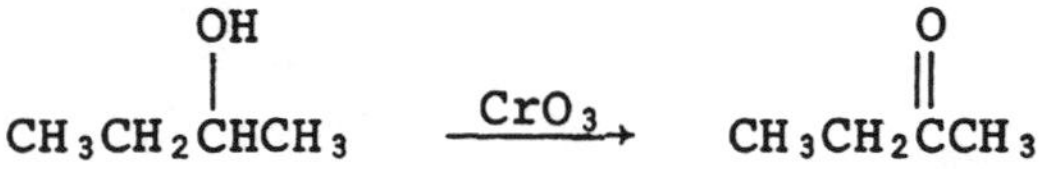 $CH_3CH_2\overset{\overset{\displaystyle OH}{|}}{C}HCH_3$ $\xrightarrow{\ CrO_3\ }$ $CH_3CH_2\overset{\overset{\displaystyle O}{\|}}{C}CH_3$

        sec-butyl alcohol   methyl ethyl ketone

To obtain the product only requires one step where sec-butyl alcohol is oxidized directly to methyl ethyl ketone with chromium trioxide.

An example of a longer synthesis in preparation of a ketone is producing acetone from a two-carbon alcohol. The only two carbon alcohol is ethanol, so this will be the precursor.

(b)

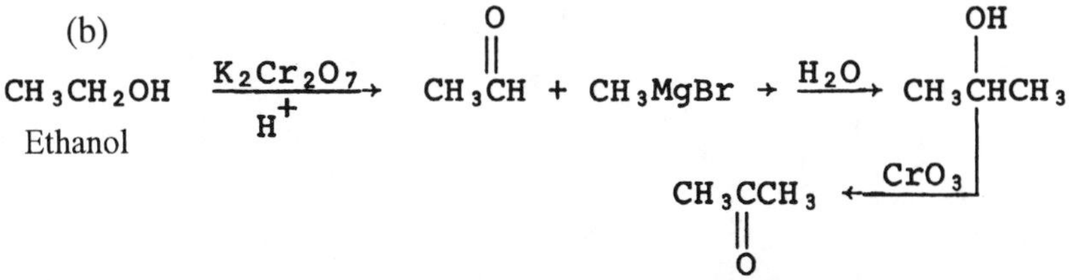

CH_3CH_2OH $\xrightarrow[H^+]{K_2Cr_2O_7}$ $CH_3\overset{\overset{\displaystyle O}{\|}}{C}H$ $+$ CH_3MgBr $\xrightarrow{H_2O}$ $CH_3\overset{\overset{\displaystyle OH}{|}}{C}HCH_3$

Ethanol

CH_3CCH_3 $\xleftarrow{\ CrO_3\ }$
    $\overset{\|}{O}$

Acetone

This synthetic reaction involves oxidation of the alcohol to yield acetaldehyde. To add another carbon atom to this first product, the Grignard reagent methyl magnesium bromide (CH_3MgBr) is added for methylation. Upon hydrolysis this yields a three-carbon compound, isopropanol, which is a secondary alcohol. As shown previously, secondary alcohols are oxidized to ketones with chromium trioxide, thus yielding the final product, acetone.

 How could the following ketones be prepared?

(a) $\overset{O}{\overset{\|}{C}} CH_2 CH_3$ on benzene ring

(b) benzene-C(=O)-benzene

(c) benzene-$\overset{O}{\overset{\|}{C}}$-$CH_2$-benzene

 All of these compounds can be prepared using a Friedel-Crafts acylation.

(a) benzene + $CH_3 CH_2 \overset{O}{\overset{\|}{C}} Cl$ $\xrightarrow{AlCl_3}$ benzene-$\overset{O}{\overset{\|}{C}} CH_2 CH_3$

(b) benzene + benzene-$\overset{O}{\overset{\|}{C}}$-Cl $\xrightarrow{AlCl_3}$ benzene-$\overset{O}{\overset{\|}{C}}$-benzene

(c)

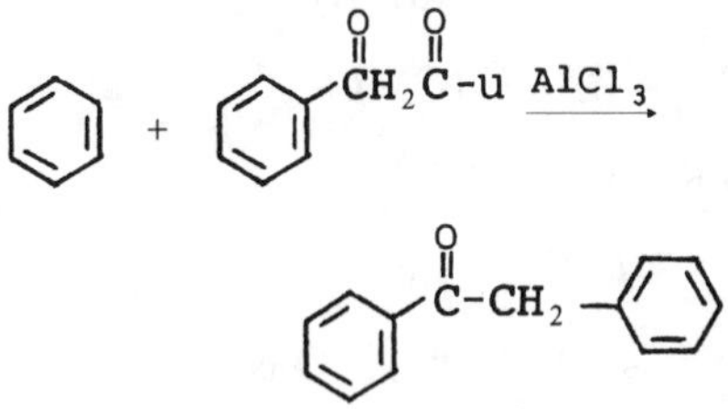

2.5 Reactions of Aldehydes and Ketones

By oxidation when treated with chromic acid or other appropriate oxidizing agents:

$$R - CHO + oxidation \rightarrow R - CO–OH \rightarrow \text{oxidized derivatives} \rightarrow \text{cleavage}$$

$$R_2C = O + oxidation \rightarrow \text{oxidized derivatives} \rightarrow \text{cleavage}$$

Example

$$3CH_3-CHO + Cr_2O_7 \xrightarrow{+8H^+} 3CH_3-CO-OH + 2Cr^{+++} + 4H_2O$$

$$CH_3-CO-CH_2-CH_2-CH_3 + Cr_2O_7 \xrightarrow{+8H^+} CH_3-CO-OH$$

$$+ CH_3CH_2CH_2\overset{O}{\overset{\|}{C}}OH$$

$$+ 2Cr^{+++} + 4H_2O$$

By reduction when treated with appropriate reducing agents (Zn/ H⁺, Na/ROH, NaHg/H₂O):

$$R - CHO + 2Na/2EtOH \rightarrow RCH_2 - OH + 2Et - O - Na$$

1° alcohol

$$R_2C = O + 2Na/2Et - OH \rightarrow R_2CH - OH + 2Et - O - Na$$

2° alcohol

By addition when treated with a Grignard reagent, hydrogen cyanide, sodium hydrogen sulfite, or ammonia:

A) $RCHO + R' MgX \rightarrow R' RCH(OMgX)$

$R_2C = O + R' MgX \rightarrow R_2R' C (OMgX)$

B) $RCHO + HCN \rightarrow RCH(OH) (CN)$

$R_2C = O + HCN \rightarrow R_2C(OH) (CN)$

C) $RCHO + HOSO_2Na \rightarrow RCH(OH) (SO_3Na)$

$R_2C = O + HOSO_2Na \rightarrow R_2C(OH) (SO_3Na)$

D) $RCHO + HNH_2 \rightarrow RCH(OH) (NH_2)$

$R_2C = O + HNH_2 \rightarrow$ complex derivatives

Substitution reactions:

A) $RCH\boxed{O + 2H}OR' \rightarrow R-CH-(OR')_2 + H_2O$

$R_2C = O + 2HOR' \xrightarrow{\;/\!/\;}$

B) $RCH\boxed{O + PCl_3}Cl_2 \rightarrow RCH \cdot Cl_2 + P \cdot O \cdot CL_3$

$R_2C = \boxed{O + PCl_3}Cl_2 \rightarrow R_2C \cdot Cl_2 + P \cdot O \cdot Cl_3$

C) $R \cdot CO\boxed{H + X}X \rightarrow R \cdot CO \cdot X + H \cdot X$

$R \cdot CO \cdot HRC\boxed{H + X}X \rightarrow R \cdot CO \cdot HRC \cdot X + H \cdot X$

Condensation reaction:

$R - CHO + R' - CH_2 - CHO(Ca(OH)_2, aq.) \rightarrow R - CHOH - CHR' - CHO$

$R-CHOH-CHR'-CHO \xrightarrow[-H_2O]{} RHC=CR'-CHO$

Polymerization of aldehydes:

$x(H-CHO)aq \xrightarrow{\text{evaporate}} (H-CHO)_x$, paraformaldehyde

Resin formation by the presence of condensing agent:

$H - CHO + C_6H_5OH(phenol) \rightarrow$ synthetic resin "bakelite"

$R-CHO + NaOH, aq., conc. \xrightarrow{\text{heat}}$ resin formation

Reaction of aldehydes and ketones with bases:

$$\overset{+}{\underset{}{\underset{\delta}{C}}} = \overset{-}{\underset{\cdot\cdot}{\underset{\delta}{O}}} \ : \quad + \quad : \overset{\cdot\cdot}{O} \diagdown^H_H \quad \underset{\longleftarrow}{\overset{H^+}{\rightleftharpoons}} \quad \diagup^{OH}_{OH} C \diagdown$$

Reaction with derivatives of ammonia:

$$\diagup C = O + (H_2N) \rightarrow \diagup C = N- + H_2O$$

carbonyl   azomethene
group   group

Cannizzaro reaction. In the presence of concentrated alkali, aldehydes containing no α-hydrogens undergo self-oxidation and reduction to yield an alcohol and a salt of a carboxylic acid.

$$2R - \overset{\overset{\displaystyle H}{|}}{C}=O \xrightarrow{\text{strong base}} + R-CH_2OH + RCOO^-$$

aldehyde
with no
α-hydrogens

(R contains no α - hydrogens)

Example $2HCHO \xrightarrow{\text{50\% NaOH}} CH_3OH + HCOO^- Na^+$

formaldehyde   methanol   sodium formate

Aldol condensation. In the presence of a dilute base or acid, two molecules of a ketone, or an aldehyde, containing α hydrogens combine to form an aldol (β-hydroxy ketone) or (β-hydroxy aldehyde).

$$\diagup C = O + \ -\overset{|}{\underset{|}{C}}-\overset{|}{\underset{H}{C}} = O \xrightarrow{\text{base or acid}} \overset{|}{\underset{OH}{C}}-\overset{|}{\underset{|}{C}}-\overset{|}{\underset{|}{C}} = O$$

an aldol

Example $2 CH_3 HO \xrightarrow{OH} CH_3 CHCH_2 CHO$

acetaldehyde
2 moles

$\underset{OH}{|}$

acetaldol

Cyanohydrin formation by the nucleophilic addition of a cyanide anion to the carbonyl group:

$$\underset{O:}{\overset{\overset{\delta+}{C}}{\|}} \,\,\overset{\delta-}{} + H^+ : \overset{-}{C} \equiv N: \,\rightarrow\, -\overset{|}{\underset{:O:_{-}}{C}}-C \equiv N + H^+ \rightarrow -\overset{|}{\underset{OH}{C}}-C \equiv N$$

Reduction to hydrocarbons:

$\xrightarrow{\text{Zn(Hg),conc. HCl}}$ $-\overset{|}{\underset{H}{C}}-H$ Clemmensen reduction for compounds sensitive to base

$\xrightarrow{\text{NH}_2\text{NH}_2,\ \text{base}}$ $-\overset{|}{\underset{H}{C}}-H$ Wolff-Kishner reduction for compounds sensitive to acid

Problem Solving Examples:

Q Write equations, naming all organic products, for the reaction (if any) of phenylacetaldehyde with:

(a) Tollens' reagent

(b) CrO_3/H_2SO_4

(c) cold dilute $KMnO_4$

(d) $KMnO_4$, H^+, heat

(e) $NaBH_4$

(f) C_6H_5MgBr, then H_2O

(g) $NaHSO_3$

(h) CN^-, H^+

(i) 2,4-dinitrophenylhydrazine

Phenylacetaldehyde $\langle\bigcirc\rangle$—$CH_2\overset{\displaystyle O}{\overset{\|}{C}}$-H undergoes those reactions that are typical of aldehydes.

(a) Tollens' reagent contains the silver ammonia ion, $Ag(NH_3)_2$. Its utility stems from the detection of aldehydes, in particular in differentiating them from ketones. This reagent oxidizes the aldehyde to a carboxylic acid; it is accompanied by reduction of silver ion to free silver (in the form of a mirror under the proper conditions). Hence,

$$\langle\bigcirc\rangle-CH_2\overset{\displaystyle O}{\overset{\|}{C}}H \ + \ Ag(NH_3)_2{}^+ \ \rightarrow \ \langle\bigcirc\rangle-CH_2COO^- \ + \ Ag$$

(Silver Mirror)

(b) Aldehydes can be easily oxidized to carboxylic acids by chromic acid. Therefore,

$$\langle\bigcirc\rangle-CH_2\overset{\displaystyle O}{\overset{\|}{C}}H \ \xrightarrow[H_2SO_4]{CrO_3} \ \langle\bigcirc\rangle-CH_2COOH$$

(c) Under the conditions of cold dilute potassium permanganate ($KMnO_4$), the aldehyde is oxidized to the carboxylic acid with the carbon skeleton number unchanged. Hence,

$$\langle\bigcirc\rangle-CH_2\overset{\displaystyle O}{\overset{\|}{C}}H \ \xrightarrow[\text{cold dilute}]{KMnO_4} \ \langle\bigcirc\rangle-CH_2COOH$$

(d) Prolonged hot treatment of a side chain on an aromatic ring with $KMnO_4$ results in its oxidation down to the ring, only a carbonyl group (–COOH) remaining to indicate the position of the original side chain.

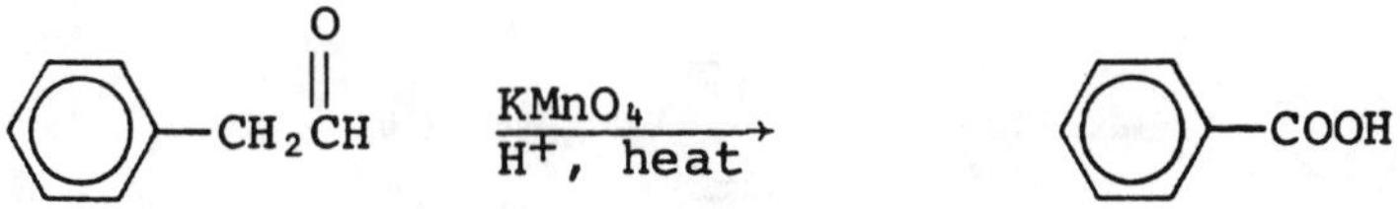

(e) Sodium borohydride, $NaBH_4$, can be used to reduce carbonyl compounds to alcohols. This reagent has the special advantage of not reducing carbon-carbon double bonds, not even those conjugated with carbonyl groups. Consequently, it is useful for the reduction of such unsaturated carbonyl compounds to unsaturated alcohols. Therefore,

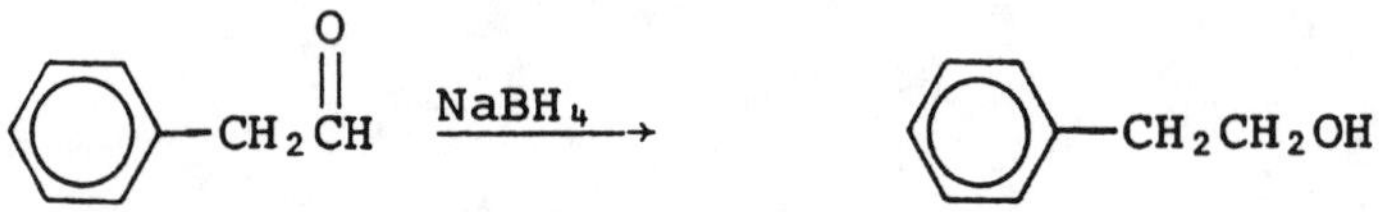

(f) The addition of Grignard reagents to aldehydes is an important route to secondary alcohols (primary alcohols in the case of formaldehyde). Thus,

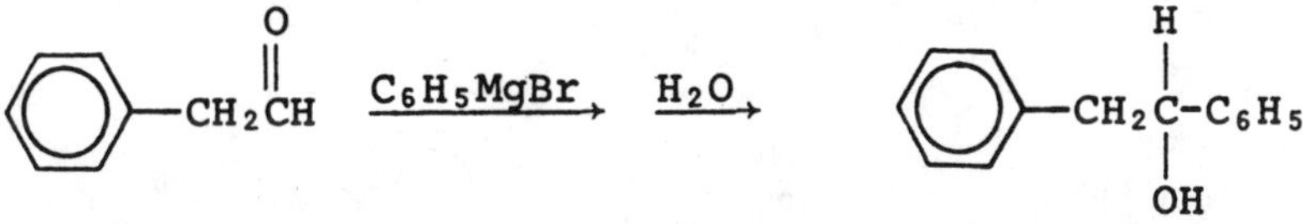

(g) Sodium bisulfite adds to aldehydes (most of them) to form bisulfite addition products. This utility is in separating a carbonyl compound from non-carbonyl compounds. The reaction is illustrated as follows:

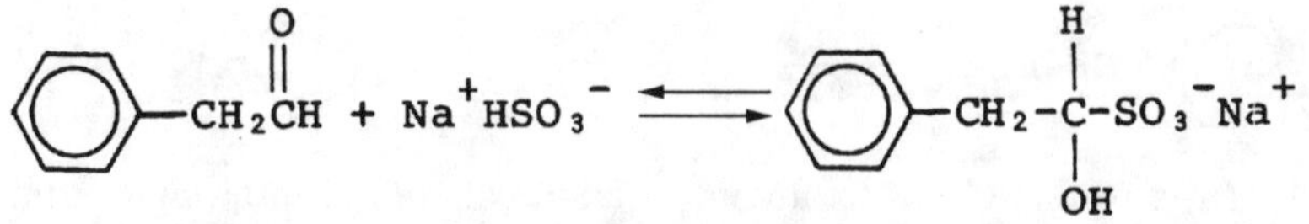

(h) Hydrogen cyanide (HCN) can be added to the carbonyl group of aldehydes and ketones to yield compounds known as cyanohydrins. The reaction is often carried out by adding mineral acid to a mixture of the carbonyl compound and aqueous sodium cyanide. Thus,

$$\text{C}_6\text{H}_5\text{—CH}_2\overset{\overset{\displaystyle O}{\|}}{\text{CH}} + \text{CN}^- \xrightarrow{\text{H}^+} \text{C}_6\text{H}_5\text{—CH}_2\overset{\overset{\displaystyle H}{|}}{\underset{\underset{\displaystyle OH}{|}}{\text{C}}}\text{—CN}$$

(i) Certain compounds that are derivatives of ammonia such as 2,4 dinitrophenylhydrazine add to the carbonyl group to form derivatives useful for the characterization and identification of aldehydes and ketones. The products possess a carbon-nitrogen double bond that was produced from elimination of a molecule of water from the initial addition products. Hence,

$$\text{C}_6\text{H}_5\text{—CH}_2\overset{\overset{\displaystyle O}{\|}}{\text{CH}} + :\text{NH}_2\text{NH—C}_6\text{H}_3(\text{NO}_2)\text{—NO}_2 \xrightarrow{\text{H}^+}$$

(2,4 dinitrophenylhydrazine)

$$\left[\text{C}_6\text{H}_5\text{—CH}_2\overset{\overset{\displaystyle OH}{|}}{\underset{\underset{\displaystyle H}{|}}{\text{C}}}\text{—NH}_2\text{NH—C}_6\text{H}_3(\text{NO}_2)\text{—NO}_2 \right] \xrightarrow{-\text{H}_2\text{O}}$$

$$\text{C}_6\text{H}_5\text{—CH}_2\overset{\overset{\displaystyle }{\underset{\underset{\displaystyle H}{|}}{\text{C}}}}{=}\text{NNH—C}_6\text{H}_3(\text{NO}_2)\text{—NO}_2$$

 Predict the product of each of the following ketone precursors after base-promoted or acid-catalyzed halogenation.

(a)

Cyclohexanone

(b) CH_3CCH_3
           ‖
           O

Acetone

(c) $CH_3\overset{\overset{\displaystyle CH_3}{|}}{C}\!\!-\!\!\overset{\overset{\displaystyle}{}}{\underset{\underset{\displaystyle}{}}{C}}\text{-}CH_3$
           | ‖
          CH_3 O

3,3-Dimenthyl-2-butanone

A   There are two ways to halogenate ketones; one is called base-promoted halogenation and the other is acid-catalyzed halogenation. In essence, acids and bases speed up the halogenation of ketones; thus, they are considered to be catalysts in these reactions. The general synthesis for both proceeds as:

$$-\overset{|}{\underset{\underset{H}{|}}{C}}\text{-}\overset{\overset{}{}}{\underset{\underset{O}{\|}}{C}}\text{-} \; + \; X_2 \quad \xrightarrow{\;H^+ \; or \; OH^-\;} \quad -\overset{|}{\underset{\underset{X}{|}}{C}}\text{-}\overset{\overset{}{}}{\underset{\underset{O}{\|}}{C}}\text{-} \; + \; HX \quad X_2 = Cl_2, \; Br_2, \; I_2$$

The kinetics of acid-catalyzed halogenation show the rate of halogenation to be independent of halogen concentration, but dependent upon ketone concentration and acid concentration.

In both types of halogenation the products are the same, except in acid-catalyzed halogenation racemization can occur; i.e., isomers of the product are formed.

To solve problems (a), (b), and, (c) the above concepts and the general equation are followed.

(a)

$+ \; Br_2 \quad \xrightarrow{\;H^+\;}$
$+ \; HBr$

Cyclohexanone   2-Bromocyclohexanone

(b)

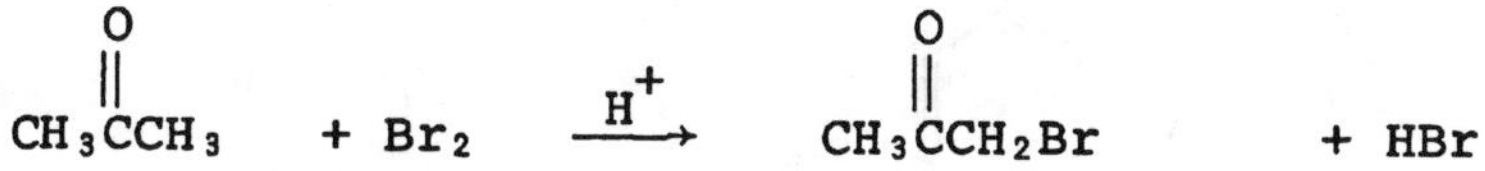

$$\underset{\text{Acetone}}{CH_3\overset{\overset{\displaystyle O}{\|}}{C}CH_3} \quad + \; Br_2 \quad \xrightarrow{\;H^+\;} \quad \underset{\text{Bromoacetone}}{CH_3\overset{\overset{\displaystyle O}{\|}}{C}CH_2Br} \qquad + \; HBr$$

(c)

$$\underset{\overset{\displaystyle |}{CH_3}}{\overset{\overset{\displaystyle CH_3}{|}}{CH_3C}} - COCH_3 \; + \; I_2 \quad \xrightarrow{\;OH^-\;} \quad \left[\; \underset{\overset{\displaystyle |}{CH_3}\; \overset{\displaystyle \|}{O}}{\overset{\overset{\displaystyle CH_3}{|}}{CH_3C}} - C\text{-}CI_3 \; \right] \longrightarrow$$

$$CHI_3 \; + \; \underset{\overset{\displaystyle |}{CH_3}}{\overset{\overset{\displaystyle CH_3}{|}}{CH_3C}} - COO^- \qquad \xleftarrow{\;OH^-\;}$$

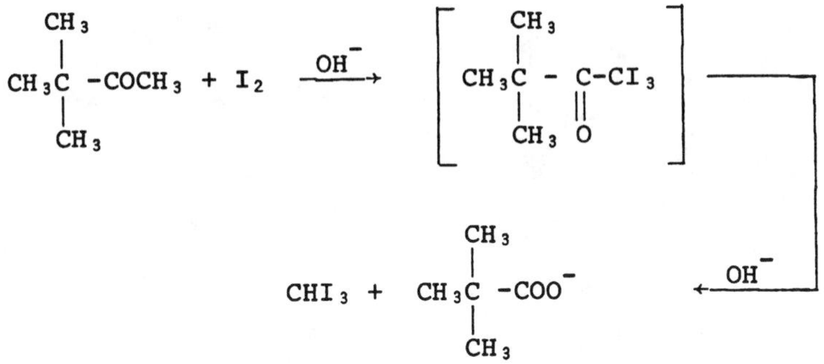

Iodoform   Trimethylacetate ion

The first product in reaction (c) is an unstable compound, so more base is added to take the reaction to equilibrium, thus forming the above product and by-product, trimethylacetate ion and iodoform, respectively.

Quiz: Arenes, Aldehydes, and Ketones

1.   Predict the product of the following reaction.

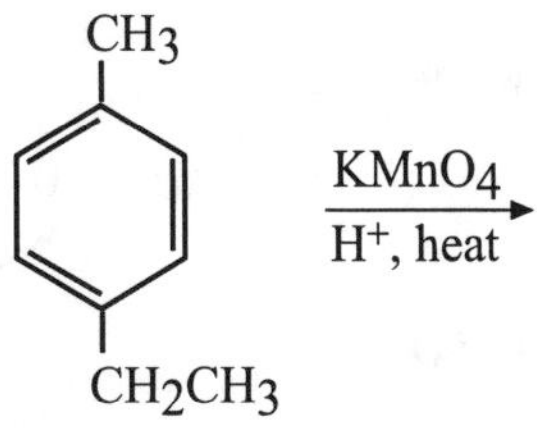

$$\xrightarrow[\text{H}^+,\ \text{heat}]{\text{KMnO}_4}$$

(A)

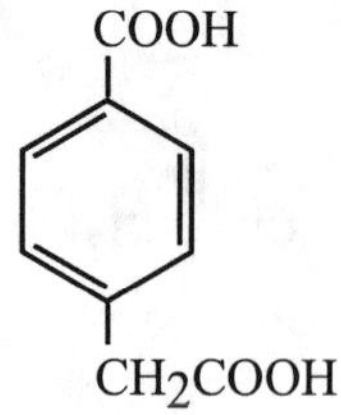

COOH

CH₂COOH

(D)

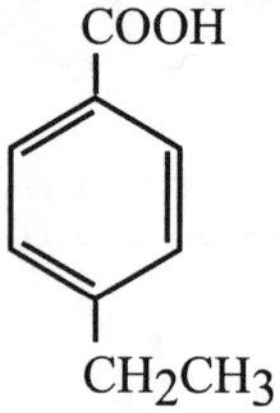

COOH

CH₂CH₃

(B)

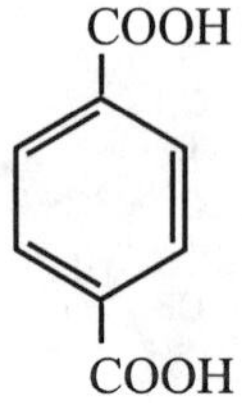

COOH

COOH

(E)

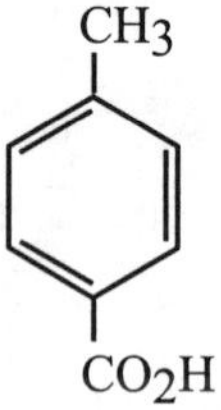

CH₃

CO₂H

(C)

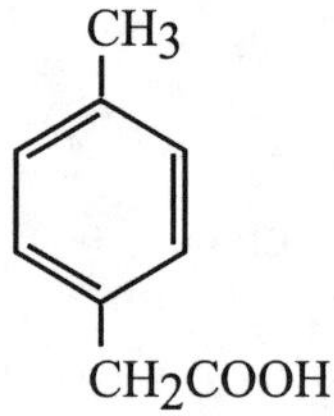

CH₃

CH₂COOH

2. What is the expected product of the following reaction?

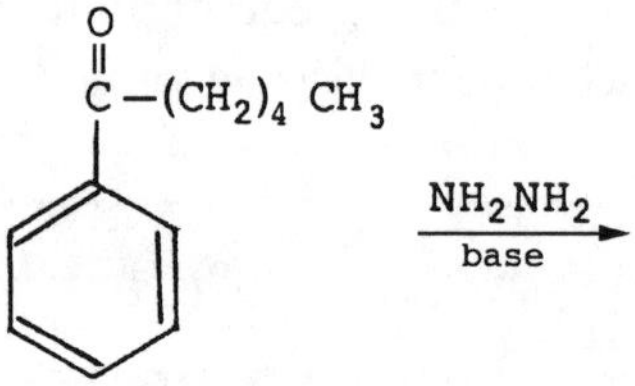

$$\overset{O}{\underset{\|}{C}}-(CH_2)_4\,CH_3$$

$$\xrightarrow[\text{base}]{NH_2NH_2}$$

(A)

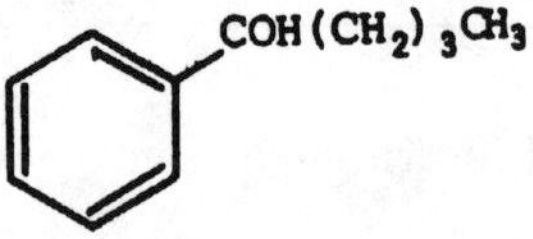

COH(CH₂)₃CH₃

(D)

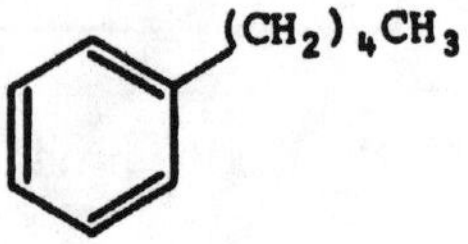

(CH₂)₄CH₃

(B)

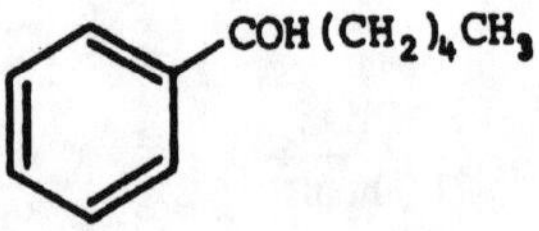

COH(CH₂)₄CH₃

(E)

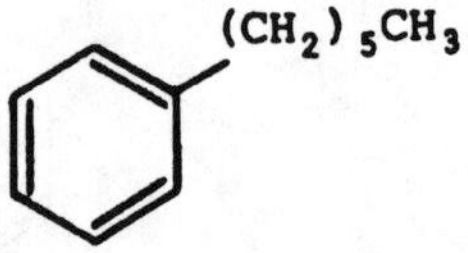

(CH₂)₅CH₃

(C)

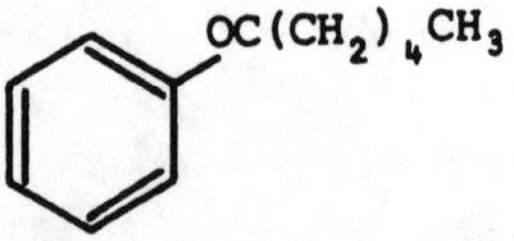

OC(CH₂)₄CH₃

3. What is the major product for the reaction below?

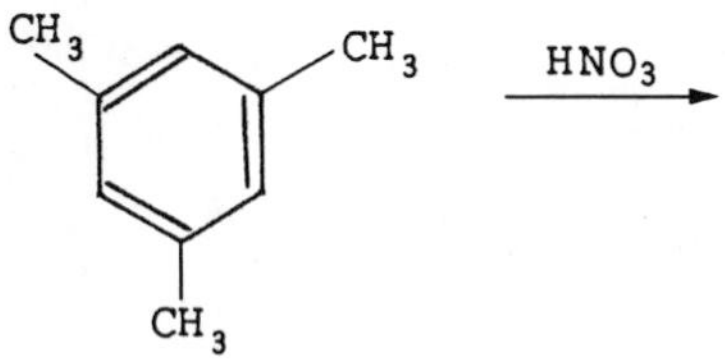

$\xrightarrow{HNO_3}$

(A)

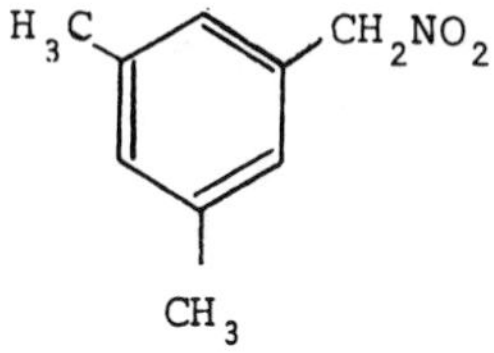

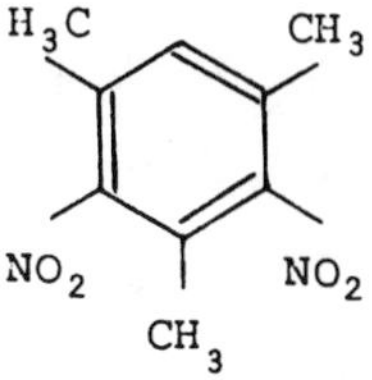

(B)

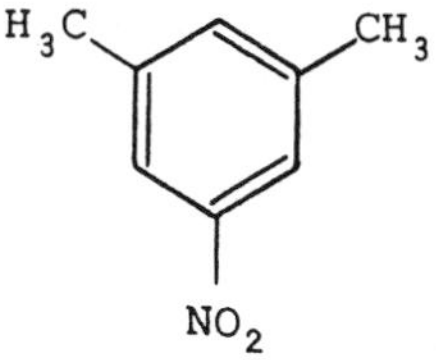

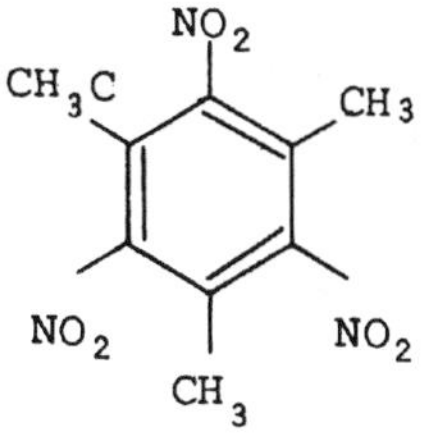

(C)

(D)

(E)

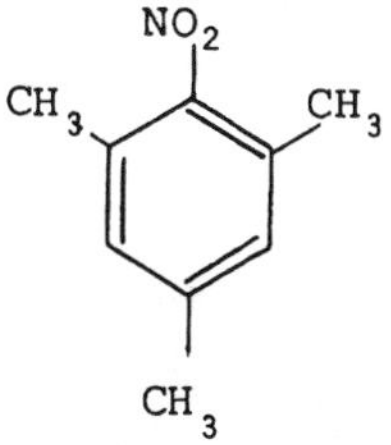

4. Which of the following compounds is expected to exhibit the highest boiling point?

(A) $CH_3CH_2CH_2OH$ (D) CH_3COOCH_3

(B) $CH_3OCH_2CH_3$ (E) $CH_3CH_2CH_3$

(C) CH_3CH_2CHO

5. What is the principal product of the following reaction?

   =O + 2CH$_3$OH $\underset{\Delta}{\overset{H^+}{\rightleftharpoons}}$

   (A) 5,5-dimethyl pentanal

   (B) Cyclopentanone dimethyl ketal

   (C) 3-methyl-1,5-pentanedione

   (D) 2-methyl cyclopentanone

   (E) Cyclopentane dimethyl acetal

6. Which of the following reactions represents an aldol condensation?

   (A)

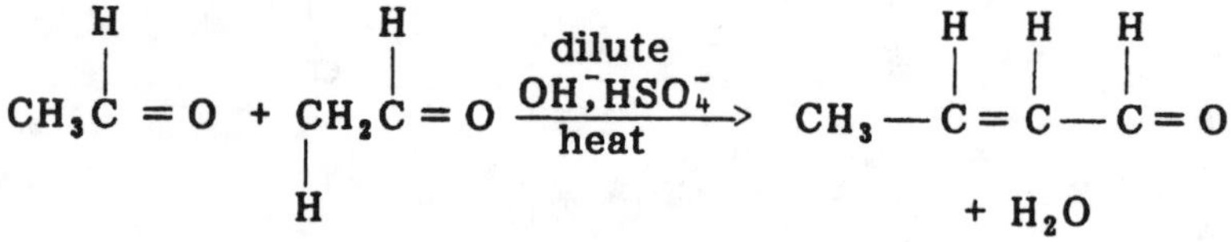

   (B)

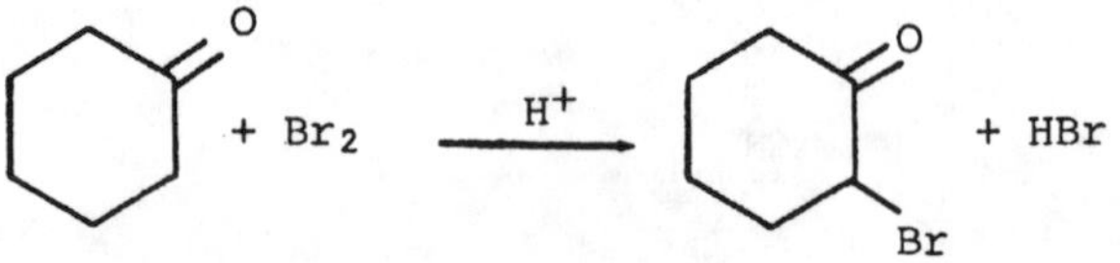

(C)

$$HOCH_2 - \overset{\overset{\displaystyle H}{|}}{N} - \overset{\overset{\displaystyle }{|}}{\underset{\underset{\displaystyle O}{\|}}{C}} - \overset{\overset{\displaystyle H}{|}}{N} - CH_2OH \xrightarrow[\text{urea}]{\text{HCHO}}$$

$$\sim\!\!\sim\!\!\sim N - CH_2 - \overset{|}{N} - \overset{\overset{\displaystyle }{|}}{\underset{\underset{\displaystyle O}{\|}}{C}} - N\!\!\sim\!\!\sim\!\!\sim$$

$$O = \overset{|}{C} \qquad\qquad \overset{|}{C}H_2$$

$$\sim\!\!\sim\!\!\sim N - CH_2 - \overset{|}{N} - \overset{\overset{\displaystyle }{|}}{\underset{\underset{\displaystyle O}{\|}}{C}} - \overset{|}{N}\!\!\sim\!\!\sim\!\!\sim$$

(D)

$$C_6H_5COOC_2H_5 + H_2O \xrightarrow{H_2SO_4} C_6H_5COOH + C_2H_5OH$$

(E)

$$(CH_3CO)_2O + \text{(trimethylbenzene)} \xrightarrow{Al\ Cl_3} CH_3-\underset{\underset{\displaystyle O}{\|}}{C}- \text{(trimethylphenyl)} + CH_3COOH$$

7. What is the product if the following compound undergoes acid-catalyzed halogenation? (X_2 is a diatomic halogen.)

$$CH_3\overset{\overset{\displaystyle CH_3}{|}}{\underset{\underset{\displaystyle CH_3}{|}}{C}} - \overset{\overset{\displaystyle O}{\|}}{C} - CH_3 + X_2 \xrightarrow{H^+} ?$$

(A) $$CH_3 - \overset{\overset{\displaystyle CH_3}{|}}{\underset{\underset{\displaystyle CH_2X}{|}}{C}} - \overset{\overset{\displaystyle O}{\|}}{C} - CH_2X$$

(B) 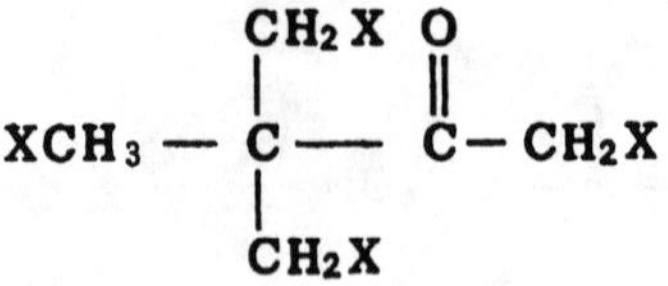 $XCH_3 - \overset{\overset{\displaystyle CH_2X}{|}}{\underset{\underset{\displaystyle CH_2X}{|}}{C}} - \overset{\overset{\displaystyle O}{\|}}{C} - CH_2X$

(C) $CH_3 - \overset{\overset{\displaystyle CH_3}{|}}{\underset{\underset{\displaystyle CH_3}{|}}{C}} - \overset{\overset{\displaystyle O}{\|}}{C} - CH_2X$

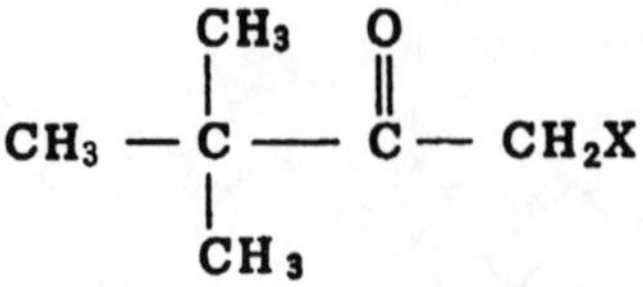

(D) 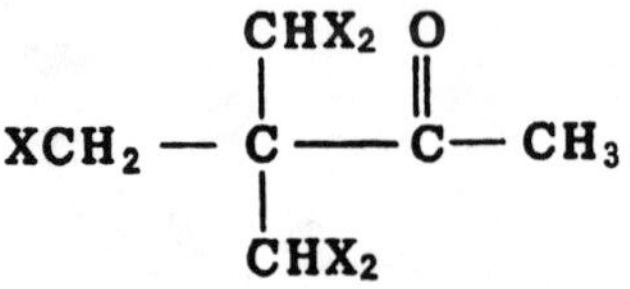 $XCH_2 - \overset{\overset{\displaystyle CHX_2}{|}}{\underset{\underset{\displaystyle CHX_2}{|}}{C}} - \overset{\overset{\displaystyle O}{\|}}{C} - CH_3$

(E) None of the above

8. Which of the following is named p-xylene?

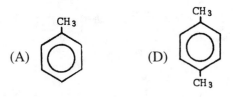

(A) [benzene ring with CH₃]

(D) [benzene ring with CH₃ top and CH₃ bottom]

(B) 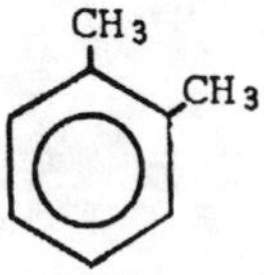 [benzene ring with CH₃ and CH₃ adjacent]

(E)  [benzene ring]

(C) 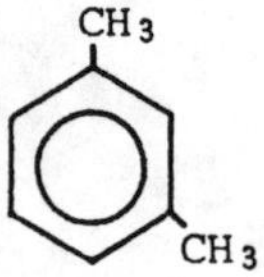 [benzene ring with CH₃ top and CH₃ meta]

9. Which of the following is least soluble in water?

(A) $H-\underset{H}{\overset{H}{C}}=O$

(D) $CH_3-\underset{CH_3}{\overset{H}{C}}=O$

(B) $CH_3\overset{H}{C}=O$

(E) $CH_3-\underset{CH_2CH_3}{\overset{H}{C}}=O$

(C) $CH_3CH_2-\overset{H}{C}=O$

10. The reaction of an alkenylbenzene below is an example of

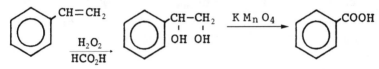

(A) catalytic hydrogenation.

(B) electrophilic addition.

(C) oxidation.

(D) ring halogenation.

(E) electrophilic aromatic substitution.

ANSWER KEY

1.	(B)	6.	(A)
2.	(E)	7.	(C)
3.	(C)	8.	(D)
4.	(A)	9.	(E)
5.	(B)	10.	(C)

CHAPTER 3

Amines

Amines are derivatives of hydrocarbons in which a hydrogen atom has been replaced by an amino group; derivatives of ammonia in which one or more hydrogen atoms have been replaced by alkyl groups are also known as amines. They are classified, according to structure, as:

$$\text{Primary} - \underset{\underset{\displaystyle H}{|}}{R-N-H}, \quad \text{Secondary} - \underset{\overset{\displaystyle H}{|}}{R-N-R},$$

$$\text{Tertiary} - \underset{\overset{\displaystyle R}{|}}{R-N-R}$$

3.1   Nomenclature (IUPAC System)

The aliphatic amine is named by listing the alkyl groups attached to the nitrogen, in alphabetical order, following them with the suffix "-amine."

CH₃—NH₂   CH₃—C̈—CH₃   [benzene ring]—CH₂—N̈—CH₂CH₃

Methyl-   tert-Butyl-   Benzyl ethylamine
amine   amine

Where the structures show:
- CH₃—NH₂, labeled Methylamine
- $CH_3-\overset{\underset{|}{NH_2}}{\underset{|}{C}(CH_3)}-CH_3$ with CH₃ on top, labeled tert-Butylamine
- Benzyl group attached to $CH_2-\overset{H}{\underset{|}{N}}-CH_2CH_3$, labeled Benzyl ethylamine

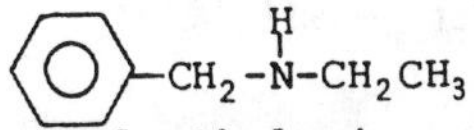

If an alkyl group occurs twice or three times on the nitrogen, the prefixes "di-" and "tri-" are used, respectively.

Example

$CH_3-NH-CH_3$ $CH_3-\overset{\underset{|}{CH_3}}{\underset{|}{N}}-CH_3$

dimethylamine   trimethylamine

If an amino group is part of a complicated molecule, it may be named by prefixing "amino" to the name of the parent chain.

Example

$NH_2CH_2CH_2OH$ $CH_3\overset{\underset{|}{NH_2}}{\underset{|}{CH}}\ CH_2\,COOH$

2-amino ethanol   3-aminobutanoic acid

An amino substituent that carries an alkyl group is named as an N-alkyl amino group.

Example

CH_3NH-CH_2COOH $CH_3-NH\overset{\underset{|}{CH_3}}{\underset{|}{CH}}(CH_2)_4CH_3$

N-methyl amino   2-(N-methylamino) heptane
acetic acid

Problem Solving Examples:

Name the following structures:

(a) $CH_3 - NH - C_2H_5$

(f) $NH_2 - CH_2 - COOH$

(b) $C_6H_5 - NH - C_2H_5$

(g) $(C_6H_5)_2NCH_3$

(c) $NH_2 - (CH_2)_3 - NH_2$

(h) $C_2H_5 - NH - CH_2C_6H_5$

(d) $(CH_3 - CH_2 - CH_2)_3N$

(i) $CH_3 - \underset{\underset{NH_2}{|}}{CH} - CH_2 - \underset{\underset{Cl}{|}}{CH} - COOH$

(e) $(CH_3)_2CH - NH_2$

(j) 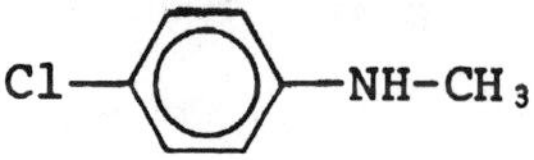 $Cl - \langle\bigcirc\rangle - NH - CH_3$

A Aliphatic amines are named by listing the alkyl groups attached to the nitrogen in alphabetical order, following them with the suffix "-amine." The entire sequence is written as one word as in the following examples:

$CH_3 - NH_2$

Methylamine

$CH_3CH_2 - NH - CH_3$

Ethylmethylamine

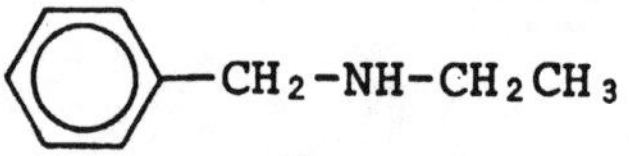

$\langle\bigcirc\rangle - CH_2 - NH - CH_2CH_3$

Benzylethylamine

$CH_3 - \underset{\underset{\underset{CH_3 \quad CH_3}{/ \quad \backslash}}{CH}}{N} - CH_2CH_3$

Ethylisopropylmethylamine

If there are two or three identical substituents on the nitrogen, the prefixes "di-" and "tri-" are used (and they are alphabetized).

$CH_3 - NH - CH_3$

Dimethylamine

$CH_3 - \underset{\underset{CH_2CH_3}{|}}{N} - CH_3$

Dimethylethylamine

If the amino group is part of a complicated molecule that contains other functional groups, it may be named as the amino substituent, as in the following examples:

$$NH_2-CH_2CH_2-OH$$

2-aminoethanol

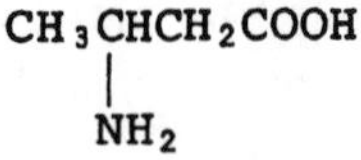

$$CH_3CHCH_2COOH$$
$$\quad\ \ |$$
$$\quad\ \ NH_2$$

β-aminobutyric acid
(or 3-aminobutanoic acid)

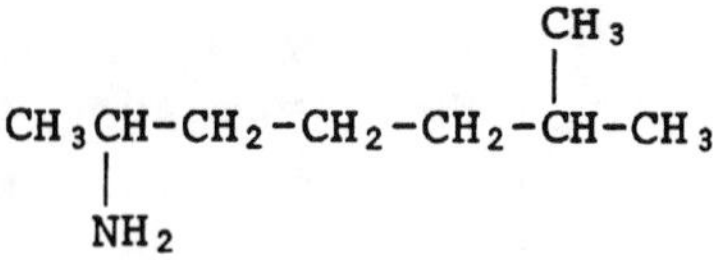

$$\qquad\qquad\qquad\ CH_3$$
$$\qquad\qquad\qquad\ |$$
$$CH_3CH-CH_2-CH_2-CH_2-CH-CH_3$$
$$\quad\ \ |$$
$$\quad\ \ NH_2$$

2-amino-6-methylheptane

An amino substituent that carries an alkyl group is named as an N-alkylamino group. For example,

$$CH_3-NH-CH_2-COOH$$

N-methylaminoacetic acid

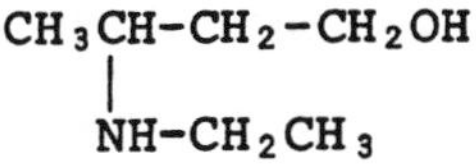

$$CH_3CH-CH_2-CH_2OH$$
$$\quad\ \ |$$
$$\quad\ \ NH-CH_2CH_3$$

3-(N-ethylamino)-1-butanol

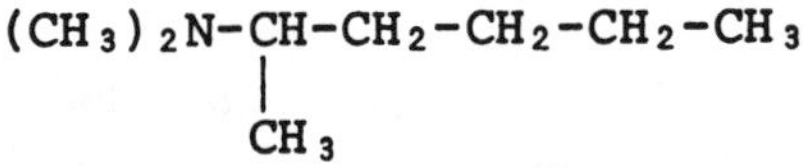

$$(CH_3)_2N-CH-CH_2-CH_2-CH_2-CH_3$$
$$\qquad\qquad |$$
$$\qquad\qquad CH_3$$

2-(N,N-dimethylamino)-hexane

Aromatic amines are usually named as derivatives of the parent compound aniline. The following examples are typical:

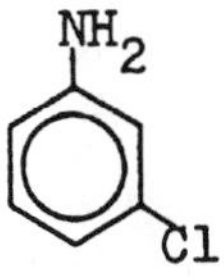

m-chloroaniline

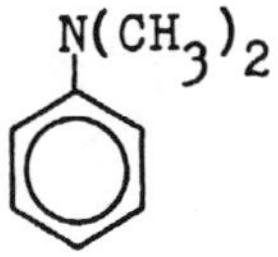

N,N-dimethylaniline

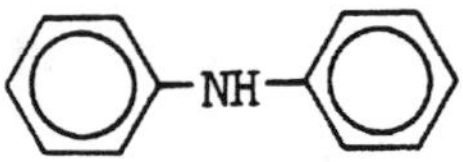

Diphenylamine

(a) $CH_3 - NH - C_2H_5$ has a methyl and an ethyl group attached to the nitrogen; thus, the name of this structure is ethylmethylamine.

(b) $C_6H_5 - NH - C_2H_5$ has a benzene ring and an ethyl group attached to the nitrogen; thus, the name of this structure is ethylphenylamine. This structure can also be named as an aniline: N-ethylaniline.

(c) $NH_2 - (CH_2)_3 - NH_2$ has two amino groups attached to the first and third carbons of a three-carbon parent; thus, the name of this structure is 1,3-diaminopropane.

(d) $(CH_3 - CH_2 - CH_2)_3N$ has three n-propyl groups attached to the nitrogen; thus, the name of this structure is tri-n-propylamine

(e) $(CH_3)_2CH - NH_2$ has an isopropyl group attached to the nitrogen; thus, the name of this structure is isopropylamine.

(f) $NH_2 - CH_2 - COOH$ has an amino group attached to the α carbon of acetic acid; thus, the name of this structure is α-aminoacetic acid (or 2-amino-ethanoic acid).

(g) $(C_6H_5)_2NCH_3$ has two benzene rings and a methyl group attached to the nitrogen; thus, the name of this structure is diphenylmethylamine.

(h) $C_2H_5 - NH - CH_2C_6H_5$ has a benzyl and an ethyl group attached to the nitrogen; thus, the name of this structure is benzylethylamine.

(i) CH₃CHCH₂CH−COOH has a chlorine atom and an amino group

       | |

    NH₂ Cl

attached to pentanoic acid at carbons 2 and 4, respectively. Thus, the name of this structure is 4-amino-2-chloropentanoic acid.

(j) Cl—⟨◯⟩—NH-CH₃ has a methyl group and p-chloroaniline fragment attached to the nitrogen; thus, the name of this structure is N-methyl-p-chloroaniline.

Q Provide an acceptable name for each of the following. Indicate whether the amino groups in these compounds are primary, secondary, or tertiary.

(a) $CH_2 = CHCH_2NH_2$

(b) $CH_3CH_2CHNHCH_3$

(c) $C_6H_5NH_2$

(d) $H_2NCH_2CO_2H$

(e) $((CH_3)_3C)_3N$

(f)

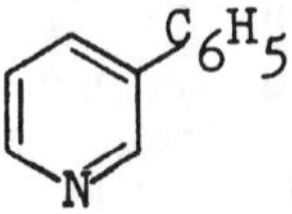

(g)

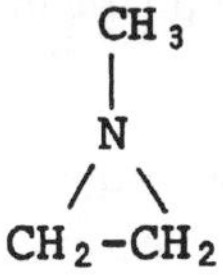

(h)

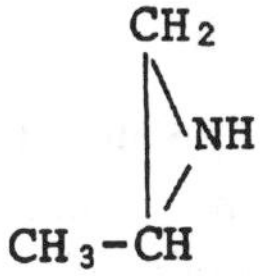

(i)

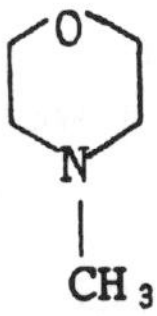

(j)

A Nitrogen appears in a variety of organic compounds. The most common of the monofunctional organic nitrogen compounds are the amines where alkyl groups and/or hydrogens are bonded directly to nitrogen. Amines are designated as primary, secondary, or tertiary 1°, 2°, or 3°, depending on the extent of substitution upon the nitrogen. For instance, an amine bearing three alkyl groups (NR_3) would be considered to be a tertiary amine. The amines are named by designating the alkyl groups attached to nitrogen. In amine nomenclature the ordering of substituents is preferably alphabetical and the entire sequence is treated as one word ending with the suffix "-amine."

(a) This amine has only the allyl group ($CH_2 = CH - CH_2 -$) substituted on the nitrogen and is a primary amine named as allylamine.

(b) Here, we now have two alkyl groups bonded to the nitrogen, and this secondary amine is called methyl-n-propylamine.

(c) This compound is most appropriately termed a primary aryl amine. It is commonly known as aniline rather than phenylamine.

(d) This substituted acetic acid can be unambiguously named as aminoacetic acid since there is only one carbon adjoining the carboxylic acid group. It is a primary amine because it has a single alkyl group derivative on the nitrogen. This compound is also known as glycine.

(e) This tertiary amine is of the general type NR_3 where here R represents a t-butyl group. Thus, the compound is tri-t-butylamine.

(f) The derivatives of pyridine (C_5NH_5) can be named by the following numbering system:

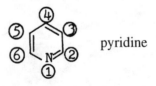 pyridine

Treating the benzene ring as a substituent, the compound is named as 3-phenylpyridine. The nitrogen has a double and single bond and is classified as a tertiary amine.

(g) Aziridine has the structure $\begin{array}{c} H \\ N \\ / \quad \backslash \\ CH_2 - CH_2 \end{array}$. We must designate substitution on the nitrogen by prefixing the substituent with the
letter N. The compound in question is named N-methyl-
aziridine and is a tertiary amine.

(h) Here the methyl group is substituted on a carbon and since
both ring carbons are equivalent, the compound is named
methylaziridine. Since the nitrogen still retains a hydrogen,
the amine is secondary.

(i) Morpholine has the structure [structure]. To show methyl substituent on the nitrogen, we would name the compound N-
methyl-morpholine. It is a tertiary amine.

(j) The only structure consistent with five consecutive methylene
groups bonded to a nitrogen would be a derivative of the six-

membered ring compound piperidine: [structure]

Again, since the substitution occurred on the nitrogen, the desired compound would be named N-methylpiperidine and is a
tertiary amine.

Name the following substances by an accepted system.

(a) $(HOCH_2CH_2)_3N$

(c) [structure] $\begin{array}{c} CH_3 \\ / \\ -N \\ \backslash \\ CH_2CH_3 \end{array}$

(b) $H_2NCH_2CH_2N^+H\text{-}3Cl$

(d) [structure] $N^+ - CH_3 \quad I^-$

A Amines are named by specifying the various substituents on nitrogen as prefixes before the word "amine," with the name written as one continuous word. Amine salts, where there is a hydrogen bonded to the nitrogen are named similarly, except the suffix is now "ammonium" with the anion being listed as a separate word.

(a) This amine has three identical groups attached to nitrogen with the substituent being 2-hydroxyethyl and the compound is named tri-(2-hydroxyethyl)-amine.

(b) This compound is more conveniently named as a derivative of an ammonium salt. Thus, naming the amine portion of the molecule amino, the compound is (2-amino-ethyl) ammonium chloride.

(c) If named as an amine, this compound is called ethylmethyl-phenylamine. Alternatively, we could name the compound as a derivative of aniline, calling it N-ethyl-N-methylaniline with the letter "N" denoting substitution on nitrogen.

(d) This compound is analogous to an ammonium ion, except that this is a derivative of pyridine instead of ammonia. Therefore, the name of the compound is N-methylpyridium iodide.

3.2 Physical Properties of Amines

A) Amines are polar compounds.

B) Primary amines form intermolecular hydrogen bonds that are weaker than those of alcohols and carboxylic acids.

C) Amines have higher boiling points than nonpolar compounds of comparable molecular size, but lower boiling points than alcohols or carboxylic acids.

D) Amines up to about six carbons are quite soluble in water and more basic than ammonia.

E) The water solubility of amines decreases with increasing size of nonpolar alkyl groups attached to nitrogen.

F) Amines are soluble in less polar solvents such as alcohol, ether, and benzene.

G) At room temperature, the lower members are gases, propylamine to dodecylamine are liquids, and the higher members are solids.

Problem Solving Examples:

Ethanolamine, $HOCH_2CH_2NH_2$, can in principle hydrogen-bond to itself in two different ways:

Actually, one of these arrangements is much more important than the other. Explain.

A A hydrogen bonded covalently to an electronegative atom can potentially be electrostatically attracted to another electronegative atom. This interaction in which hydrogen acts as a bridge between two electronegative atoms is called a hydrogen bond. Of the two possible hydrogen bond arrangements in ethanolamine, the first arrangement in which the hydrogen involved in the hydrogen bond is covalently bonded to the oxygen is the important one. This is because a proton is more apt to be transferred from oxygen as in the first structure than a transfer from nitrogen as is true in the latter arrangement. This difference in acidity is due to the fact that oxygen is more electronegative than nitrogen and can better support the incipient negative charge. By similar reasoning, nitrogen can better accept a proton and the subsequent positive charge.

Explain the feature of amine basicity.

 Ammonia in aqueous solution is weakly basic, and its reaction with water is usually represented by the following equations:

$$NH_3 + H_2O \leftrightarrow NH_4^+ + OH^-$$

The basic character of ammonia depends on the fact that its unshared pair of electrons can react with a proton from water to form the ammonium ion and liberate the hydroxide ion. This reaction is actually a displacement reaction.

$$NH_3 + H_2O \rightleftharpoons NH_4^+ + OH^-$$

$$NH_4OH \rightleftharpoons NH_4^+ + OH^-$$

In fact, any molecule containing an unshared pair of electrons may show basic properties, as in the above example. Like their parent compound ammonia, amines exhibit significant basicity. Their reaction with water is analogous to the reaction with ammonia, as seen in the following figure:

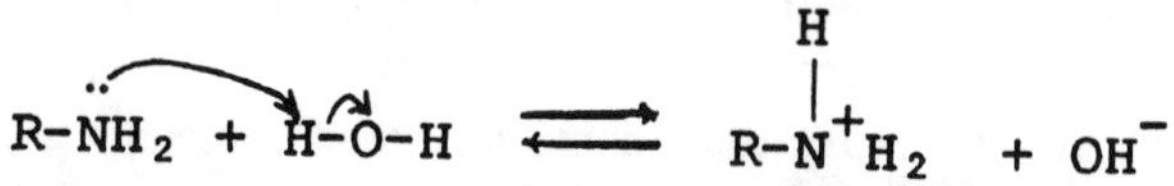

$$R-\overset{..}{N}H_2 + H-O-H \rightleftharpoons R-\overset{\overset{H}{|}}{\underset{..}{N}}{}^+H_2 + OH^-$$

This figure was written for a primary amine $R-\overset{..}{N}H_2$, but both secondary amines $R_2\overset{..}{N}H$ and tertiary amine $R_3\overset{..}{N}$ have an unshared pair of electrons and show basic behavior in water.

The strength of a base is expressed as the equilibrium constant K_b for the protonation reaction. Thus, for the reaction

$$RNH_2 + H_2O \leftrightarrow R\overset{+}{N}H_3 + OH^-$$

the equilibrium constant is:

$$K_b = \frac{\left[R\overset{+}{N}H_3 \right]\left[OH^- \right]}{\left[RNH_2 \right]}$$

The more the reaction is shifted to the right (toward products), the larger the K_b value; a large K_b indicates a stronger base. For ammonia $K_b = 1.8 \times 10^{-5}$, while for methylamine $K_b = 4.4 \times 10^{-4}$. Methylamine is thus a stronger base than ammonia.

The differences in base strength are due to the availability of the unshared electron pair on the nitrogen of an amine. It is known that methyl groups are electron-donating relative to hydrogen. Thus, when a methyl group is present on an amine, electrons are closer to the nitrogen as compared to when nitrogen is bonded to hydrogen.

Thus, there are more electrons close to the nitrogen of the methyl-substituted amine, and the electron pair on this nitrogen is more available to react with the proton from water.

A similar argument predicts correctly that dimethylamine is a stronger base than methylamine.

The base strength of the aromatic amines is significantly lower than that of the aliphatic amines. The K_b of aniline is 4.2×10^{-10}, while that of ethylamine is 4.7×10^{-4}. The relatively low basicity of the aromatic amines can be understood by keeping in mind the notion of electron availability while examining the resonance forms of aniline.

Several of the resonance forms of aniline have the unshared electron pair of the nitrogen in the ring. The electron pair of aniline is thus less available for protonation by water. Since aliphatic amines are not capable of resonance, their electron pair is more available than that of aromatic amines, and the lower basicity of aromatic amines is accounted for.

Similarly, there are even more resonance forms for diphenylamine in which the electron pair on the nitrogen is delocalized within the ring. Thus, diphenylamine is a weaker base than aniline.

 Without referring to tables, arrange the compounds of each set in order of basicity:

(a) ammonia, aniline, cyclohexylamine

(b) ethylamine, 2-aminoethanol, 3-amino-l-propanol

(c) aniline, p-methoxyaniline, p-nitroaniline

In order to compare the relative basicities of various compounds, one must understand what factors influence basicity. First of all, aromatic amines such as aniline are less basic than aliphatic amines such as ammonia. This can be seen in terms of aniline's various resonance structures:

The Kekulé forms where the double bonds are delocalized around the aromatic ring are found also in the anilinium ion.

However, while the nitrogen can delocalize its free pair of electrons in aniline to positions ortho and para to the amino group, this delocalization does not occur in the anilinium ion. Therefore, aniline is stabilized by resonance unlike the anilinium ion, and the activation energy for protonation is increased by this resonance. In comparison, neither ammonia nor the ammonium ion is stabilized by resonance. Thus, aromatic amines are less basic than aliphatic amines. An additional factor that determines basicity is the substituents in the amine. Summarizing, we may say that any substituent that stabilizes the positive charge of the ammonium ion increases the amine's basicity, and anything that destabilizes it, lowers the amine's basicity.

(a) We know that the alkyl groups of cyclohexylamine would stabilize the subsequent anilinium ion compared to the unsubstituted ammonia. Also, we already know acylic amines are better bases than aromatic amines. Thus, the order of basicity is:

(b) The inductive effect of the hydroxy functionality is that of an electron-withdrawing group while an alkyl group is electron donating. We also note that a group's inductive effect is dependent on its distance from the functionality it is affecting (in this case, the functionality is the amine group). Thus, the basicity sequence is:

$$CH_3CH_2NH_2 > HOCH_2CH_2CH_2NH_2 > HOCH_2CH_2NH_2$$

(c) When comparing the basicities of substituted anilines, we must consider both the substituent's inductive effect and its interaction with aniline's delocalization into the aromatic ring. For instance, the structure,

is more stable than

because of the electron-withdrawing effect of the nitro group. Therefore, the resonance stabilization would be greater for p-nitroaniline and it would be a weaker base than aniline. The methoxyl group ($-OCH_3$) is also electron-withdrawing, but we recall that it also stabilizes positive charge by resonance delocalization and in a similar way it would stabilize the anilinium ion. Therefore, the order of basicity is:

3.3   Preparation of Amines

Alkyl halides undergo nucleophilic substitution by ammonia-yielding ammonium salts. Subsequent treatment with a base liberates the free amine.

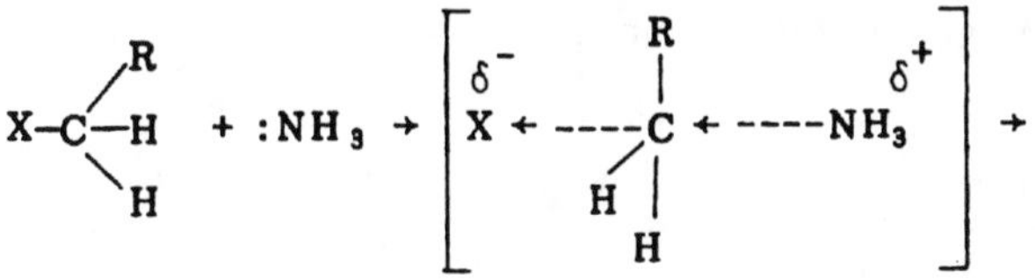

$$RCH_2N^+H_3X^-$$

$$RCH_2N^+H_3X^- + OH \rightarrow RCH_2NH_2 + H_2O + X^-$$

The reduction of appropriate compounds by catalytic hydrogenation or use of certain reducing agents:

A) a) $RCN + 4Na/4CH_3CH_2OH \rightarrow RCH_2NH_2 + 4CH_3CH_2ONa$

 b) $RC(=NOH)H + 4Na/4CH_3CH_2OH \rightarrow RCH_2NH_2 + 4CH_3CH_2ONa + H_2O$

 c) $RC(=NNH_2)H + 4Na/4CH_3CH_2OH \rightarrow RCH_2NH_2 + 4CH_3CH_2ONa + NH_3$

 d) $4RCH_2NO_2 + 9Fe/(FeCl_2,H+) + 4H_2O \rightarrow 4RCH_2NH_2 + 3Fe_3O_4$

B) $RNC + 4Na/4CH_3CH_2OH \rightarrow RNHCH_3 + 4CH_3CH_2-ONa$

Hofmann reaction. Reduction of amides by bromine and alkali to give primary amines. The steps of this reaction are as follows.

$$R-CO-NH_2 + Br_2/NaOH, aq \rightarrow [R-CO-NHBr] + NaBr + H_2O$$

$$R-CO-NHBr + NaOH, aq \rightarrow R-N=C=O + NaBr + H_2O$$

alkyl isocyanate

$$R-N=C=O + 2NaOH, aq \rightarrow R-NH_2 + Na_2CO_3$$

primary amine

In summary, the Hofmann reaction results in a primary amine with one or less carbons than the parent amide.

$$RC{\overset{O}{||}}NH_2 \longrightarrow RNH_2$$

Gabriel synthesis. Reaction of alkyl halides and alkali with potassium phthalimide to give primary amines.

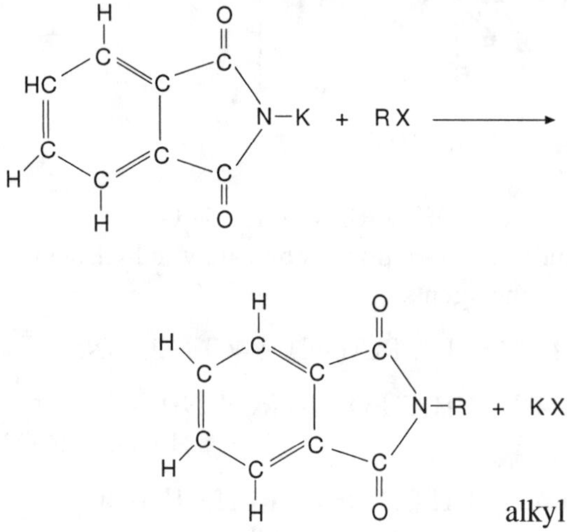

alkyl phthalimide

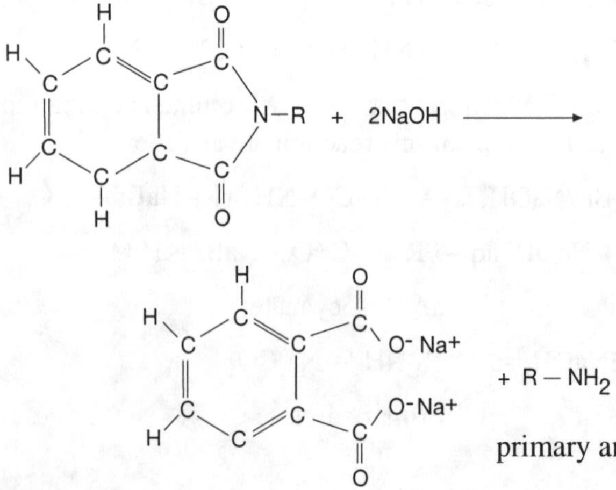

primary amine

Reaction of methanol with ammonia in the presence of a catalyst, to give a mixture of methyl amines:

$CH_3OH + NH_3/catalyst \rightarrow CH_3 - NH_2 + H_2O$

$CH_3OH + CH_3 - NH_2/catalyst \rightarrow (CH_3)_2NH + H_2O$

$CH_3OH + (CH_3)_2 - NH/catalyst \rightarrow (CH_3)_3N + H_2O$

Ethanolamines are prepared by the action of ammonia on ethylene oxide:

$$H_2C\!\!-\!\!CH_2 + NH_3 \rightarrow HO-CH_2-CH_2-NH_2 \quad (ethanolamine)$$

The reduction of nitriles by hydrogen and a catalyst to produce primary amines:

$$RC \equiv N \xrightarrow{2H_2, Catalyst} R-CH_2-NH_2$$
$$\text{primary amine}$$

The reduction of nitro compounds to give primary amines:

$$\begin{matrix} ArNO_2 \\ or \\ RNO_2 \end{matrix} \xrightarrow[H_2, catalyst]{metal, H^+; or} \begin{matrix} ArNH_2 \\ or \\ RNH_2 \end{matrix}$$

A) Catalytic hydrogenation

Example

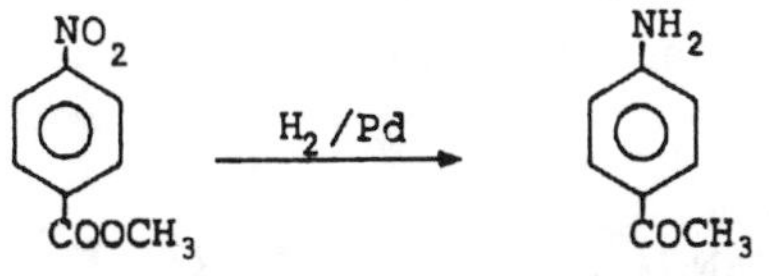

Methyl-p-
nitrobenzoate

Methyl-p-
aminobenzoate

B) Metal-acid reduction

Example

$$CH_3CH_2CH_2-NO_2 \xrightarrow{Fe/HCl} CH_3CH_2CH_2NH_2$$
$$\text{1-nitropropane} \qquad\qquad \text{n-propylamine}$$

C) Lithium aluminum hydride

Example

$$CH_3CH_2\underset{\underset{NO_2}{|}}{CH}\ CH_3 \longrightarrow CH_3CH_2\underset{\underset{NH_2}{|}}{CH}CH_3$$

2-nitrobutane   2-butylamine

Problem Solving Examples:

 Synthesize n-pentylamine from any alcohol using the process of

(a) Hofmann degradation,

(b) nitrile reduction, and

(c) reductive amination.

In this problem we are asked to synthesize n-pentylamine $(CH_3(CH_2)_4NH_2)$ from alcohols in a variety of ways. The general method for solving a synthesis problem is to begin with the product and work backwards to the reactants. In this case, we are told to achieve the transformation by a particular route, so we need to use specific intermediates in each case.

(a) The Hofmann degradation involves the conversion of an amide,

$$Ar(R)\overset{\overset{O}{\|}}{C}-NH_2,$$ to an amine, $Ar(R)NH_2$ with one less carbon. The amine is treated with alkaline solutions of bromine, chlorine, or iodine. The mechanism is controversial but an N-haloamide, an isocyanate, and an unstable carbamic acidate are known to be intermediates.

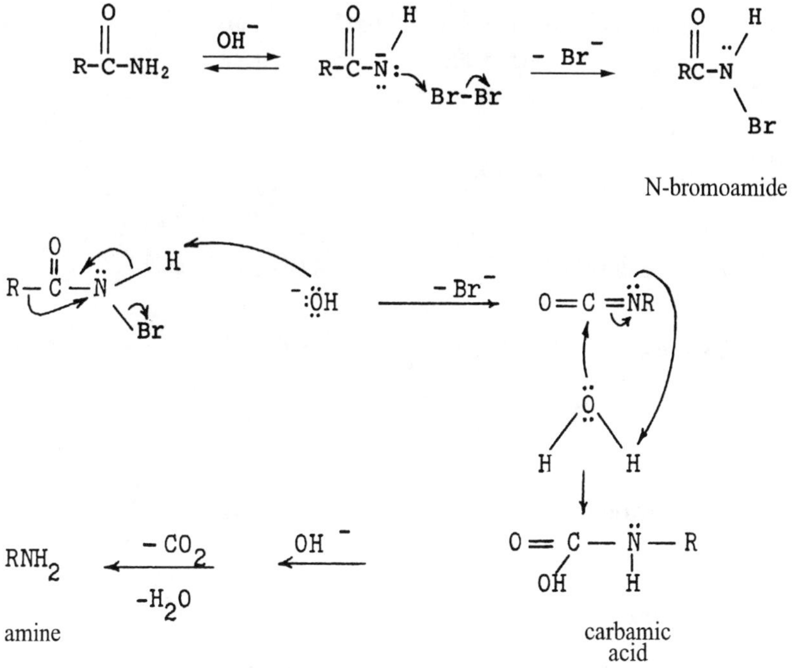

N-bromoamide

This rearrangement proceeds with complete retention of configuration in relation to the chiral center of the migrating moiety. Also, this "migration" does not involve the generation of a free particle such as a carbanion. This concurs with the stereochemistry of the process in that a free particle would mean the loss of configuration and an optically inactive product would be formed, which is not the case. It has been proposed that the carbonyl carbon attaches itself to the nitrogen before it leaves the chiral center. Various groups have different migratory aptitudes with a phenyl group migrating faster than an alkyl group. Returning to our problem of synthesizing n-pentylamine via the Hofmann degradation, and realizing that this process reduces a carbon chain by one carbon, we would start with an amide of six carbons. The amide can be derived from an acid chloride of equal length, which would in turn be synthesized from its corresponding carboxylic acid. Oxidation of our starting material, n-hexanol, would yield the carboxylic acid. Summarizing this synthetic route:

$$CH_3(CH_2)_5OH \quad \xrightarrow[OH^-/H_2O]{KMnO_4/\Delta} \quad \xrightarrow{H^+} \quad CH_3(CH_2)_4COOH$$

$$CH_3(CH_2)_4\overset{O}{\underset{//}{C}}OOH \quad \xrightarrow{PCl_3} \quad CH_3(CH_2)_4\overset{O}{\underset{//}{C}}-Cl$$

$$CH_3(CH_2)_4\overset{O}{\underset{//}{C}}-Cl \; + \; NH_3 \; (excess) \; \rightarrow \; CH_3(CH_2)_4\overset{O}{\underset{//}{C}}-NH_2$$

$$CH_3(CH_2)_4\overset{O}{\underset{//}{C}}-NH_2 \quad \xrightarrow[Br_2]{OH^-} \quad CH_3(CH_2)_4NH_2$$

Phosphorus trichloride is just one agent that would achieve our desired transformation to the acid chloride ($SOCl_2$, PCl_5 are others). Also, we used an excess of ammonia because the subsequently produced HCl would destroy the amine.

(b) Nitriles with the general formula RCN can be converted to a primary amine, $R - CH_2NH_2$, by a variety of ways, including hydroboration, use of lithium aluminum hydride, and catalytic hydrogenation. The last method has the disadvantage that the newly formed amine will react with the intermediate imine, $RCH = NH$ to give a symmetrical secondary amine, $RCH_2-NH - CH_2R$. This loss of yield can be circumvented by converting the amine to an amide and its salt while carrying out the reduction. The final product is liberated by alkaline hydrolysis. The cyanide anion has nucleophilic capabilities, and we use this to produce the five-carbon nitrile, $CH_3(CH_2)_3CN$, from the four-carbon alkyl halide via a nucleophilic displacement of the S_N2 type. Subsequent reduction of the nitrile will yield the desired amine while treatment of n-butanol, our starting material, with the brominating agent, PBr_3, will yield the alkyl halide. Summarizing this sequence:

$$CH_3(CH_2)_3OH \xrightarrow{PBr_3} CH_3(CH_2)_3Br$$

$$CH_3(CH_2)_3Br \xrightarrow{CN^-} CH_3(CH_2)_3CN$$

$$CH_3(CH_2)_3CN \xrightarrow[\text{dioxane}]{B_2H_6} CH_3(CH_2)_3CH_2NH_2$$

(c) The process of reductive amination involves reacting a carbonyl compound in the presence of an amine and a reducing agent. Reductive animation has the synthetic advantage of being able to produce 1°, 2°, and 3° amines by picking the appropriate amine-carbonyl compound combination. Since treating a carbonyl compound with ammonia does not affect the number of carbons, we would start with n-pentanol and oxidize it selectively to its corresponding aldehyde, which would be treated with ammonia in the presence of hydrogen and a catalyst. The sequence would be as follows:

$$CH_3(CH_2)_4OH \xrightarrow[\text{pyridine}]{CrO_3} CH_3(CH_2)_3\overset{\displaystyle O}{\overset{\displaystyle \|}{C}}-H$$

valeraldehyde

$$CH_3(CH_2)_3\overset{\displaystyle O}{\overset{\displaystyle \|}{C}}-H \xrightarrow[\text{Pt/H}_2]{NH_3} \left[CH_3(CH_2)_3\overset{\displaystyle H}{\underset{\displaystyle \|}{\overset{\displaystyle |}{\overset{\displaystyle N}{C}}}}H \right]$$

reduction ↓

$$CH_3(CH_2)_3\ CH_2NH_2$$

Q Isopropylamine and n-pentylamine are to be synthesized. In the laboratory synthesis, outline all the steps from alcohols of four carbons or less, using any needed inorganic reagents.

A These organic compounds can be prepared by employing reductive amination and reduction of nitriles, two ways of preparing amines.

Isopropylamine Synthesis: Many aldehydes and ketones can be converted to amines (of the same carbon number) by reductive amination: reduction in the presence of ammonia. This reduction can be accomplished catalytically or by use of sodium cyanohydridoborate, $NaBH_3CN$. The reaction can be generalized as follows:

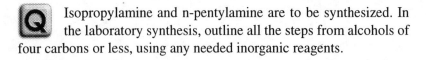

Note: A similar sequence can be written for an aldehyde, with a 1° amine being formed.

This reaction means that if acetone $\left(CH_3-\overset{\displaystyle O}{\underset{\displaystyle \|}{C}}-CH_3 \right)$ were present,

isopropylamine $\left(CH_3-\overset{\displaystyle H}{\underset{\displaystyle NH_2}{\overset{\displaystyle |}{\underset{\displaystyle |}{C}}}}-CH_3 \right)$ could be synthesized by reductive

amination. Acetone can be obtained from isopropyl alcohol $\left(CH_3-\overset{\displaystyle H}{\underset{\displaystyle OH}{\overset{\displaystyle |}{\underset{\displaystyle |}{C}}}}-CH_3 \right)$ by oxidation with CrO_3, in glacial acetic acid. CrO_3 oxidizes 2° alcohols (such as isopropyl alcohol) to the ketone. Note that isopropyl alcohol is allowed as the starting material since it is less than four carbons, as the problem demands. The sequence of reactions can therefore be outlined:

$$CH_3-\underset{\underset{OH}{|}}{\overset{\overset{H}{|}}{C}}-CH_3 \xrightarrow{CrO_3} CH_3-\underset{\underset{O}{||}}{C}-CH_3 + NH_3 \xrightarrow{H_2, Ni} CH_3-\underset{\underset{NH_2}{|}}{\overset{\overset{H}{|}}{C}}-CH_3$$

(Isopropyl   (Acetone)   (Isopropyl
alcohol)   amine)

N-Pentylamine Synthesis: N-pentylamine's structure ($CH_3CH_2CH_2$ $CH_2CH_2NH_2$) indicates it is composed of five carbon atoms. Since the starting material cannot have over four carbon atoms, the synthesis of the amine must include an increase in the carbon skeleton of the starting material. Synthesis via reduction of nitriles ($-C \equiv N$) has the special feature of increasing the length of a carbon chain, producing a primary amine that has one more carbon atom than the alkyl halide from which the nitrile was made.

Reduction of a nitrile will be used in the synthesis of N-pentylamine. This necessitates, therefore, that an alkyl halide of four carbon length be produced. (Addition of the nitrile ($-C \equiv N$) will increase the length to the desired number of five carbon atoms.)

Since the structure of N-pentylamine is a straight-chain carbon sequence, the alkyl halide should be a straight chain. The alkyl halide can be made from an alcohol, using a catalyst, PX_3:

$$R - OH \quad + \quad PX_3 \quad \rightarrow \quad RX + H_3PO_3$$

(Alcohol)   (X = I, Cl, Br)   (Alkyl halide)

Hence, one needs an alcohol of four carbons that possesses a straight-carbon chain. N-butyl alcohol fits this description. N-butyl alcohol is reacted with a phosphorus trihalide, say PBr_3.

$$CH_3CH_2CH_2CH_2OH + PBr_3 + CH_3CH_2CH_2CH_2Br + H_3PO_3$$

(N-butyl alcohol)   (N-butyl bromide)

The nitrile is then formed by a nucleophilic displacement.

$$CH_3CH_2CH_2CH_2Br + NaCN \rightarrow CH_3CH_2CH_2CH_2CN$$

(N-butyl nitrile)

Reduction of this compound produces the desired product:

$$CH_3CH_2CH_2CH_2CN \xrightarrow[\text{catalyst}]{2H_2} CH_3NH_2CH_2CH_2CH_2NH_2$$

(N-pentylamine)

The overall synthesis can be written as:

$$n - BuOH \xrightarrow{PBr_3} n - BuBr \xrightarrow{CN^-} n - Bu - C \equiv N \xrightarrow{H_2, Ni}$$
$$n - BuCH_2NH_2$$

3.4   Reactions of Amines

Formation of Addition Products

A) 1) $RNH_2 + HX \rightarrow RNH_3^+ X^-$

2) $RNH_2 + RX \rightarrow R_2NH_2^+ X^-$

B) 1) $R_2NH + HX \rightarrow R_2NH_2^+ X^-$

2) $R_2NH + RX \rightarrow R_3NH^+ X^-$

C) 1) $R_3N + H X \rightarrow R_3NH^+X^-$

2) $R_3N + R X \rightarrow R_4N^+X^-$

Conversion into Amides

A) RNH_2 ⎯⎯⎯⎡ $\xrightarrow[-HCl]{R'COCl}$ **R'CONH·R**   (N-substituted amide)

⎣ $\xrightarrow[-HCl]{ArSO_2Cl}$ **ArSO_2NHR**   (N-substituted sulfonamide)

B)

$$R_2NH-\begin{cases}\xrightarrow[-HCl]{R'COCl} & R'CONR_2 \quad \text{(N,N—disubstituted amide)}\\[2em]\xrightarrow[-HCl]{ArSO_2Cl} & ArSO_2NR_2 \quad \text{(N,N—disubstituted sulfonamide)}\end{cases}$$

C) Tertiary amines, R_3N, do not react in this manner.

Formation of Ammonium Bases

A) $R\text{-}NH_3^+ + OH^-$

B) $R_2\text{-}NH_2^+ + OH^-$

C) $R_3\text{-}NH^+ + OH^-$

Halogenation of amines by hypochlorous acid or tert-butyl-hypo-bromite in alkaline solution:

A) $R\text{-}NH_2 + 2Cl_2 \xrightarrow{NaOH/Cl_2} R\text{-}N\begin{smallmatrix}Cl\\\\Cl\end{smallmatrix} + 2HCl$

$$\text{N,N-dichloroalkylamine}$$

B) $R_2NH + Cl_2 \xrightarrow{NaOH/Cl_2} R_2N\text{-}Cl + HCl$

$$\text{N- chlorodialkylamine}$$

Basicity of amines: Since ammonia and amines contain a nitrogen with an unshared electron pair, they act as bases, accepting protons and forming ammonium ions and alkyl ammonium ions.

A) $\ddot{N}H_3 + H^+ \rightarrow NH_4^+,$ Ammonium ion

B) $R\text{-}\ddot{N}H_2 + H^+ \rightarrow R\text{-}NH_3^+,$ Alkylammonium ions

Ammonia and amines are stronger bases than water, forming ammonium salts in aqueous mineral acids.

Reaction of Amines with Nitrous Acid

A) Primary aliphatic and aromatic amines are converted into diazonium salts.

a) $ArNH_2 \xrightarrow{\text{HONO}} ArN \equiv N^+$

   aryldiazonium salt

b) $RNH_2 \xrightarrow{\text{HONO}} [R-N \equiv N^+] \xrightarrow{H_2O} N_2 +$ mixture of alcohols an alkenes

   alkyldiazonium salt
   unstable

B) Secondary amines react to yield N-nitrosoamines.

$$\begin{array}{c} ArNHR \\ \text{or} \\ R_2NH \end{array} \xrightarrow{\text{HONO}} \begin{array}{c} R \\ | \\ Ar-N-N = O \\ \text{or} \\ R_2N-N = O \end{array}$$

C) Tertiary aliphatic amines are oxidized to yield N-nitrosodialkylamines and a mixture of ketones and aldehydes.

$R_3N \xrightarrow{\text{HONO}} R_2N-N =O +$ mixture of aldehydes and ketones

D) Tertiary aromatic amines undergo electrophilic nitrosation at the benzene ring.

p-nitroso compound

Electrophilic substitution of aromatic amines. Amino groups ($-NH_2$, $-NHR$, $-NR_2$), but not ammonium groups in aryl ammonium ions, activate the benzene ring to which they are attached for electrophilic substitution. They release electrons due to their unshared electron pairs.

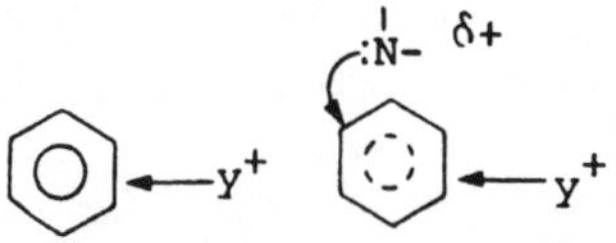

Problem Solving Examples:

Q Write balanced equations, naming all organic products, for the following reactions:

(a) n-butyryl chloride + methylamine

(b) acetic anhydride + N-methylaniline

(c) tetra-n-propylammonium hydroxide + heat

(d) tetramethylammonium hydroxide + heat

(e) N,N-dimethylacetamide + boiling dilute HCl

(f) benzanilide + boiling aqueous NaOH

(g) $m\text{-}O_2NC_6H_4NHCH_3$ + $NaNO_2$ + H_2SO_4

(h) m-toluidine + Br_2(aq) in excess

(i) p-toluidine + $NaNO_2$ + HCl

(j) $p\text{-}CH_3C_6H_4NHCOCH_3$ + HNO_3 + H_2SO_4

(k) benzanilide + Br_2 + Fe

A Nitrogen takes a variety of forms in organic chemistry, many of which are biologically important. This problem involves the reactions of amines and their related compounds. It is best to examine the reactants and see what type of reaction is involved. For example, if given the reactants ammonia and acetic acid, we know that this is an acid-base reaction involving a proton transfer. The reaction would be designated as:

$$NH_3 + CH_3COOH \rightarrow CH_3COO^- \overset{+}{N}H_4$$

Other general reactions of amines would be nucleophilic displacement, electrophilic aromatic substitution in anilines, and substitution via diazonium salts.

(a) The reactants are an acid chloride and an amine. The reaction may be seen as nitrogen's free pair of electrons, attacking the electropositive carbon of the carbonyl group in an addition-elimination reaction with the chloride anion leaving.

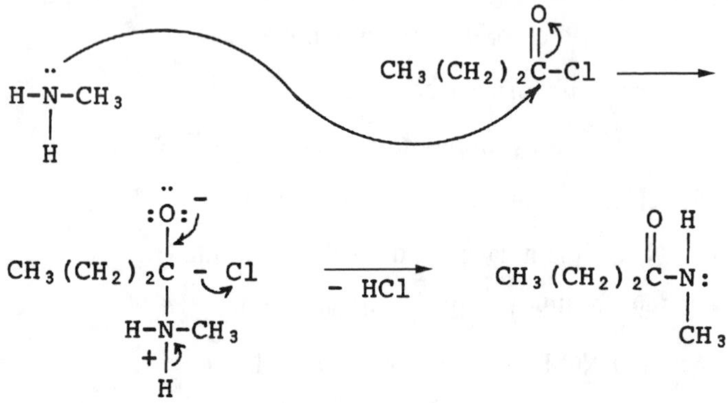

The HCl that is generated will protonate the unreacted amine to give a salt.

$$HCl + NH_2CH_3 \rightarrow \overset{+}{N}H_3CH_3 \quad Cl^-$$

The balanced equation for the entire reaction is:

$$2NH_2CH_3 + CH_3(CH_2)_2\overset{\displaystyle O}{\overset{\displaystyle \|}{C}}-Cl \rightarrow CH_3(CH_2)_2\overset{\displaystyle O}{\overset{\displaystyle \|}{C}}-NHCH_3 + \overset{+}{N}H_3CH_3Cl^-$$

(b) In this reaction, the amine will attack the anhydride to yield an amide with the carboxylate anion acting as a leaving group. The other product is the salt of the amide.

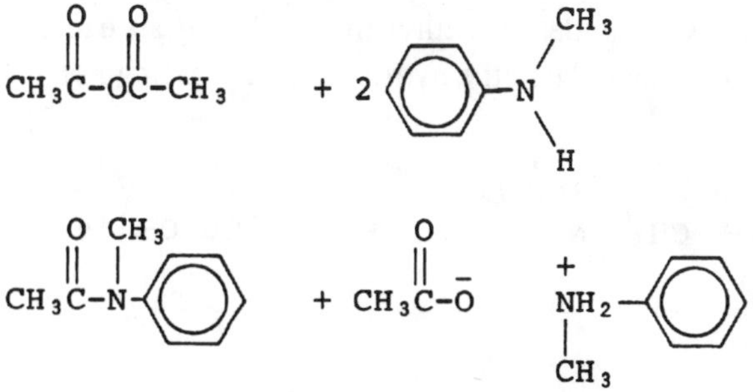

(c) In this problem, we have tetra-n-propylammonium hydroxide as a reactant. This salt belongs to the general class of compounds known as quaternary ammonium hydroxides with the formula $R_4N^+OH^-$. When these compounds are heated strongly, they undergo an elimination reaction of the $E2$ variety in which the hydroxide ion acts as a base and a tertiary amine leaves. In this case:

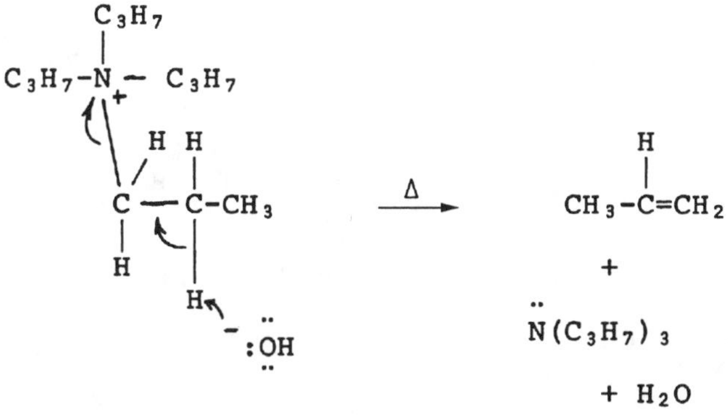

This is an example of a Hofmann elimination and follows an anti-orientation of leaving groups with the less-substituted alkene product predominating.

(d) Tetramethylammonium hydroxide is a quaternary ammonium hydroxide that cannot possibly undergo an elimination reaction. Hofmann eliminations usually have substitution reactions as side reactions, but in this case substitution is the only feasible route. The substitution is of an S_N2 variety with the relatively unhindered methyl moiety being an excellent site for such attack. Here the hydroxide ion is acting as a nucleophile.

$$CH_3 \overset{+}{-} \overset{\overset{\displaystyle CH_3}{|}}{\underset{\underset{\displaystyle CH_3}{|}}{N}} - CH_3 \quad \rightarrow \quad CH_3OH \quad + \quad CH_3 - \overset{\overset{\displaystyle \cdot\cdot}{}}{\underset{\underset{\displaystyle CH_3}{|}}{N}} - CH_3$$

$$^-:\!OH$$

(e) When acid derivatives such as esters, amides, and acid chlorides are subject to aqueous acid, they undergo a hydrolysis reaction, forming a carboxylic acid and another molecule. In this case, hydrolysis of the amide yields the acid and a protonated amine by the following mechanism:

$$CH_3 - \overset{\overset{\displaystyle O}{||}}{C} - N \overset{\displaystyle CH_3}{\underset{\displaystyle CH_3}{\big\langle}} \quad \xrightarrow{H_3O^+} \quad CH_3\overset{\overset{\displaystyle OH}{|}}{\underset{\underset{\displaystyle +}{}}{C}} - N \overset{\displaystyle CH_3}{\underset{\displaystyle CH_3}{\big\langle}} \quad \rightarrow$$

$$\overset{O}{\underset{H \quad H}{\diagup \cdot\cdot \diagdown}}$$

$$CH_3\overset{\overset{\displaystyle OH}{|}}{\underset{\underset{\displaystyle O}{|}}{C}} - \overset{\cdot\cdot}{N} \overset{\displaystyle CH_3}{\underset{\displaystyle CH_3}{\big\langle}} \quad \rightarrow \quad CH_3\overset{\overset{\displaystyle OH}{|}}{\underset{\underset{\displaystyle O}{|}}{C}} - \overset{\overset{\displaystyle H}{|}}{\underset{\underset{\displaystyle +}{}}{N}} - CH_3 \quad \xrightarrow{-H_2O}$$

$$\overset{+}{\underset{H \quad H}{\diagup}} \qquad\qquad\qquad \underset{CH_3}{\big\langle}$$

$$\overset{O}{\underset{\underset{H \quad H}{\diagup \cdot\cdot \diagdown}}{\cdot\cdot \rightarrow H}}$$

$$CH_3\overset{\overset{\displaystyle O}{||}}{\underset{\underset{\displaystyle OH}{\diagdown}}{C}} \quad + \quad \overset{+}{N}H_2(CH_3)_2$$

The net reaction is:

$$H_3O^+ + CH_3\overset{\displaystyle O}{\overset{\|}{C}}-N(CH_3)_2 \longrightarrow CH_3\overset{\displaystyle O}{\overset{\|}{C}}-OH + \overset{+}{N}H_2(CH_3)_2$$

(f) Hydrolysis of a carboxylic acid derivative may be undertaken under alkaline conditions as well as acidic conditions. Benzanilide, which is the condensation product of benzyl chloride and aniline, will yield aniline and the anion of benzoic acid when subjected to alkaline hydrolysis. The mechanism involves the nucleophilic attack of the hydroxide ion upon the carbonyl group.

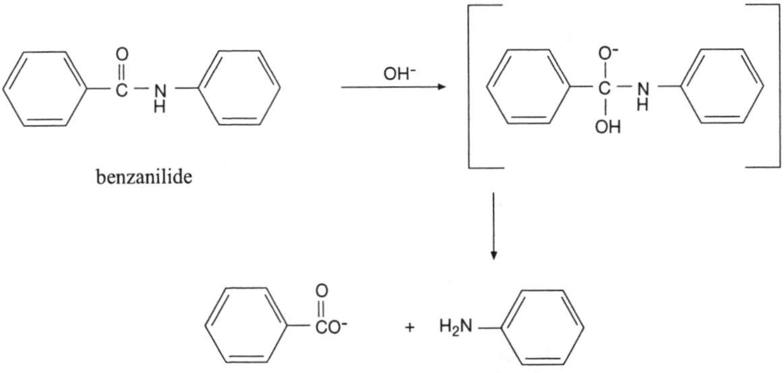

benzanilide

benzoic acid anion   aniline

(g) When sodium nitrite ($NaNO_2$) and H_2SO_4 are mixed, nitrous acid is generated. This short-lived reagent reacts with secondary amines, both alkyl and aryl, to form an N-nitroso compound. We note that this reagent is specific to the amino group and the nitro group is unaffected.

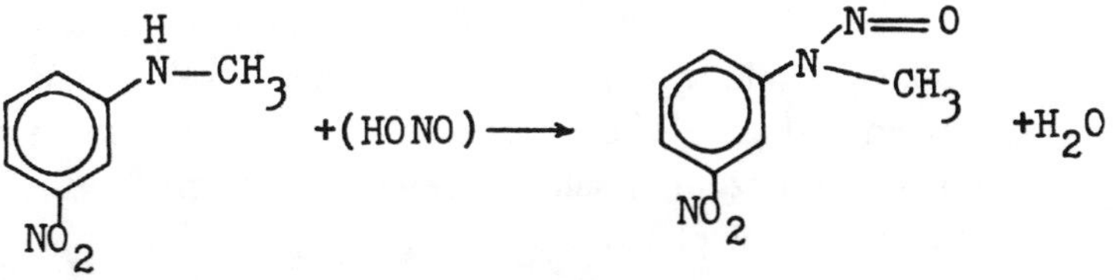

(h) When located directly on an aromatic ring, the amino group is an activating ortho-para director. This activation and orientation in electrophilic aromatic substitution can be seen in terms of aniline's resonance structures, which include forms where the negative charge is delocalized to positions ortho and para to the amino group.

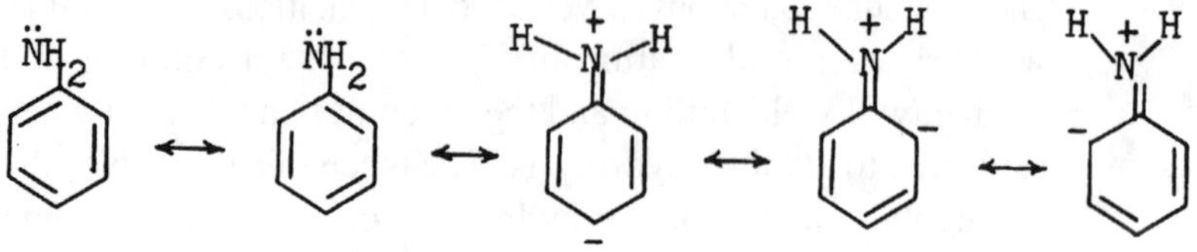

m-toluidine, 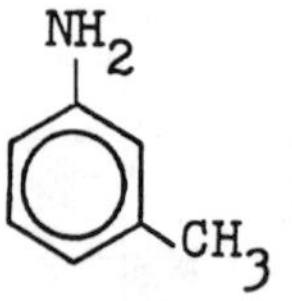 contains both the amino

group and the methyl group, both of which are activators. However, the amino group is the stronger activator of the two, and the incoming bromines are directed ortho and para to it. We can see the aniline's great reactivity by the fact that no catalyst was needed and that all positions ortho and para to the amino group were brominated.

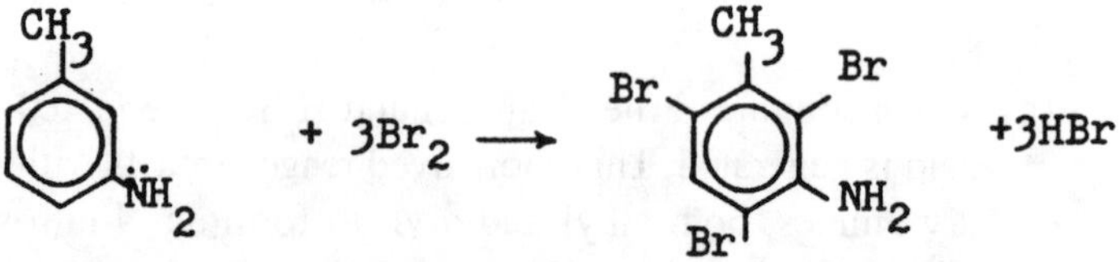

(i) When an aromatic or aliphatic amine is reacted with nitrous acid, they form a diazonium salt $R(Ar) - N_2^+X^-$. While the alkyl diazonium salt is unstable and loses nitrogen to form a variety of alkenes and alcohols, the aryl type is stable and has synthetic importance. When an aniline such as p-toluidine is converted to its diazonium salt, the temperature is kept low to avoid formation of the phenol.

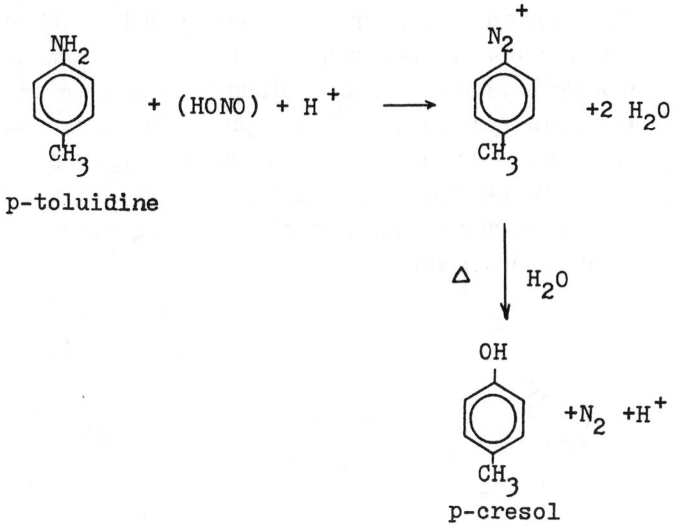

p-toluidine

$+ (HONO) + H^+ \longrightarrow$

$+2 H_2O$

$\Delta \downarrow H_2O$

$+N_2 +H^+$

p-cresol

(j) We note that aceto-p-toluidide contains an acetyl group attached to the amino group. The carbonyl carbon is electropositive and lessens the adjacent nitrogen's tendency to delocalize the electrons within the aromatic ring. Hence, the reactivity to electrophilic substitution of aceto-p-toluidide is less than its corresponding aniline, p-toluidine. This can be seen in the comparable bromination reactions of these compounds in which only the aniline will undergo polysubstitution.

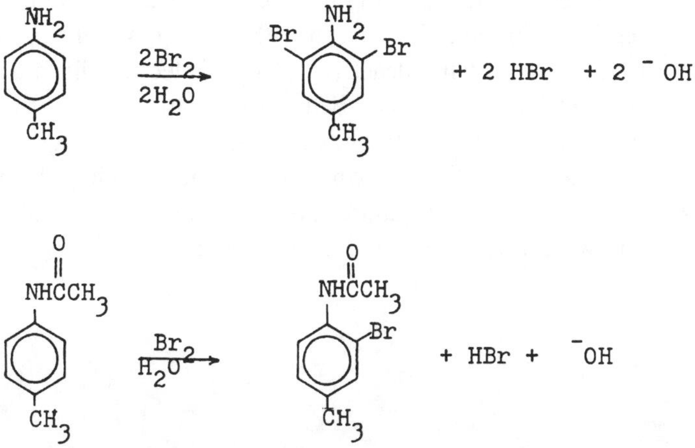

$\xrightarrow[2H_2O]{2Br_2}$

$+ 2 \ HBr \ + \ 2 \ {}^-OH$

$\xrightarrow[H_2O]{Br_2}$

$+ \ HBr \ + \ {}^-OH$

Thus, when aceto-p-toluidide is treated with the nitrating mixture, nitration occurs only at a single ortho position. We note that we may remove the acetyl group by alkaline hydrolysis. It is a useful synthetic procedure to protect the amino group by acetylating it and removing it after undertaking the desired reaction. Besides reducing the activity of the aniline, acetylation protects the amino group from protonation or oxidation during the nitration reaction.

(k) Benzanilide, contains two aromatic rings; one of which is bounded by the amide nitrogen and the other is bounded by a carbonyl group. The nitrogen can delocalize its free electron pair less than an amino group can but it activates the adjacent ring nevertheless. The carbonyl group, on the other hand, is electron withdrawing, and the aromatic ring bounded by it is deactivated toward electrophilic aromatic substitution. Thus, when benzanilide undergoes a bromination reaction, the electrophile is more likely to attack the more activated aromatic ring, which is the one bounding the nitrogen. We note that the amide nitrogen is an ortho-para director, and we get both ortho and para products.

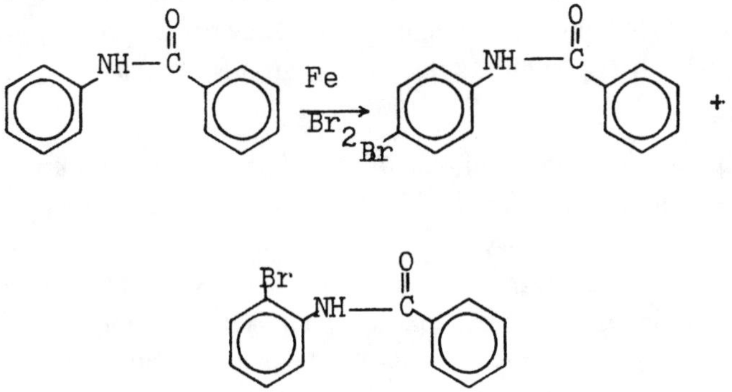

Q Complete the following:

(a) $CH_3CH_2-\underset{\underset{CH_3}{|}}{N}-Ph + CH_3I \longrightarrow$

(b) 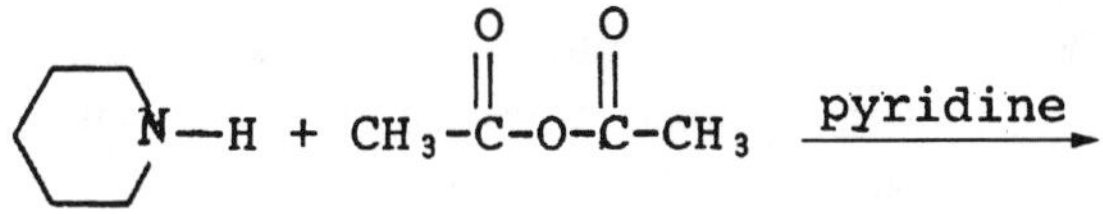 $N-H + CH_3-\underset{\underset{O}{\|}}{C}-O-\underset{\underset{O}{\|}}{C}-CH_3 \xrightarrow{\text{pyridine}}$

(c) $(CH_3)_3N + CH_3-\underset{\underset{Cl}{\diagdown}}{\overset{\overset{O}{\diagup\!\!\diagup}}{C}} \longrightarrow$

(d) 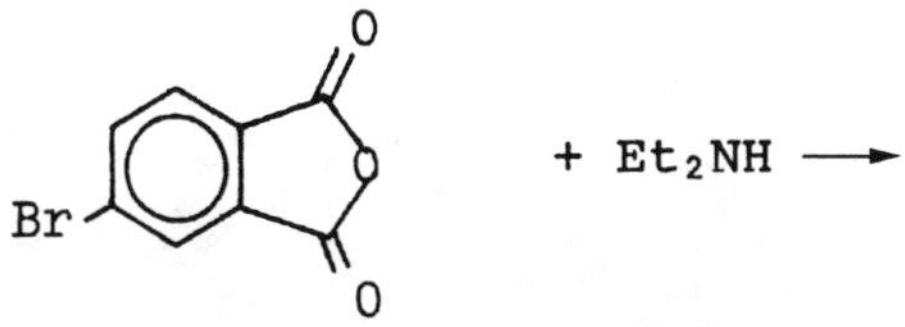 $+ Et_2NH \longrightarrow$

A Amines contain a nitrogen with a free pair of electrons that make them potent nucleophiles. They can attack various sites of positive charge, such as the carbonyl groups of carboxylic acid derivatives.

(a) Amines may be alkylated by an $S_N 2$ displacement upon a primary or secondary halide. This, however, is a poor way of converting a primary amine into a secondary one, or a secondary amine into a trisubstituted type. This is because the product amine is more nucleophilic than the reactant amine and further substitution may take place. The action of an amine may become synthetically feasible if a large excess of the amine is used or, as in the present case, a trisubstituted amine is converted to a quaternary ammonium salt.

$$CH_3CH_2 \overset{..}{N} - Ph \quad CH_3 \frown I \longrightarrow CH_3CH_2 \overset{+}{N} Ph \quad I^- $$
$$\underset{CH_3}{|} \qquad\qquad\qquad \overset{CH_3}{|}\underset{CH_3}{|}$$

(b) When an amine is reacted with an anhydride, an amide and its salt is formed. To increase the concentration of the amide, a "scavenger" base is also put in the reaction mixture to combine with the subsequently formed acid which would otherwise react with the amine. In this case, we would form the acetamide of piperidine and a pyridium salt.

(c) When a tertiary amine is placed together with an acyl halide, it appears not to react in that the amine is recovered unchanged after workup with aqueous base. Actually, they do react to form salts, which are effective acylating agents. In the presence of water, they form an acid and the protonated amine.

$$(CH_3)_3N + CH_3\text{-}\overset{\displaystyle O}{\overset{\|}{C}}\text{-}Cl \rightarrow CH_3\overset{\displaystyle O}{\overset{\|}{C}}\overset{+}{N}(CH_3)_3 \quad \overset{-}{Cl}$$

$$\downarrow H_2O$$

$$CH_3\overset{\displaystyle O}{\overset{\|}{C}}\text{-}OH$$

$$+$$

$$\overset{-}{Cl}\ \overset{+}{HN}(CH_3)_3$$

(d) When an amine is reacted with an unsymmetrical anhydride, it will attack the anhydride carbonyl group that is most electron-deficient. In this case, we know that bromine can delocalize its electrons onto the aromatic ring and put a negative charge on the carbon para to the bromine, making the adjacent carbonyl carbon less electropositive. Thus, the nucleophilic amine will attack the carbonyl (which is next to the ring carbon meta) to the bromine substituent.

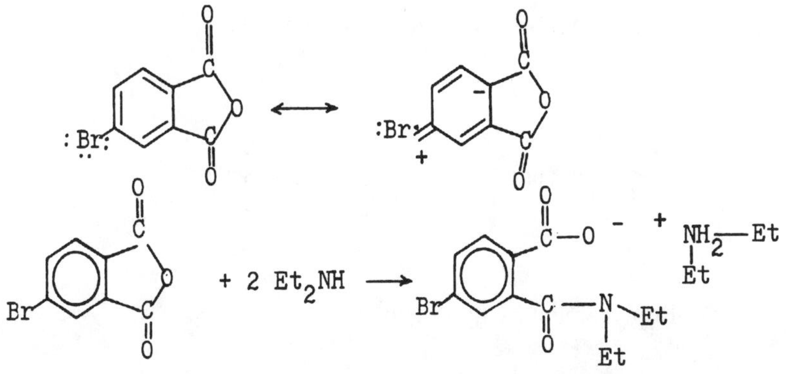

3.5 Aromatic Diazonium Salts

Aliphatic diazonium salts are very unstable and decompose spontaneously after generation. Aromatic diazonium salts, on the other hand, are more stable and are very reactive. They are important intermediates and are seldom isolated or purified.

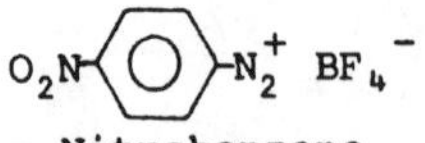

O_2N
N_2^+ BF_4^-

p-Nitrobenzene diazonium fluoroborate

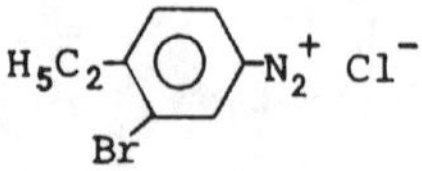

H_5C_2
N_2^+ Cl^-
Br

3-Bromo-4-ethyl benzene diazonium chloride

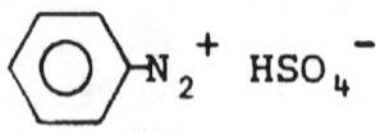

N_2^+ HSO_4^-

Benzene diazonium hydrogen sulfate

Nomenclature

Compounds containing the group $-N \equiv N^+$ are called diazonium salts. When naming aromatic diazonium salts, the word "diazonium" is added to the name of the aryl group, followed by the name of the anion.

Reactions of Aromatic Diazonium Salts

A) Nucleophilic displacement of nitrogen.

$$ArN_2^+ + :Z \rightarrow ArZ + N_2$$

a) Replacement by hydroxyl.

$$ArN_2^+ + H:OH \xrightarrow{H^+} \underset{\text{phenol}}{ArOH} + N_2$$

CH_3
N_2^+ HSO_4^-
+ H_2O
H^+
heat
CH_3
OH
+ N_2
+ H_2SO_4

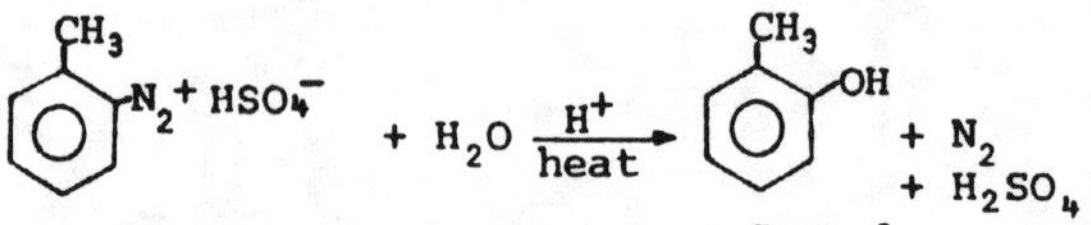

o-Cresol

b) Replacement by hydrogen.

$$ArN_2^+ + HO:H \xrightarrow{H_3PO_2} Ar\,H + H_3PO_3 + N_2$$

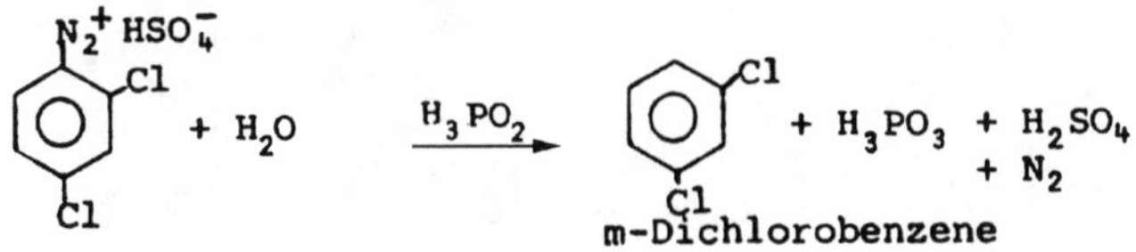

m-Dichlorobenzene

c) Replacement by iodine.

$$ArN_2^+ + :I^- \rightarrow ArI + N_2$$

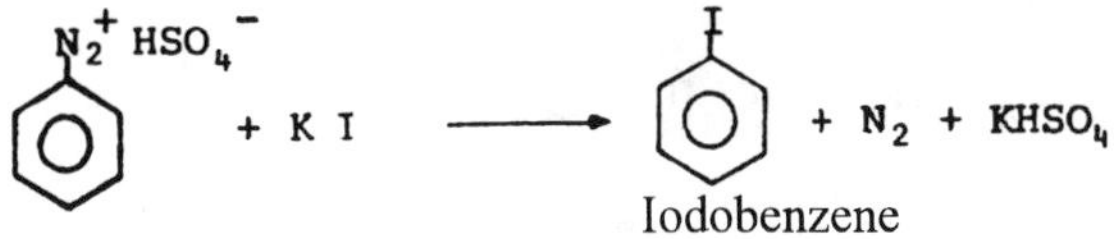

Iodobenzene

d) Replacement by fluorine.

$$ArN_2^+ \; BF_4^- \xrightarrow{\text{heat}} ArF + N_2 + BF_3$$

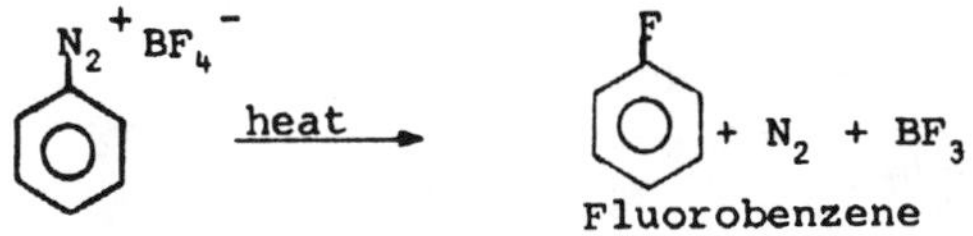

Fluorobenzene

e) Replacement by chlorine, bromine, or the nitrile group. Sandmeyer reaction.

1) $ArN_2^+ + Cu:Cl \rightarrow ArCl + N_2$

2) $ArN_2^+ + Cu:Br \rightarrow ArBr + N_2$

3) $ArN_2^+ + Cu:CN \rightarrow ArCN + N_2$

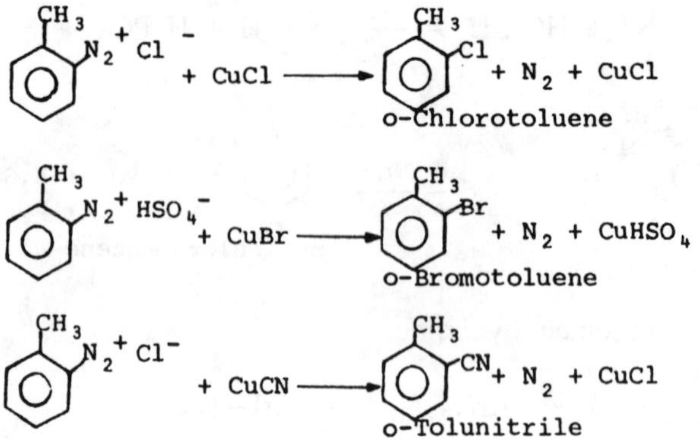

$$\text{o-Chlorotoluene}$$

$$\text{o-Bromotoluene}$$

$$\text{o-Tolunitrile}$$

f) Replacement by the thiol group (–SH).

Example

$$\underset{\text{potassium ethyl xanthate}}{C_6H_5N_2^+SO_4^- + EtO-\overset{\overset{S}{\|}}{C}-SK} \rightarrow C_6H_5S-\overset{\overset{S}{\|}}{C}-OEt \xrightarrow{\text{KOH,hyd}}$$

$$EtOH + COS + C_6H_5S^-K^+ \xrightarrow{H_3O^+} C_6H_5SH$$

g) Replacement by the nitro group.

Example

$$C_6H_5N_2^+BF_4^- + NaNO_2 \xrightarrow{Cu} C_6H_5NO_2 + NABF_4 + N_2$$

h) The Gatterman reaction. Replacement by aryl group.

Example

$$2C_6H_5N_2^+ \xrightarrow{Cu} C_6H_5-C_6H_5$$
$$\text{Biphenyl}$$

B) Reduction to hydrazines

Example

$$C_6H_5N_2^+ Cl^- \xrightarrow{Na_2SO_3} C_6H_5NHNH_2$$

C) Coupling

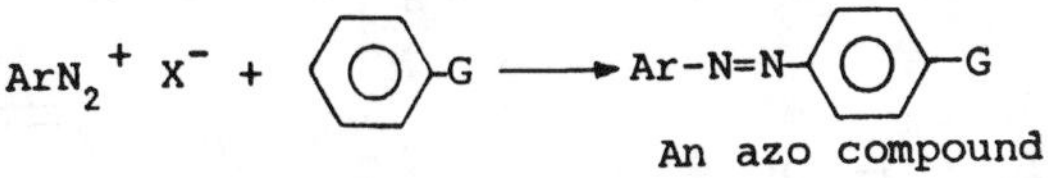

$$ArN_2^+ \ X^- \ + \ \langle O \rangle{-}G \longrightarrow Ar{-}N{=}N{-}\langle O \rangle{-}G$$

An azo compound

G must be a strong electron-releasing group:

OH, NR$_2$, NHR, NH$_2$

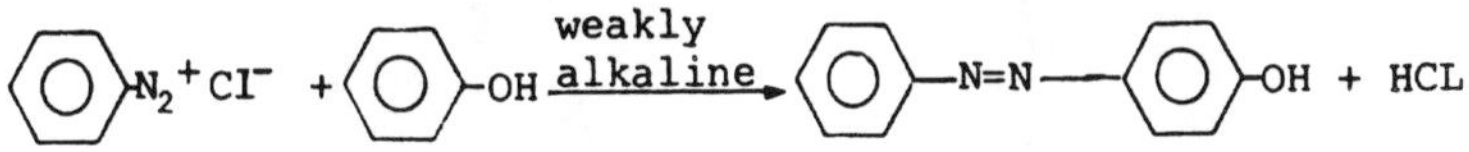

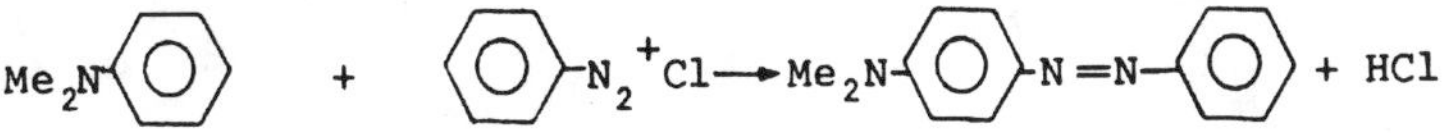

$$\langle O \rangle{-}N_2^+Cl^- \ + \ \langle O \rangle{-}OH \xrightarrow[\text{alkaline}]{\text{weakly}} \langle O \rangle{-}N{=}N{-}\langle O \rangle{-}OH \ + \ HCL$$

$$Me_2N\langle O \rangle \quad + \quad \langle O \rangle{-}N_2^+Cl \longrightarrow Me_2N{-}\langle O \rangle{-}N{=}N{-}\langle O \rangle \ + \ HCl$$

Problem Solving Examples:

Benzenediazonium chloride couples with phenol, but not with the less reactive anisole. 2,4-dinitrobenzenediazonium chloride, however, couples with anisole; 2,4,6-trinitrobenzenediazonium chloride even couples with the hydrocarbon mesitylene (1,3,5-trimethylbenzene).

(a) How can you account for these differences in behavior?

(b) Would you expect p-toluenediazonium chloride to be more or less reactive as a coupling reagent than benzene-diazonium chloride?

(a) Diazo coupling results from the attack of a diazonium salt on a benzene ring containing highly activating substituents (e.g., –O⁻, –N(CH₃)₂, –OH). The coupling reaction proceeds through an electrophilic aromatic substitution mechanism whereby the electron-releasing activating group serves to stabilize the positive charge placed on the ring by the reaction with the electrophilic diazonium salt.

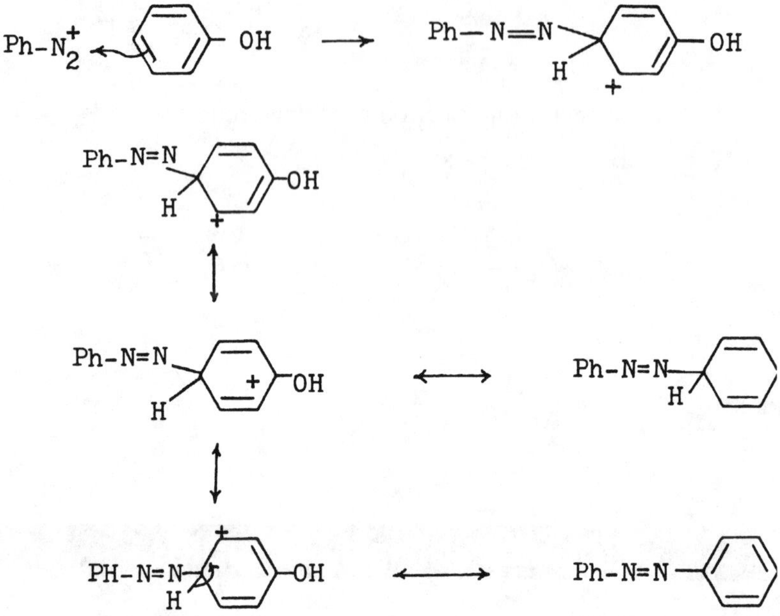

The methoxy group on anisole is not as strong an electron-releasing group as the hydroxy group on phenol, hence, it will not activate the aromatic ring sufficiently to allow a reaction with benzene-diazonium chloride.

The presence of the nitro groups on the diazonium salt, however, serves to increase the electrophilicity of the diazo group, due to its electron-withdrawing properties.

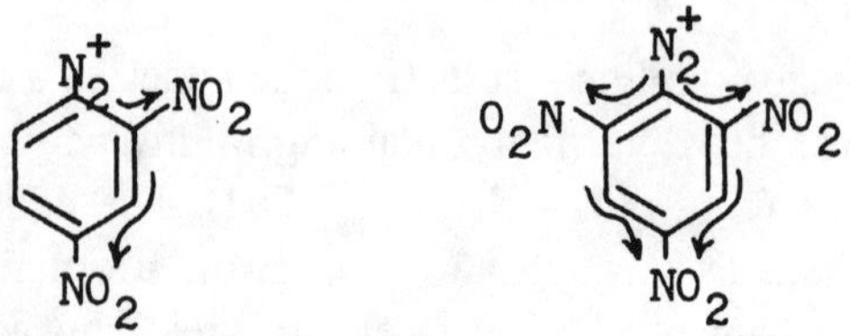

The increase in the electrophilicity of the diazo group permits it to couple with benzene rings substituted with substituents of relatively low activating energy. Thus, two nitro groups on a benzene-diazonium salt will allow it to react with anisole, three with 1,3,5-trimethylbenzene.

(b) Conversely, the presence of an electron-releasing group on the diazonium salt would serve to decrease the electrophilicity of the azo group. This, in turn, would decrease the diazo coupling

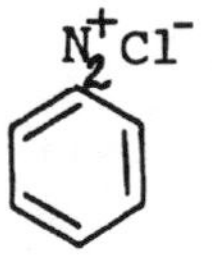

benzene
diazonium chloride

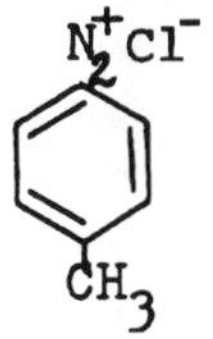

p-toulene
diazonium chloride

ability of a particular compound. Hence, one would expect benzene-diazonium chloride to be more reactive than p-toluene diazonium chloride, which possesses an electron-releasing methyl.

Q Diazotization of 2,4-dinitroaniline in aqueous solution is accompanied by some conversion to phenols in which a nitro group is replaced by a hydroxyl group.

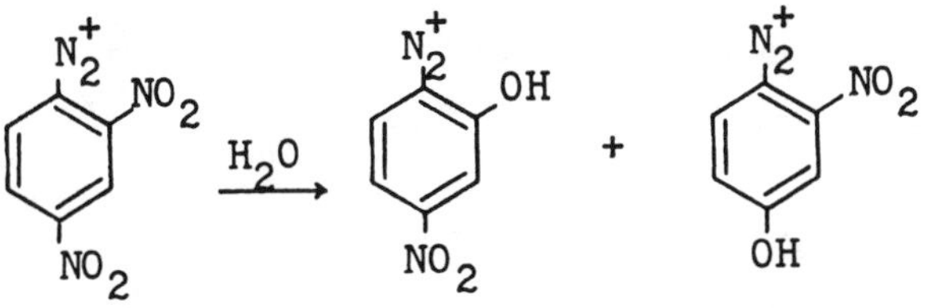

Give a reasonable mechanism for this reaction.

A The conversion of small amounts of 2,4-dinitroaniline to phenol can be accounted for by a nucleophilic aromatic substitution mechanism in which the diazo group acts as an activating group.

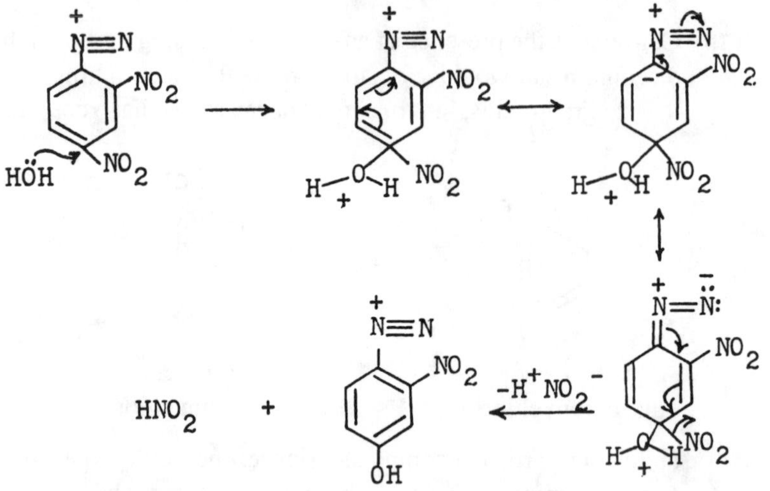

The production of the ortho phenol will proceed by an identical mechanism.

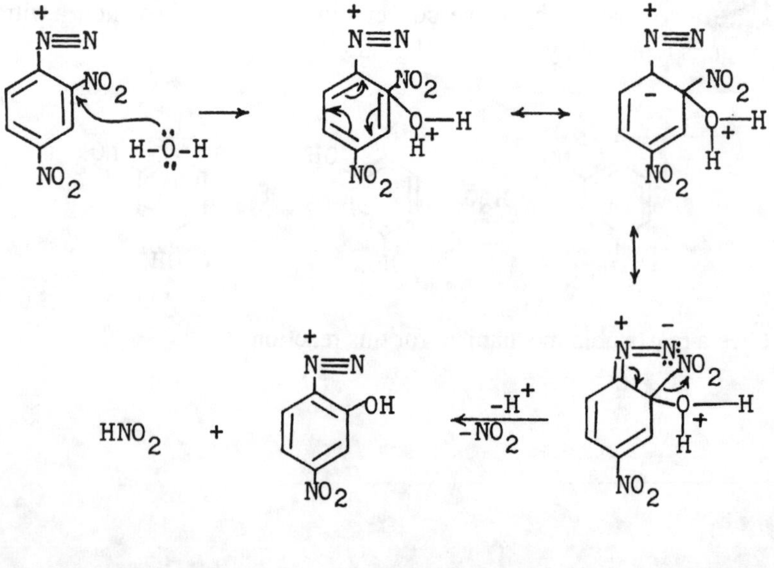

Quiz: Amines

1. The evolution of nitrogen gas upon the addition of sodium nitrite in mineral acid solution identifies the presence of

   (A) an aromatic secondary amine.

   (B) a tertiary amine.

   (C) an aliphatic primary amine.

   (D) an aromatic primary amine.

   (E) an aliphatic tertiary amine.

2. Which of the following is designated dimethylaminoethane under IUPAC rules of nomenclature?

   (A) $CH_3CH_2N(CH_3)_2$

   (D) $(CH_3)_2CH-C\equiv N$

   (B) $(CH_3)_2CH-CH_2NH_2$

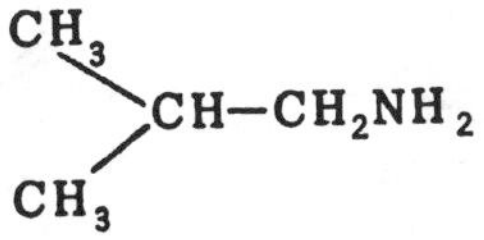

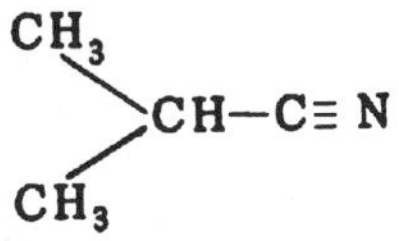

   (E) None of the above.

   (C) $(CH_3)_3CNH_2$

3. Heating of nitrobenzene, in iron and dilute hydrochloric acid, in the presence of a catalyst, yields which one of the following structures as the final product?

(A)

(D)

(B)

(E)

(C)

4. Amines CANNOT be formed by reduction of

   (A) nitriles.   (D) imines.

   (B) oximes.   (E) amides.

   (C) thiols.

5. Which one of the following reactions is unlikely to be correct?

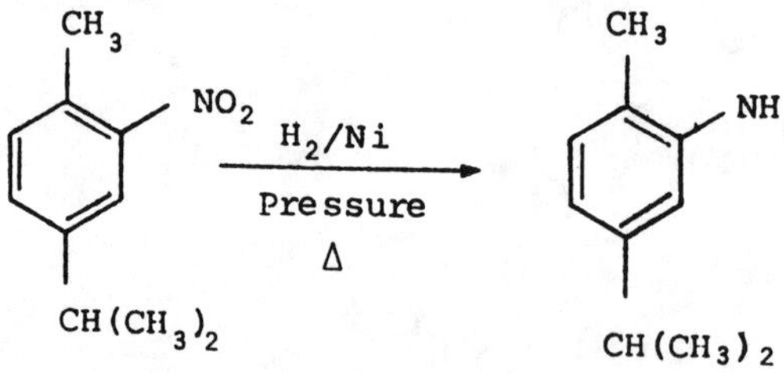

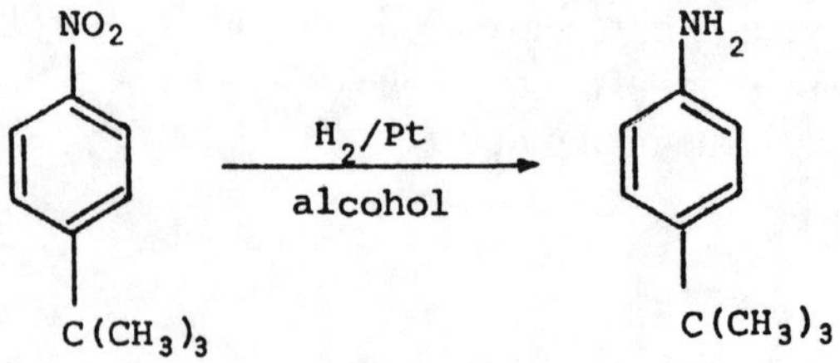

(C)

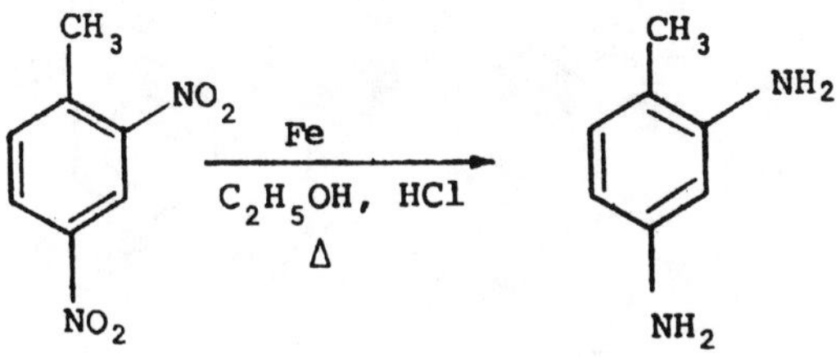

$$\xrightarrow[\substack{C_2H_5OH, \ HCl \\ \Delta}]{Fe}$$

(D)

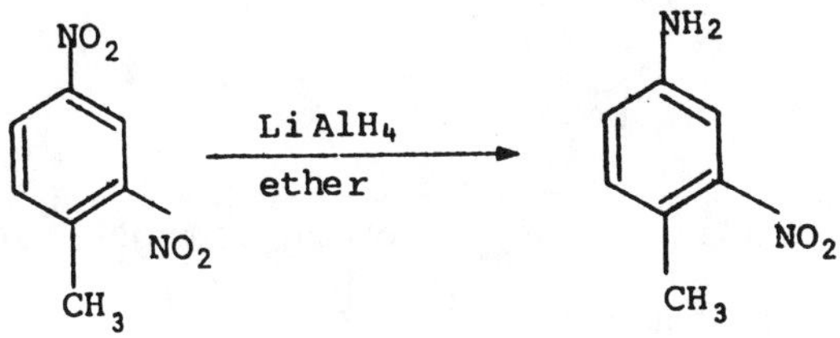

$$\xrightarrow[ether]{Li \ AlH_4}$$

(E)

$$\xrightarrow[alc. \ Na \ OH]{Zn}$$

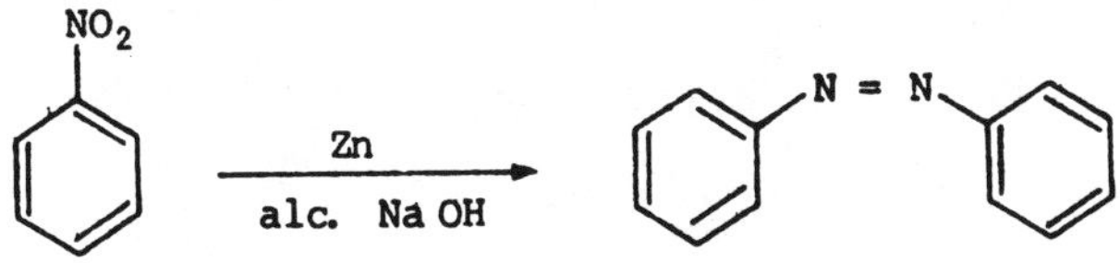

6. The reaction below takes place by nucleophilic substitution; which of the following is the product for this reaction?

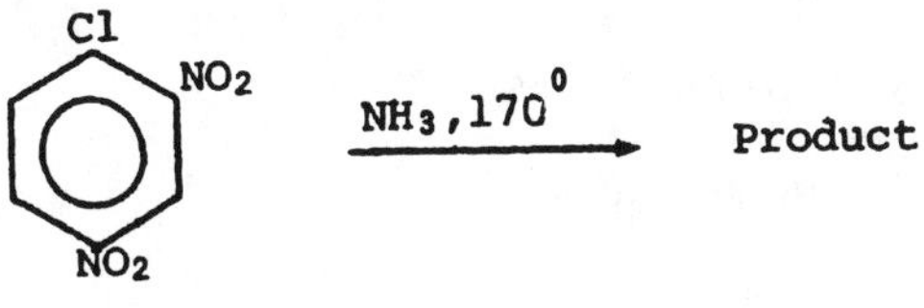

$$\xrightarrow{NH_3, 170^0} \quad Product$$

(A)

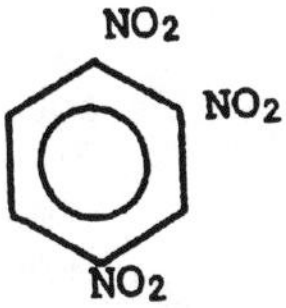

(B)

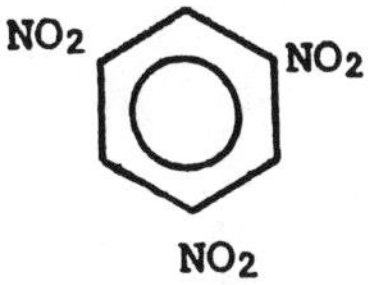

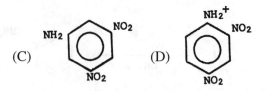

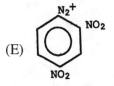

7. All of the following are physical properties of amines EXCEPT

   (A) nonpolar.

   (B) hydrogen bonding capability.

   (C) lower boiling point than alcohols.

   (D) soluble in ether.

   (E) may exist as gases.

8. All of the following are products of reactions of amines EXCEPT

   (A) amides.

   (B) diazonium salts.

   (C) ammonium acids.

   (D) ketones.

   (E) aldehydes.

9. Aromatic diazonium salts contain which one of the following groups?

   (A) $ArNH_2$ (D) $N \equiv N^+$

   (B) NH_2 (E) NO_2

   (C) NH_3^+

10. Aromatic diazonium salts may undergo replacement by all of the following EXCEPT

    (A) thiol groups.   (D) hydrogen.

    (B) iodine.   (E) phenol.

    (C) nitro groups.

ANSWER KEY

1.	(C)		6.	(D)
2.	(A)		7.	(A)
3.	(A)		8.	(C)
4.	(C)		9.	(D)
5.	(D)		10.	(E)

CHAPTER 4

Phenols and Quinones

4.1 Nomenclature of Phenols

Phenols have the general formula ArOH. The –OH group in phenols is attached directly to the aromatic ring.

Phenols are named as derivatives of phenol, which is the simplest member of the family. Methyl phenols are given the name cresols. Phenols are also called "hydroxy-" compounds.

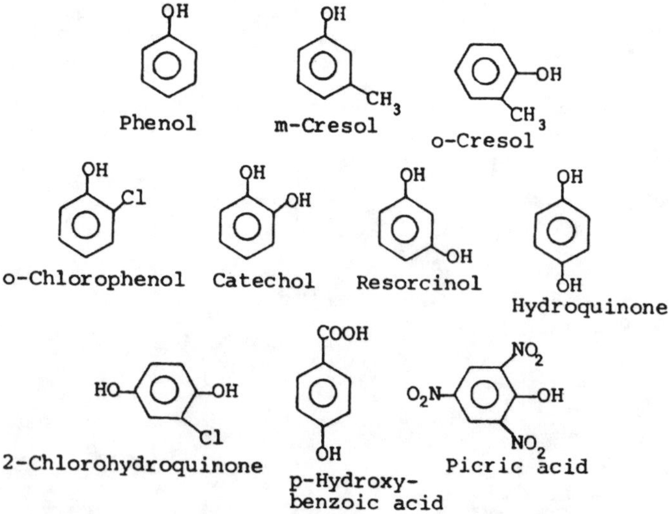

Phenol

m-Cresol

o-Cresol

o-Chlorophenol   Catechol   Resorcinol   Hydroquinone

2-Chlorohydroquinone

p-Hydroxy-
benzoic acid

Picric acid

Salicylic acid   Phloroglucinol

Vanillin   3,4-Xylenol   β-Naphthol (or 2-Naphthol)   α-Naphthol

Problem Solving Example:

Write structures for each of the following names:

(a) m-cresol   (e) p(t-tolyl)azophenol

(b) 3-hydroxybenzenesulfonamide   (f) benzyl phenyl ether

(c) 3-chloro-1,2-benzoquinone   (g) 3-(o-hydroxyphenyl)

(d) o-methoxyphenol   pentanoic acid

     (h) 2-methoxy-1,4-naphthoquinone

(a) Cresols are phenols with methyl substituents. Meta-cresol has a methyl group in the "3" position with the hydroxyl group in the "1" position. The structure of this compound is

(b) This compound, 3-hydroxybenzenesulfonamide, is a derivative of benzenesulfonic acid: . 3-hydroxybenzenesulfonic

acid has an alcohol functionality in the "3" position

The amide derivative of this substituted sulfonic acid is the

compound we desire:

SO₂NH₂ ... OH structure (benzene ring with SO₂NH₂ and OH substituents)

(c) 1,2-benzoquinone is a dicarbonyl compound, derived from the oxidation of catechol:

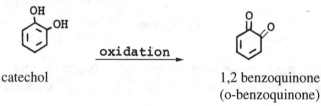

catechol   1,2 benzoquinone
                        (o-benzoquinone)

The numbering scheme for 1,2 benzoquinone is: A chlorine substituent in the "3" position gives us the compound 3-chloro-1,2-benzoquinone

(d) The methoxy group is –OCH₃. In this substituted phenol, the methoxy group is ortho or adjacent to the alcohol functionality. The structure of this compound is:

OH / OCH₃ (benzene ring structure)   o-methoxy phenol

(2-methoxy phenol)

(e) The substituent, p-tolyl, is derived from toluene: (benzene ring with CH₃). It is in the para position where the attachment is made to the azo group (– N = N –) . This compound is a derivative of azobenzene;

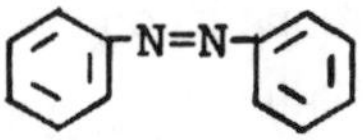

, with the two substituted benzene rings being

p-tolyl: 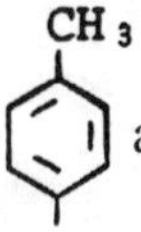 and phenol with the substitution occurring at the

para position:  . The compound that is named as a substi-

tuted phenol has the structure: 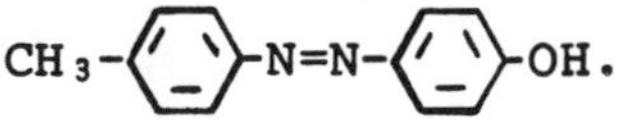 .

(f) As seen from its name, this is an ether with the general formula ROR'. Here, R is a benzyl substituent

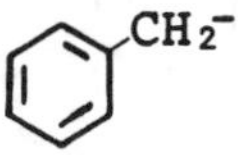

and R' is a phenyl group .

The structure is:

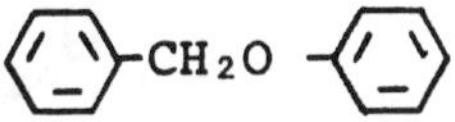

(g) This compound is a substituted carboxylic acid. The acid contains five carbons (pentanoic acid) and is substituted on the
third carbon from the carboxyl group. The substituent on the
carboxylic acid is a phenol with substitution occurring ortho
to the hydroxyl group. The structure is therefore:

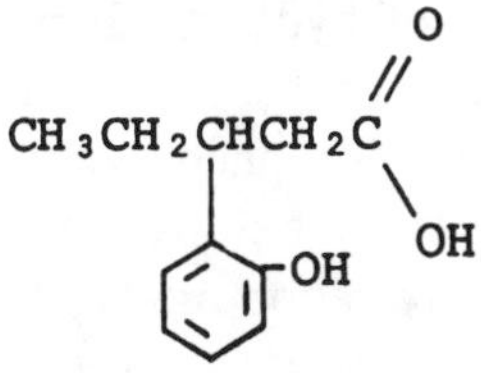

(h) 1,4 naphthoquinone is a dicarbonyl aromatic compound. It is derived from the oxidation of α-naphthol.

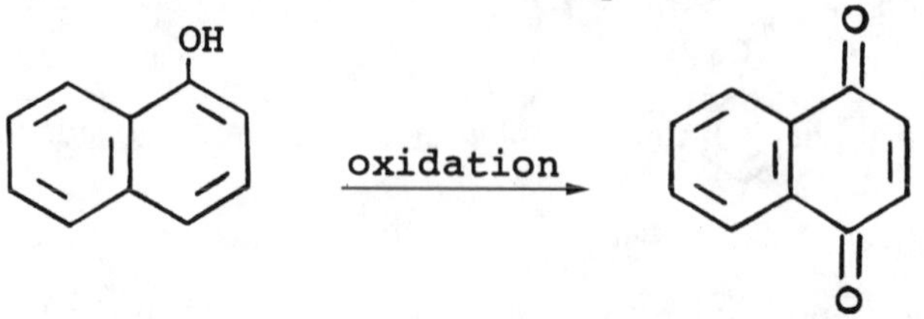

α-naphthol

1,4 naphthoquinone
(α-naphthoquinone)

The numbering scheme for 1,4 naphthoquinone is:

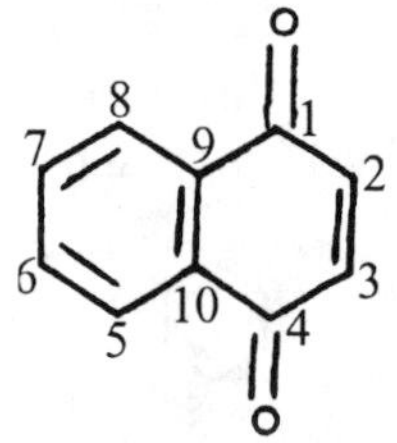

Here the substituent is the methoxyl group in the "2" position. The structure is therefore:

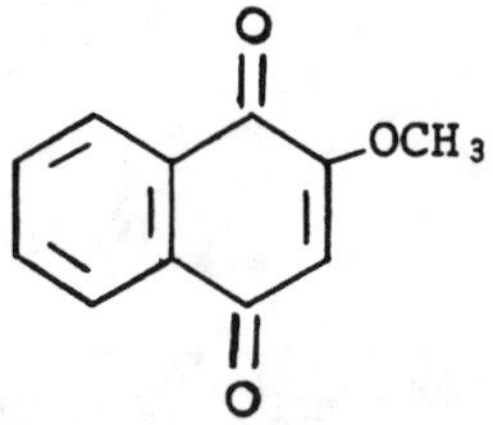

4.2   Physical Properties of Phenols

Simple phenols are low-melting solids or liquids. They have high boiling points because of hydrogen bonding. Phenol (C_6H_5OH) is soluble in water to some extent because of the hydrogen bonding present in both phenol and water. Most phenols are insoluble in water. Phenols are colorless. The are easily oxidized, except when purified. Many phenols are colored by oxidation products.

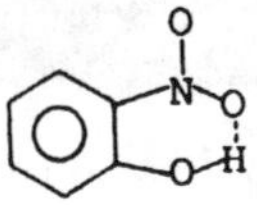

Intramolecular hydrogen bonding:

Chelation

The intramolecular hydrogen bonding within a single molecule, takes the place of intermolecular hydrogen bonding with other phenol molecules and with water molecules.

The holding of a hydrogen or metal atom between two atoms of a single molecule is called chelation.

Acidity of Phenols

$$ArOH \underset{H^+}{\overset{OH^-}{\rightleftharpoons}} ArO^-$$

a phenol   a phenoxide ion
(acid)   (salt)

insoluble in   soluble in water
water

The solubility properties of phenols and their salts are opposite; the salts are insoluble in organic solvents and soluble in water.

Phenols are stronger acids than alcohols because the negative charge created by deprotonation can be distributed in the phenol by resonance. Deprotonation lowers the energy level of the compound and also stabilizes it.

Because phenols dissolve in aqueous hydroxide, but not in aqueous bicarbonate, they are more acidic than water, but less acidic than carboxylic acid.

Problem Solving Examples:

Q In which of the following compounds would you expect intramolecular hydrogen bonding to occur: o-nitroaniline, o-cresol, o-hydroxybenzoic acid (salicylic acid), o-hydrobenzaldehyde (salicylaldehyde), o-fluorophenol, o-hydroxybenzonitrile?

A A hydrogen bond is an attraction between a hydrogen atom covalently bonded to an electronegative element and a second electronegative element. These electronegative elements are usually nitrogen, oxygen, and fluorine. If both electronegative elements and the proton are in the same molecule, it is called intramolecular hydrogen bonding. O-cresol cannot have intramolecular hydrogen bonding because it has only one electronegative element. (Carbon is not electronegative enough for hydrogen bonding.)

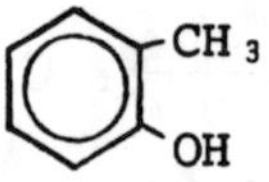

o-cresol

O-hydroxybenzonitrile cannot have hydrogen bonding either but for a different reason: the linear geometry of the nitrile group prevents significant interaction between the nitrogen and the hydroxyl proton.

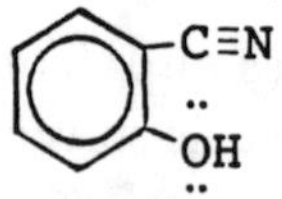

o-hydroxybenzonitrile

In both salicylic acid

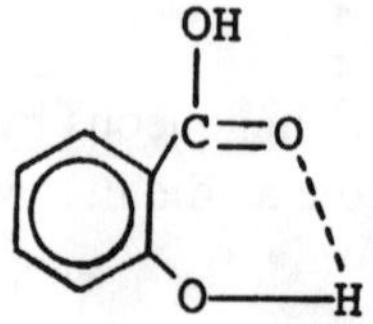

and salicaldehyde: the proton involved in hydrogen bond-

ing is between two oxygen atoms. In o-nitroaniline: it is

between nitrogen and oxygen.

In o-fluorophenol the proton is between fluorine and oxygen.

Q (a) Why is phenol C_6H_5OH a stronger acid than an alcohol? (b) Would para-acetylphenol be a stronger or weaker acid than phenol? (c) Would 2,4-dinitrophenol be a stronger or weaker acid than p-nitrophenol? Draw contributing structures to support the answers.

A Acidity, the tendency of a compound to release a proton, depends on the compound's surrounding environment and its stability after deprotonation. If it is in a nonpolar environment, therefore, not having a compound ready to accept the proton, it tends to be less acidic. If it is in a basic environment, surrounded by proton-acceptors, it tends to release the proton more readily.

(a) Phenol is more acidic than simple alcohols because the negative charge created by deprotonation can be distributed in phenol by resonance. This lowers the energy level of the deprotonated compound, thus stabilizes it. For an alcohol, deprotonation leads to RO–. For phenol, the negative charge can be shared by the resonating structures:

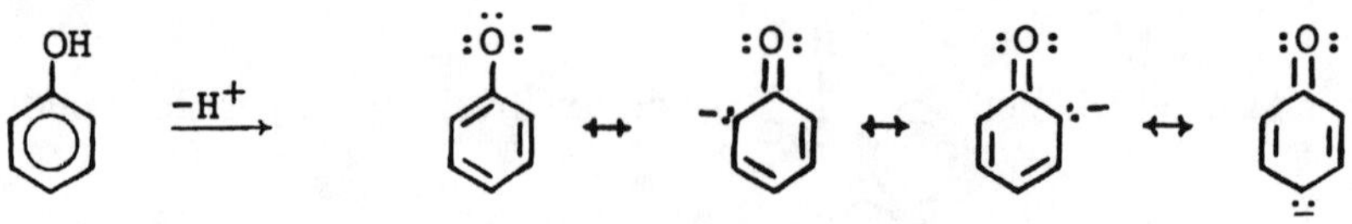

Therefore, phenol can release its proton more readily, and is thus a stronger acid than an alcohol.

(b) Para-acetylphenol, $CH_3C(=O)$ —⟨O⟩— OH , is more acidic than phe-

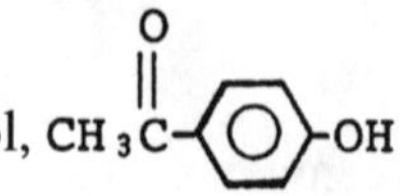

nol. The acetyl group, $CH_3C(=O)$—, can delocalize the negative charge. It adds an extra contributing structure, thus further stabilizes the deprotonated compound. The contributing structures are:

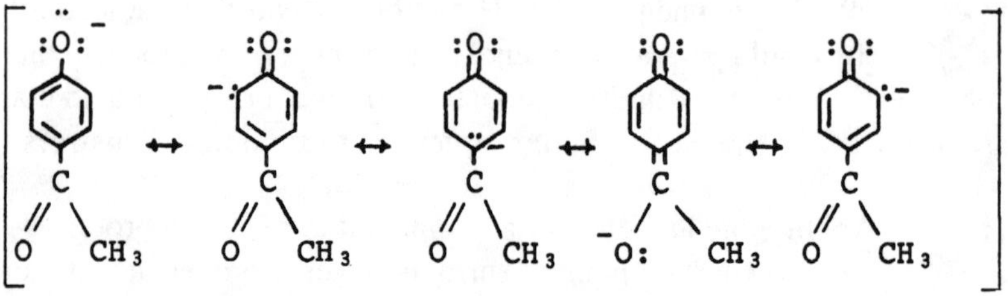

(c) 2,4-dinitrophenol and p-nitrophenol have the following structures, respectively.

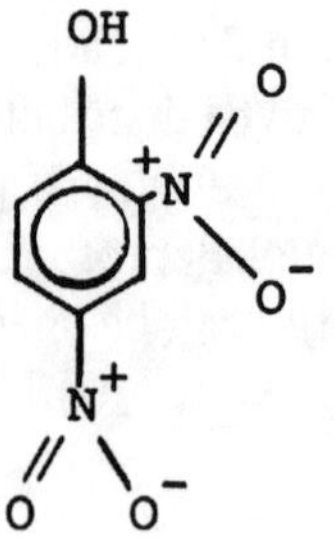

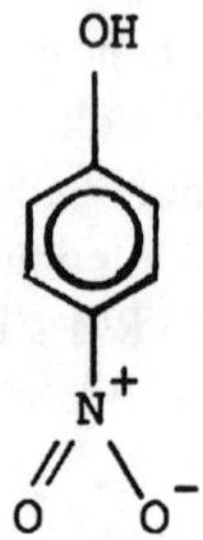

As can be seen from the previous answers, conjugated double bonds are essential for charge delocalization, which is the essence of extra stabilization. From the structures shown, one can see that 2,4-dinitrophenol has one more conjugated double bond than p-nitrophenol. This implies that 2,4-dinitrophenol has more contributing structures than p-nitrophenol; therefore, it is more acidic. The contributing structures for both deprotonated compounds are shown.

(1) p-nitrophenol (five major contributing structures)

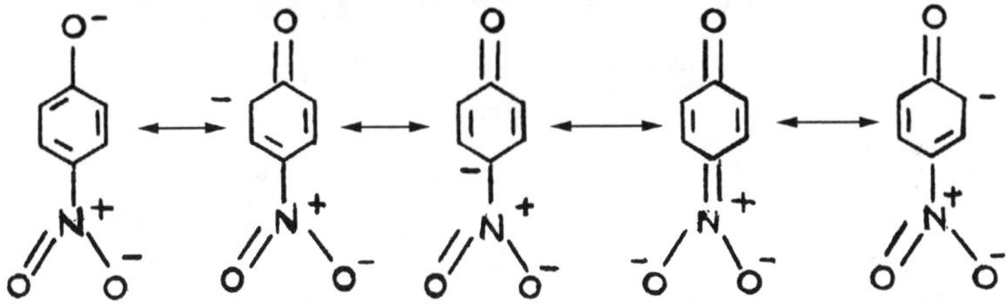

(2) 2,4-dinitrophenol (six major contributing structures)

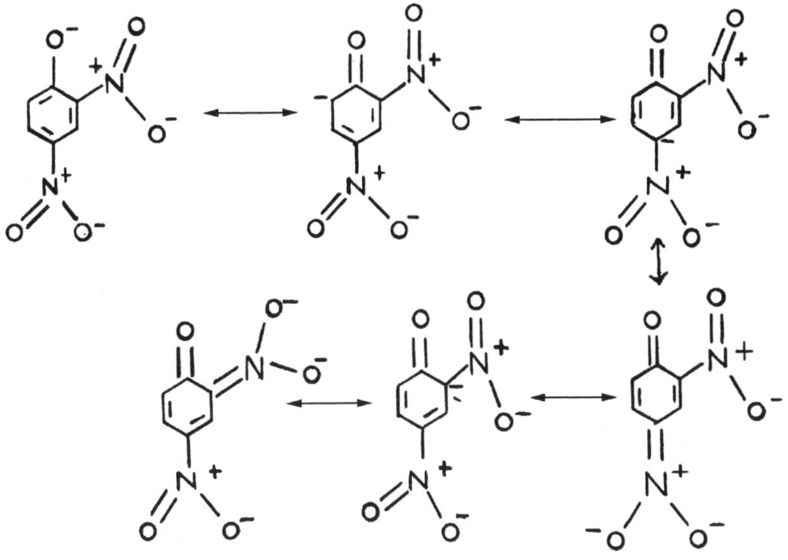

4.3   Preparation of Phenols

A) Nucleophilic displacement of halides

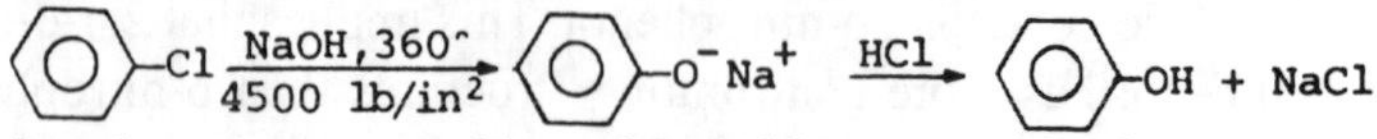

Chlorobenzene   Sodium phenoxide   Phenol

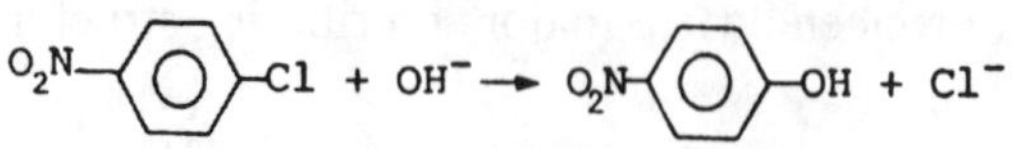

p-Chlorophenol   p-Nitrophenol

B) Oxidation of cumene

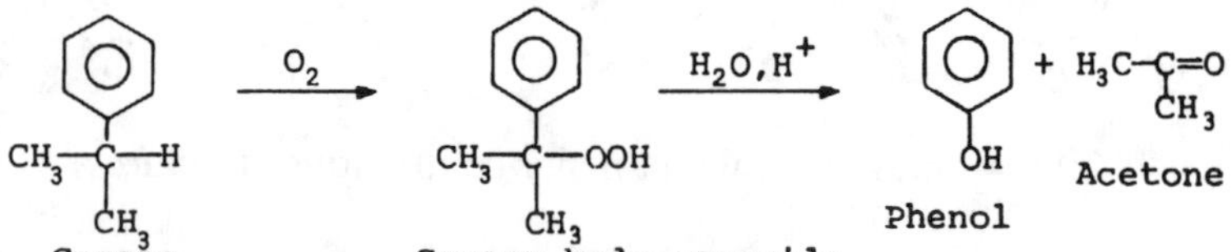

CH$_3$—C—H   CH$_3$—C—OOH   OH   Acetone

CH$_3$ CH$_3$ Phenol
Cumene   Cumene hydroperoxide

C) Hydrolysis of diazonium salts

$$ArN_2^+ + H_2O \rightarrow ArOH + H^+ + N_2$$

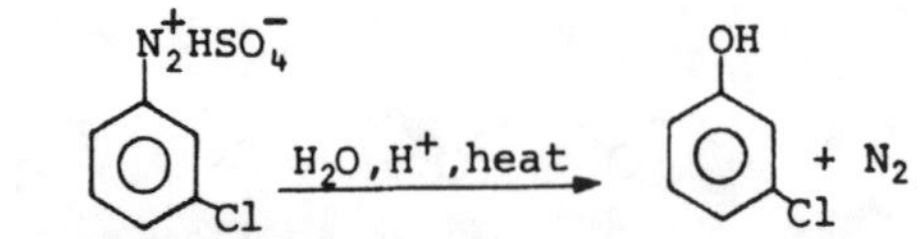

m-Chlorobenzene-   m-Chlorophenol
diazonium hydrogen
sulfate

D) Oxidation of organothallium compounds

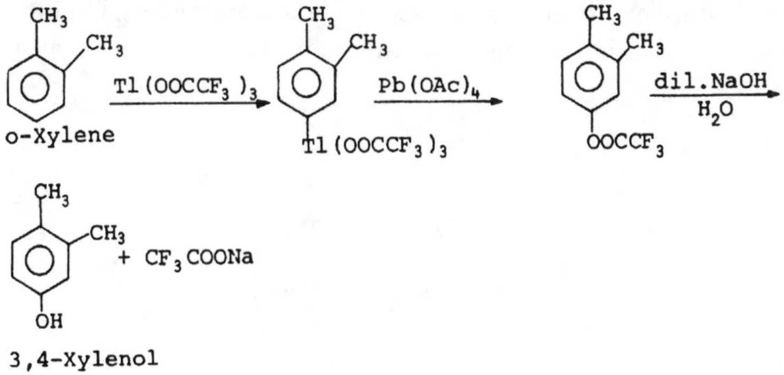
CH3
CH3
o-Xylene
Tl(OOCCF3)3
Tl(OOCCF3)3
Pb(OAc)4
OOCCF3
dil.NaOH
H2O
OH
+ CF3COONa
3,4-Xylenol

E) Oxidation of arylthallium compounds

$$ArH \xrightarrow{Tl(OOCCF_3)_3} ArTl(OOCCF_3)_2 \xrightarrow{Pb(OAc)_4}$$

arylthallium trifluoroacetate

$$ArOOCCF_3 \xrightarrow[heat]{H_2O, OH^-} ArO^- \xrightarrow{H^+} ArOH$$

trifluoroacetate

$$\uparrow Tl(OOCCF_3)_3$$

ArH

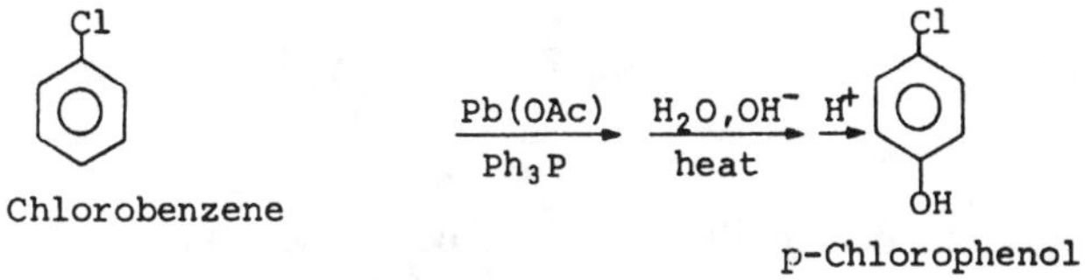
Cl
Chlorobenzene
Pb(OAc)
Ph3P
H2O,OH−
heat
H+
OH
p-Chlorophenol

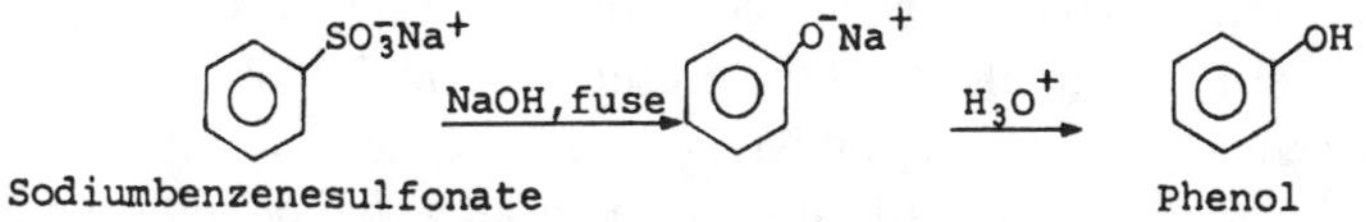
SO3−Na+
NaOH, fuse
O−Na+
H3O+
OH
Sodiumbenzenesulfonate
Phenol

Problem Solving Examples:

 4-n-hexylresorcinol is used in certain antiseptics. Outline its preparation starting with resorcinol and any aliphatic reagents.

OH

$CH_2(CH_2)_4CH_3$

4-n-hexylresorcinol

 4-n-hexylresorcinol has the structure:

OH

OH

$CH_2(CH_2)_4CH_3$

To synthesize this compound from the

polyhydric alcohol resorcinol $\left[\begin{array}{c} OH \\ \\ OH \end{array} \right]$, one would have to add a

straight six-carbon chain to the aromatic ring by an electrophilic addition. This is best done through a Friedel-Crafts acylation reaction. Here the hexanoyl cation

$$\left(\begin{array}{c} :O: \\ \parallel \\ CH_3(CH_2)_4C^+ \end{array} \right)$$

acts as an electrophile upon the electron system of an aromatic ring. Following this addition, the carbonyl group is reduced specifically by the Clemmensen reduction, which involves treating the carbonyl compound with zinc amalgam in the presence of acid. It should be noted that Friedel-Crafts alkylation would be inadequate for this procedure because an alkyl cation (desired from an alkyl halide, etc.) will undergo rearrangement unlike the acylium ion and also in Friedel-Crafts alkylation, there is the significant danger of polysubstitution upon the aromatic ring. This is particularly relevant in this case since resorcinol, even to a greater extent than phenol itself, is activated toward electrophilic attack. This can be seen in resorcinol's resonance structures if oxygen donates a pair of electrons to the benzene ring to give a negative charge within the ring:

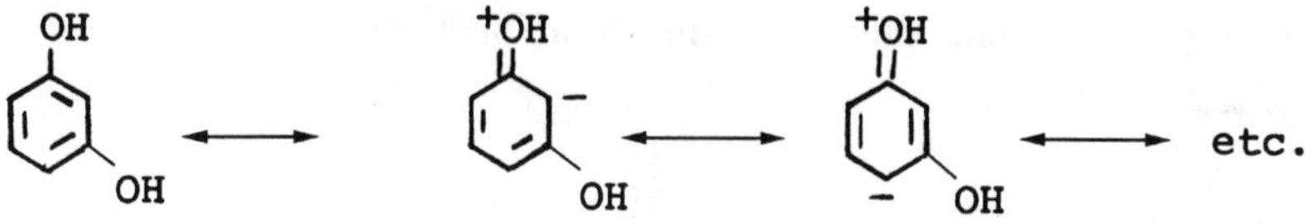

The carbonyl group acts as a deactivating group when substituted directly on a ring, and therefore, polyacylation is rare. The synthesis of 4-n-hexylresorcinol proceeds as follows:

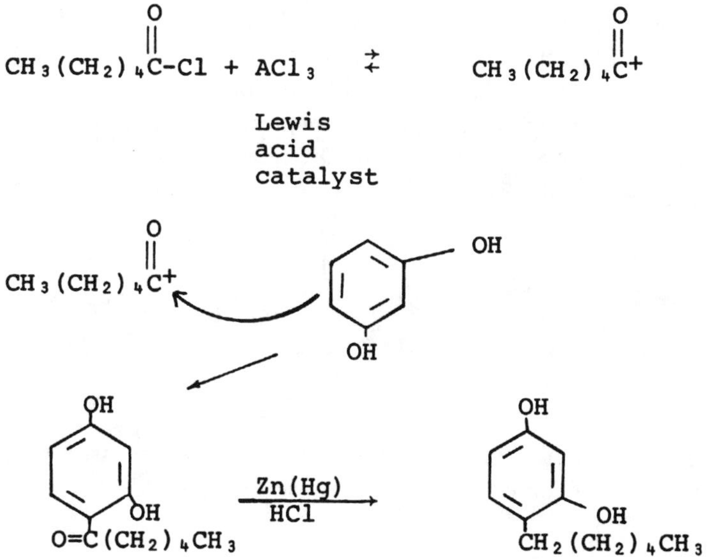

It should be noted that the hydroxyl group is an ortho-para director. Also, there is very little substitution between the hydroxyl groups of resorcinol for steric reasons.

 Suggest a mechanism for the steps in the synthesis of phenolphthalein.

 Phenolphthalein, a pH indicator, has the structure:

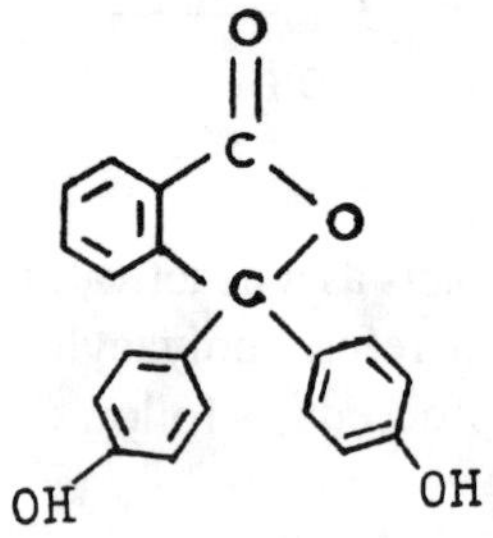

It is synthesized by reacting one mole of phthallic anhydride with two moles of phenol in the presence of a strong acid (i.e., H_2SO_4). The mechanism involves the sequential formation of two resonance stabilized cations. The first is formed by the protonation of one of phthallic anhydride's carbonyl groups. The subsequent cation is resonance stabilized by the adjacent oxygen atom:

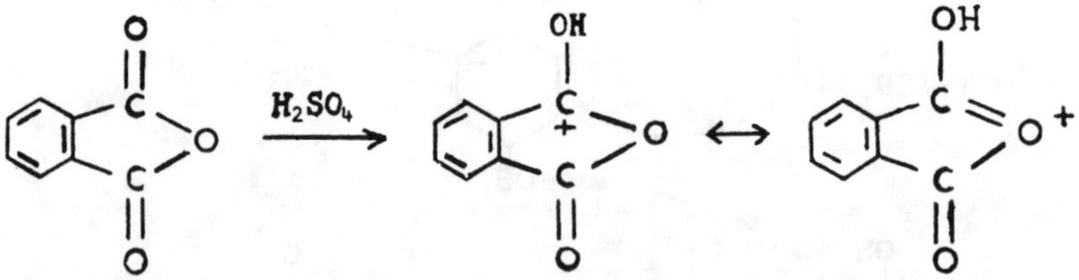

This cation is an electrophile that may attack the electron-rich aromatic ring of phenol to give a para substituted phenol.

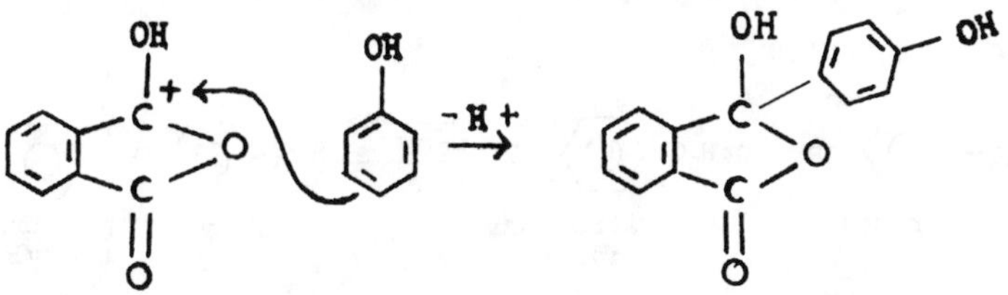

This substituted phenol still has a hydroxyl group that may be protonated to give a resonance stabilized cation. Protonation occurs at this hydroxyl because of the dibenzylic cation produced. This may act as an electrophile upon phenol's aromatic system to give phenolphthalein, containing two parasubstituted phenols.

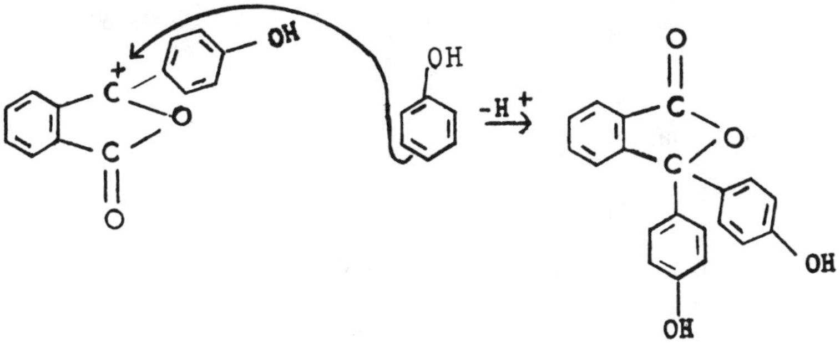

4.4   Reactions of Phenols

A) Salt formation

$$ArOH + H_2O \leftrightharpoons ArO^- + H_3O^+$$

Phenol   Sodiumphenoxide

B) Ether formation—Williamson synthesis

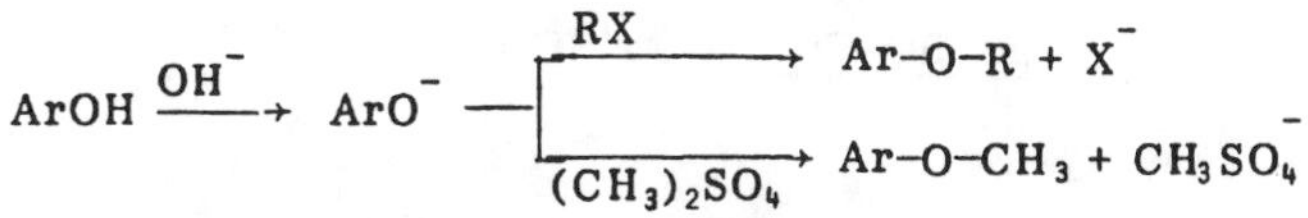

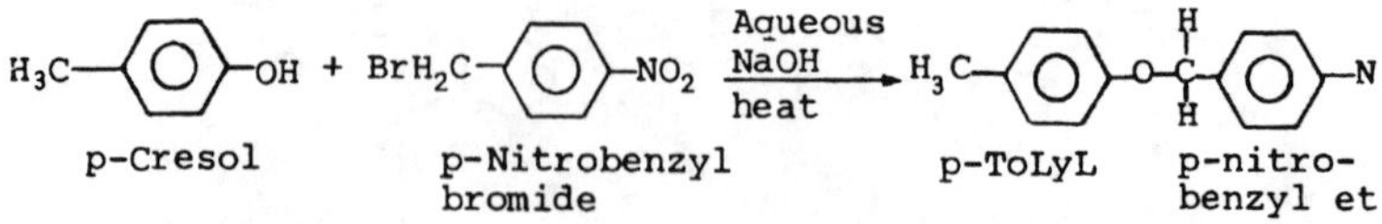

$$H_3C-\text{(ring)}-OH + BrH_2C-\text{(ring)}-NO_2 \xrightarrow[\text{heat}]{\text{Aqueous NaOH}} H_3C-\text{(ring)}-O-\overset{H}{\underset{H}{C}}-\text{(ring)}-N$$

p-Cresol   p-Nitrobenzyl bromide   p-ToLyL   p-nitro-benzyl et

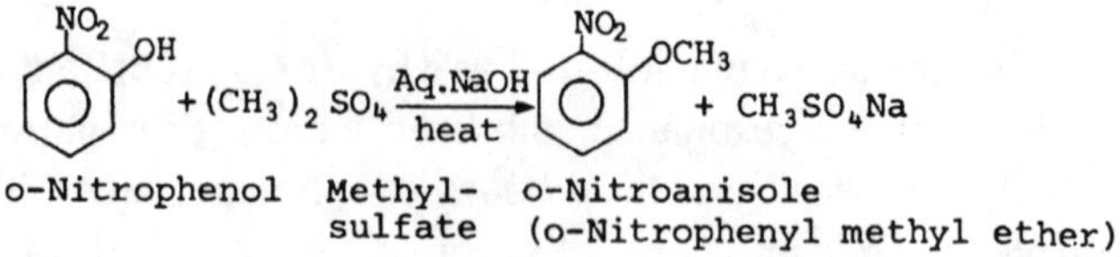

o-Nitrophenol   Methyl-sulfate   o-Nitroanisole (o-Nitrophenyl methyl ether)

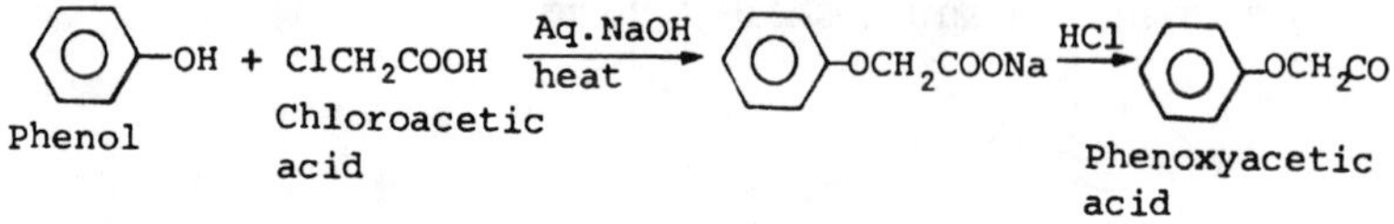

Phenol   Chloroacetic acid   Phenoxyacetic acid

C) Ester formation

$$HOAc + ArO\overset{O}{\overset{||}{C}}CCH_3 \xleftarrow{Ac_2O} ArOH \xrightarrow[\text{ArCOCl}]{\text{RCOCl}}$$

$$R\overset{O}{\overset{||}{C}}OAr\ (Ar\overset{O}{\overset{||}{C}}OAr) + HCl$$

$$TlCl + ArOTs \xleftarrow{TsCl} ArOTl \xrightarrow[\text{ArCOCl}]{\text{RCOCl}}$$

$$R\overset{O}{\overset{||}{C}}OAr\ (Ar\overset{O}{\overset{||}{C}}OAr) + TlCl$$

$$ArOH \begin{cases} \xrightarrow[\text{(ArCOCl)}]{\text{RCOCl}} & RCOOAr(ArCOOAr) \\ \\ \xrightarrow{Ar'SO_2Cl} & Ar'SO_2OAr \end{cases}$$

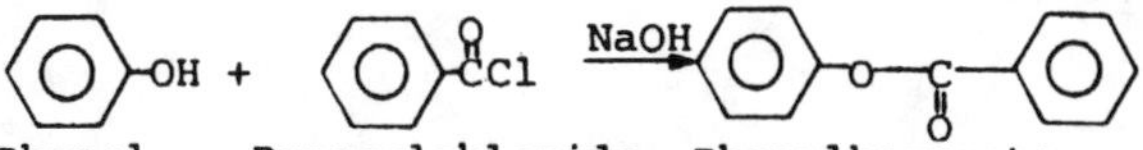

Phenol + Benzoylchloride → (NaOH) Phenylbenzoate

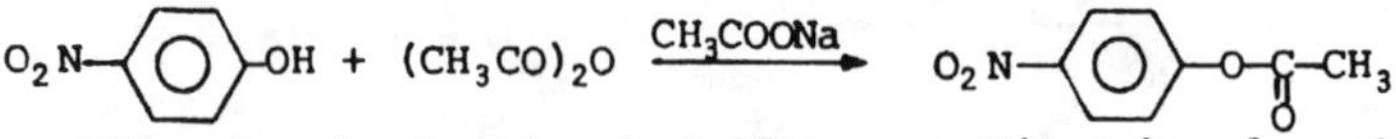

p-Nitrophenol + Aceticanhydride → (CH₃COONa) p-Nitrophenyl acetate

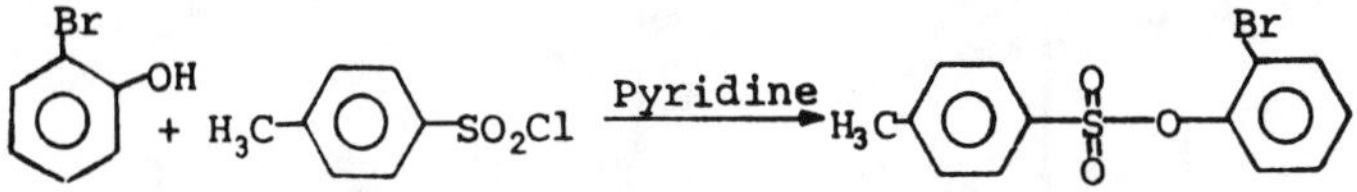

o-Bromophenol + p-Toluenesulfonyl chloride → (Pyridine) o-Bromophenyl-p-Toluene sulfonate

D) Ring substitution

-OH   Activate powerfully, and direct ortho, para in

-O⁻   Electrophilic aromatic substitution

-OR: Less powerful activator than -OH.

a) Nitration

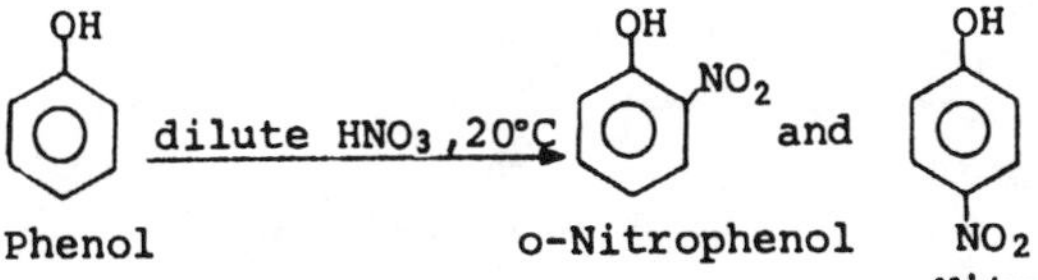

Phenol → (dilute HNO₃, 20°C) o-Nitrophenol   and   p-Nitrophenol

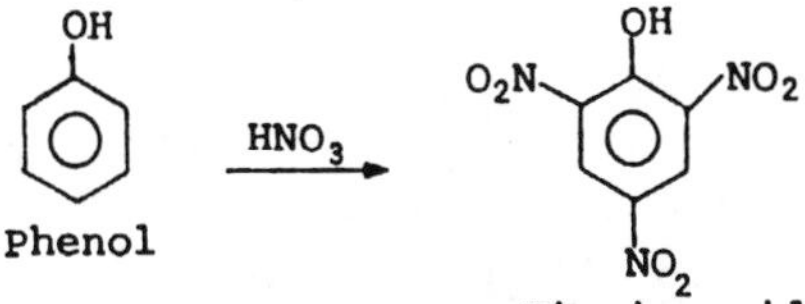

Phenol → (HNO₃) Picric acid

b) Sulfonation

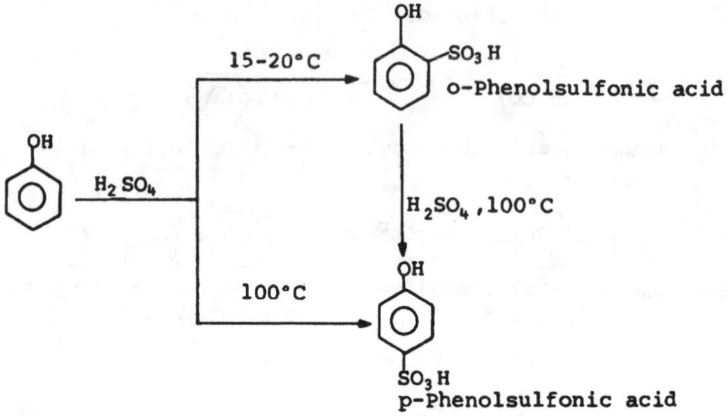
OH
H_2SO_4
15-20°C
OH
SO_3H
o-Phenolsulfonic acid
H_2SO_4, 100°C
100°C
OH
SO_3H
p-Phenolsulfonic acid

c) Halogenation

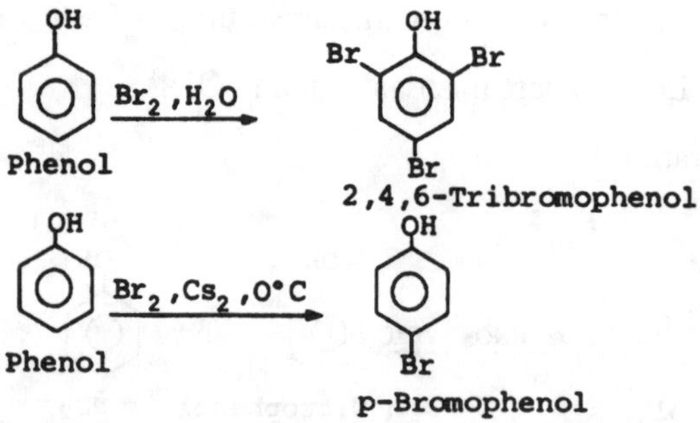
OH
Br_2, H_2O
Phenol
OH
Br
Br
Br
2,4,6-Tribromophenol
OH
Br_2, Cs_2, 0°C
Phenol
OH
Br
p-Bromophenol

d) Nitrosation

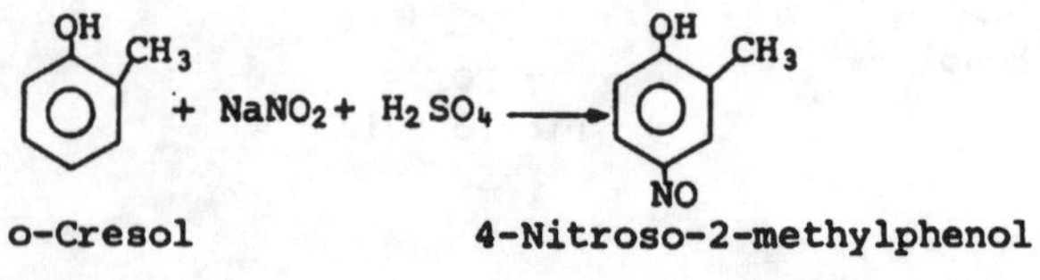
OH
CH_3
+ $NaNO_2$ + H_2SO_4
o-Cresol
OH
CH_3
NO
4-Nitroso-2-methylphenol

e) Friedel-Crafts alkylation

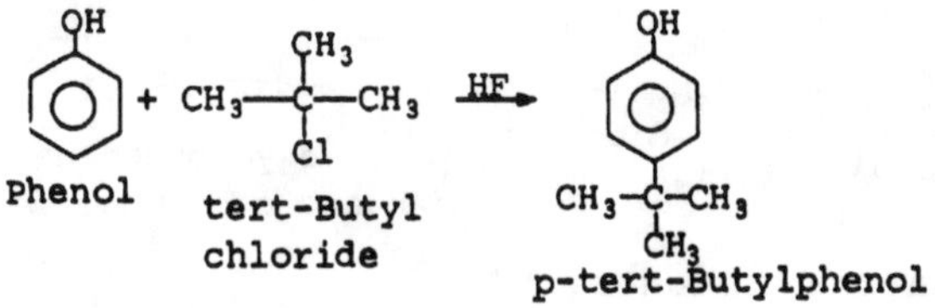

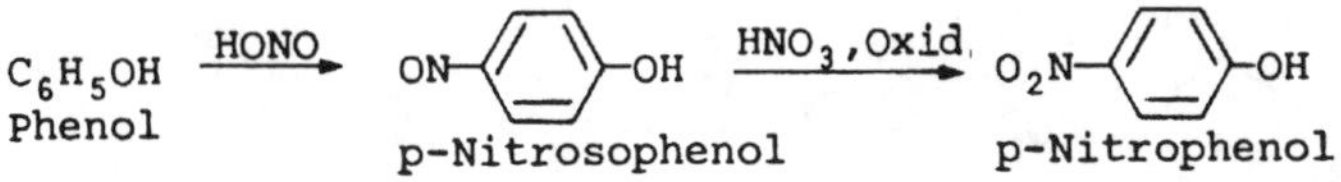

Phenol + tert-Butyl chloride →(HF) p-tert-Butylphenol

C_6H_5OH (Phenol) →(HONO) p-Nitrosophenol →(HNO_3, Oxid) p-Nitrophenol

f) Friedel-Crafts acylation.

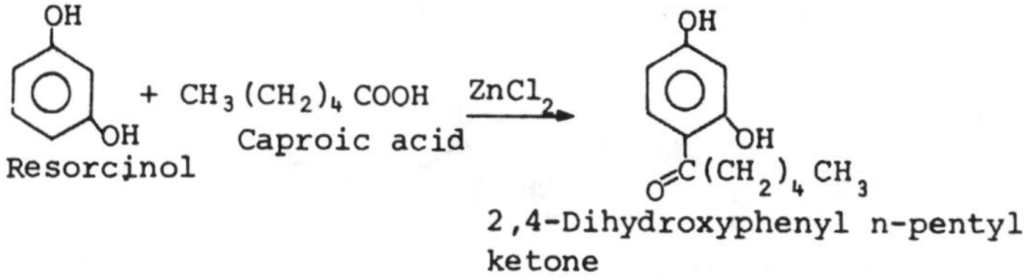

Resorcinol + $CH_3(CH_2)_4COOH$ (Caproic acid) →($ZnCl_2$) 2,4-Dihydroxyphenyl n-pentyl ketone

1) Fries Rearrangement

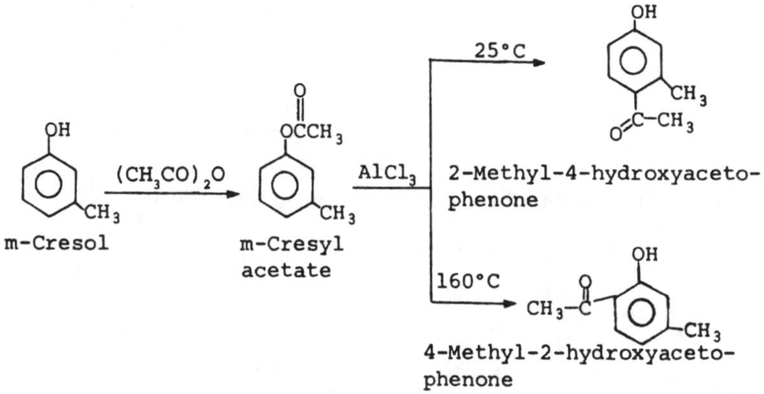

m-Cresol →$(CH_3CO)_2O$ m-Cresyl acetate →($AlCl_3$)

25°C → 2-Methyl-4-hydroxyaceto-phenone

160°C → 4-Methyl-2-hydroxyaceto-phenone

g) Coupling with diazonium salts

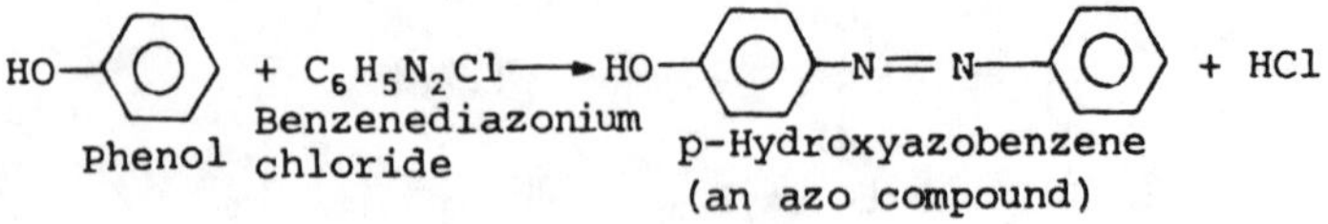

HO—⟨O⟩ + C$_6$H$_5$N$_2$Cl ⟶ HO—⟨O⟩—N=N—⟨O⟩ + HCl

Phenol   Benzenediazonium chloride   p-Hydroxyazobenzene (an azo compound)

h) Carbonation. Kolbe reaction.

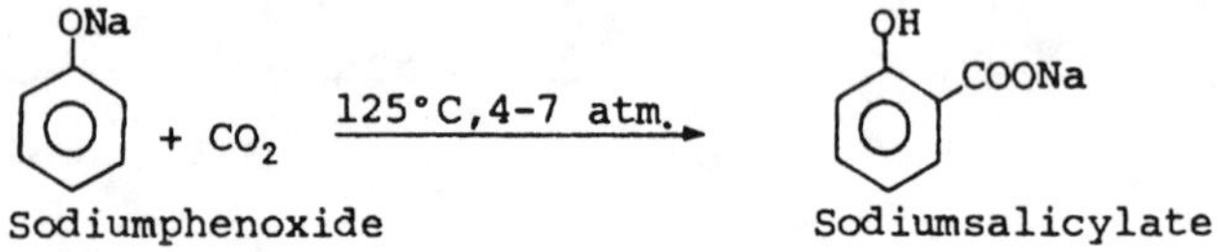

ONa

⟨O⟩ + CO$_2$ $\xrightarrow{125°C, 4-7 \text{ atm.}}$

Sodiumphenoxide

OH
⟨O⟩ COONa

Sodiumsalicylate

i) Reaction with formaldehyde

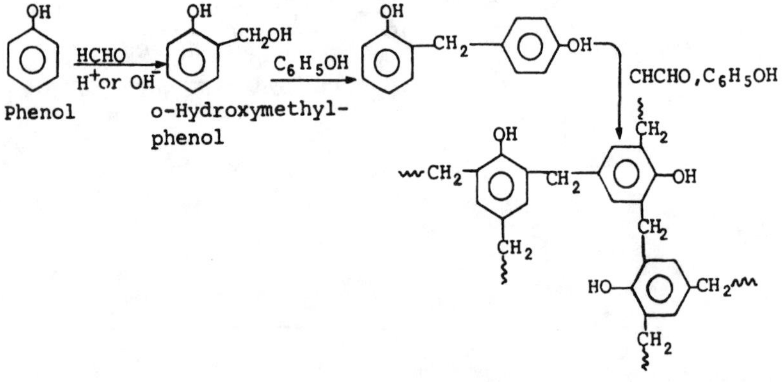

OH
⟨O⟩ $\xrightarrow[H^+ \text{or } OH^-]{HCHO}$

Phenol

OH
⟨O⟩ CH$_2$OH $\xrightarrow{C_6H_5OH}$

o-Hydroxymethyl-phenol

OH
⟨O⟩ CH$_2$—⟨O⟩—OH—

CHCHO, C$_6$H$_5$OH

∿CH$_2$—⟨O⟩ CH$_2$—⟨O⟩—OH

A phenol-formaldehyde resin

j) Introduction of the –COR group. Ketone formation.

1) Fries rearrangement

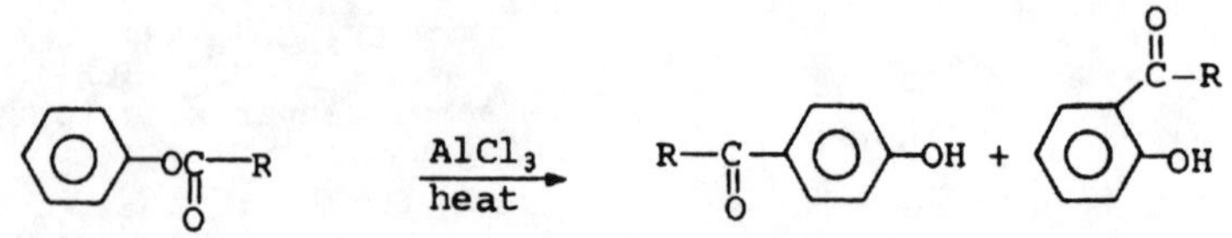

⟨O⟩—O—C—R $\xrightarrow[\text{heat}]{AlCl_3}$ R—C—⟨O⟩—OH + ⟨O⟩—OH
        ‖   ‖
        O   O

Phenyl ester   Acyl derivatives (ketones)

2) Houben-Hoesch synthesis

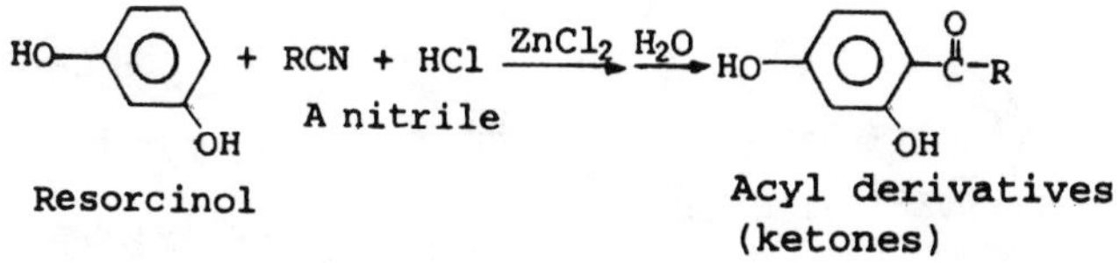

HO—⟨benzene⟩—OH + RCN + HCl $\xrightarrow{\text{ZnCl}_2}$ $\xrightarrow{\text{H}_2\text{O}}$ HO—⟨benzene⟩—C(=O)—R

A nitrile

Resorcinol

Acyl derivatives (ketones)

k) Introduction of the –CHO group (formylation). Aldehyde formation.

1) Reimer-Tiemann reaction

C_6H_5OH Phenol $\xrightarrow{\text{CHCl}_3,\text{NaOH(aq.)}}$ ⟨benzene⟩—OH, CHO

Salicylaldehyde

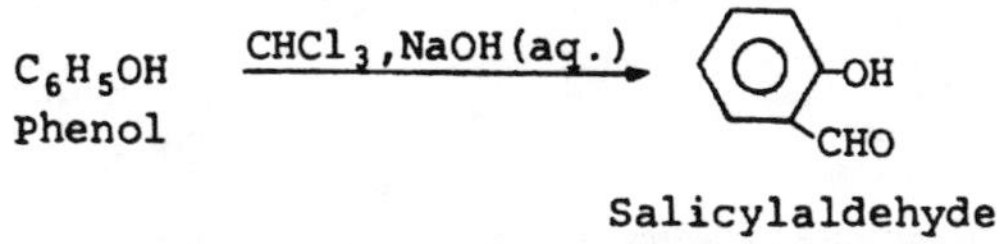

2) Gatterman reaction. Special case of the Houben-Hoesch reaction.

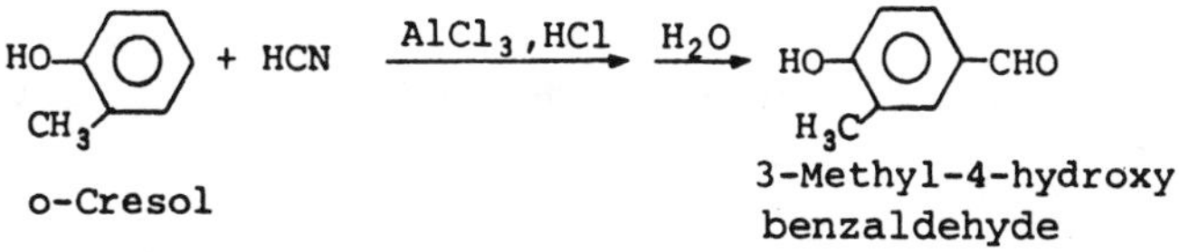

HO—⟨benzene⟩—CH₃ + HCN $\xrightarrow{\text{AlCl}_3,\text{HCl}}$ $\xrightarrow{\text{H}_2\text{O}}$ HO—⟨benzene⟩—CHO, H₃C

o-Cresol

3-Methyl-4-hydroxy benzaldehyde

l) Introduction of the –COOH group (carboxylation). Acid formation.

1) Kolbe synthesis

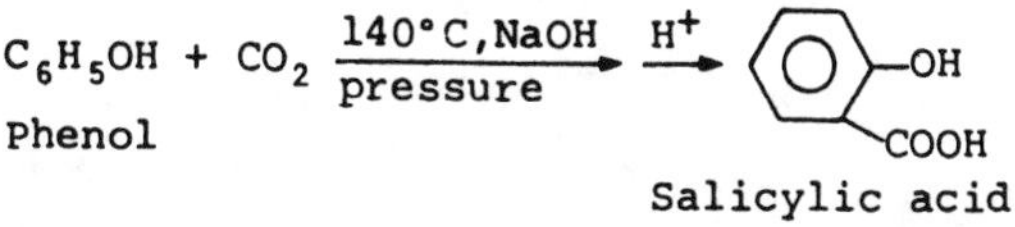

C_6H_5OH + CO_2 Phenol $\xrightarrow[\text{pressure}]{140°C,\text{NaOH}}$ $\xrightarrow{\text{H}^+}$ ⟨benzene⟩—OH, COOH

Salicylic acid

2) Use of CCl_4 in the Reimer-Tiemann reaction.

C_6H_5OH Phenol $\xrightarrow{\text{CCl}_4,\text{NaOH}}$ $\xrightarrow{\text{H}^+}$ ⟨benzene⟩—OH, COOH

Salicylic acid

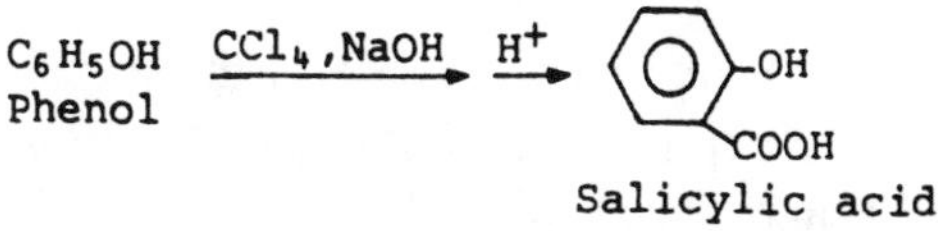

m) Claisen rearrangement. Allyl ethers of phenol yield o-allyl phenol upon heating.

n) Oxidation

Problem Solving Examples:

   Give structures of all compounds below:

(a) p-nitrophenol + C_2H_5Br + NaOH(aq) $\rightarrow$ A($C_8H_9O_3N$)

A + Sn + HCl $\rightarrow$ B($C_8B_{11}ON$)

B + $NaNO_2$ + HCl, then phenol $\rightarrow$ C($C_{14}H_{14}O_2N_2$)

C + ethyl sulfate + NaOH (aq) $\rightarrow$ D($C_{16}H_{18}O_2N_2$)

D + $SnCl_2$ $\rightarrow$ E($C_8H_{11}ON$)

E + acetyl chloride $\rightarrow$ phenacetin ($C_{10}H_{13}O_2N$), an analgesic ("pain-killer") and antypyretic ("fever killer")

(b) β-(-o-hydroxyphenyl) ethyl alcohol + HBr $\rightarrow$ F(C_8H_9OBr)

F + KOH $\rightarrow$ coumarane (C_8H_8O), insoluble in NaOH

A   (a) Treating a phenol with base in the presence of an alkyl halide is representative of the Williamson synthesis. The reaction produces phenolic ethers by an S_N2 (substitution nucleophilic-bimolecular) with the phenoxide anion acting as a nucleophile and the halide as the leaving group. In this case, the displacement results in the formation of p-nitrophenylethylether.

$$O_2N-\underset{\substack{\text{p - nitrol}\\\text{phenol}}}{\bigcirc}-OH \xrightarrow{\substack{OH^-\\H_2O}} O_2N-\bigcirc-\ddot{O}^- \overset{H}{\underset{H}{\underset{|}{CH_3\overset{|}{C}-Br}}} \xrightarrow{-Br^-}$$

$O_2N-\underset{A}{\bigcirc}-OC_2H_5$ In a useful reaction, aromatic nitro com-

pounds are reduced to their corresponding aniline derivatives. The reduction involves treating the nitro compound with a metal catalyst. Usually base is needed to isolate the final product.

$$(A) \quad \underline{\textbf{reduction}}\xrightarrow{} (B) H_2N-\bigcirc-OC_2H_5$$

When sodium nitrite ($NaNO_2$) is treated with acid, nitrous acid (HNO_2) is produced. This is the reagent in the mixture that converts the aniline derivative to its diazonium salt. In this case:

$$\underset{OC_2H_5}{\overset{NH_2}{\bigcirc}} \xrightarrow{HNO_2} \underset{OC_2H_5}{\overset{N_2^+}{\bigcirc}}$$

This diazonium cation is a very weak electrophile that may react with aromatic rings but with only very highly activated ones, such as phenol. The final product is an azo compound.

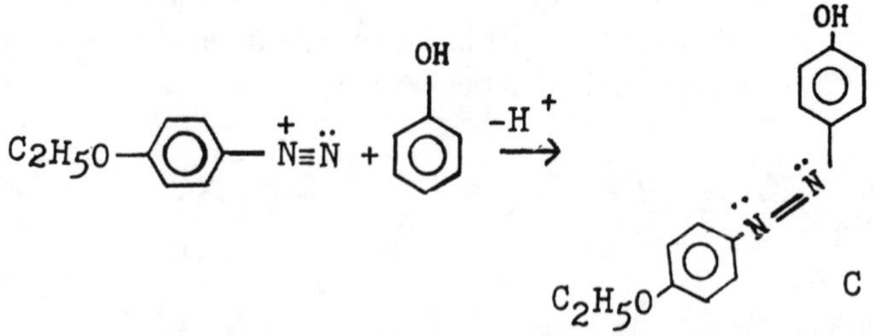

$$C_2H_5O-\!\!\!\bigcirc\!\!\!-\overset{+}{N}\!\equiv\!\ddot{N} \;+\; \bigcirc\!\!\!-OH \;\xrightarrow{-H^+}\; $$

This azo compound contains a phenolic hydroxyl group and therefore it may undergo a Williamson synthesis. Ethyl sulfate is an alkylating reagent that reacts with the phenoxide ion to produce an ethyl ether.

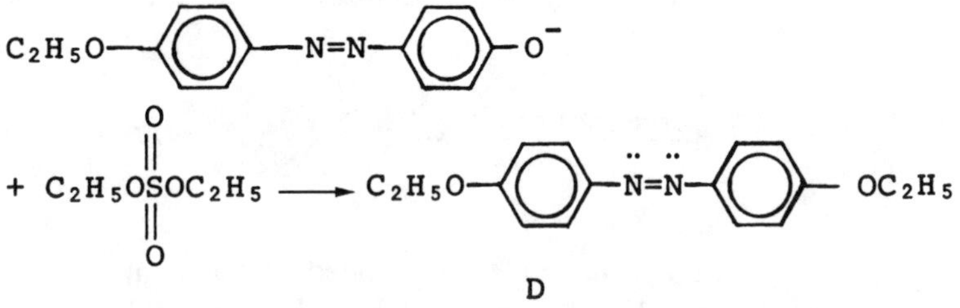

$$C_2H_5O-\!\!\!\bigcirc\!\!\!-N\!=\!N-\!\!\!\bigcirc\!\!\!-O^-$$

$$+ \; C_2H_5O\overset{O}{\underset{O}{\overset{\|}{\underset{\|}{S}}}}OC_2H_5 \;\longrightarrow\; C_2H_5O-\!\!\!\bigcirc\!\!\!-\ddot{N}\!=\!\ddot{N}-\!\!\!\bigcirc\!\!\!-OC_2H_5$$

$$D$$

The final product retains its azo linkage, which can be cleaved into two aniline derivatives, which in this case are equivalent.

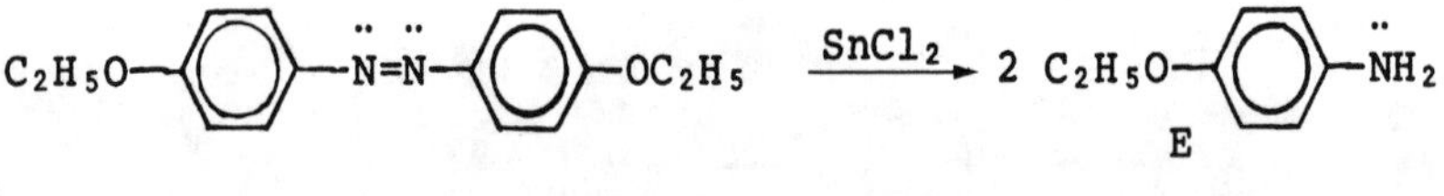

$$C_2H_5O-\!\!\!\bigcirc\!\!\!-\ddot{N}\!=\!\ddot{N}-\!\!\!\bigcirc\!\!\!-OC_2H_5 \;\xrightarrow{SnCl_2}\; 2 \; C_2H_5O-\!\!\!\bigcirc\!\!\!-\ddot{N}H_2$$

$$E$$

This compound, like any amine, can undergo an acylation reaction in which the pair of electrons on nitrogen attacks the partially positively charged carbonyl carbon of an acylating agent (i.e., acetyl chloride) to give an amide.

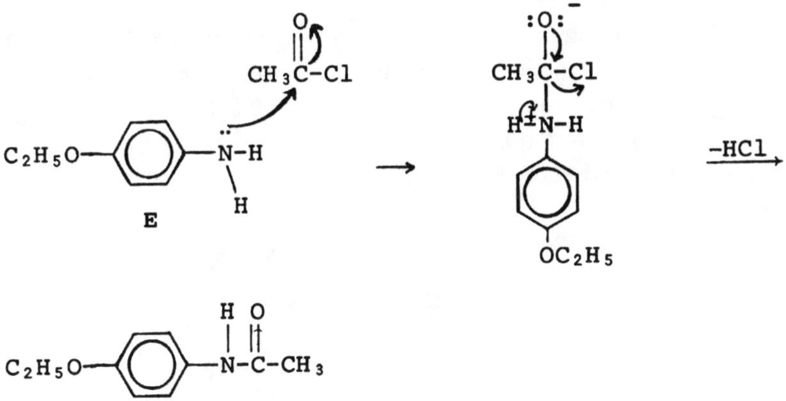

phenacetin

(b) β-(o-hydroxyphenyl) ethyl alcohol has the structural formula:

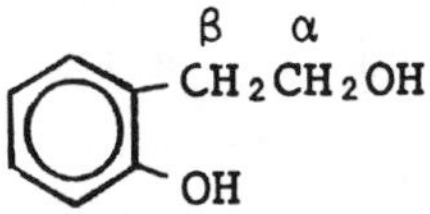

with the β carbon of ethanol being the one

that does not bear the hydroxyl group. The question arises whether the bromine would be on the α or β carbon upon treat ment with HBr. Any mechanism that involves a cation would

rearrange to give the stable benzylic cation,

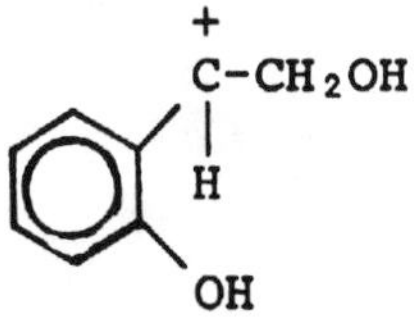

and therefore give the β bromide.

On the other hand, a concerted mechanism (S_N2 displace ment) would have the α bromide formed by the following mechanism:

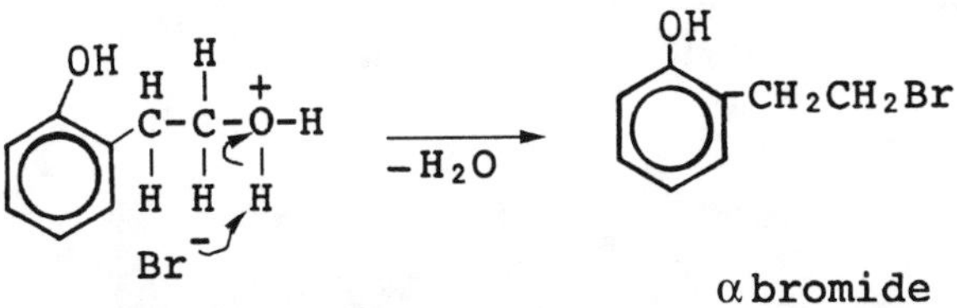

α bromide

Whichever compound is formed will react with base to form another compound with the loss of HBr. This is typical of a Williamson synthesis but since the alcohol and the halide are in the same molecule, this is an intramolecular reaction. Looking at the possible products if the a and b bromide undergo an intramolecular Williamson synthesis, we see that the ethereal ring contains five and four carbons, respectively.

α bromide $\xrightarrow{\text{OH}^-}$

β bromide $\xrightarrow{\text{OH}^-}$

The ether containing the five-membered ring is more stable entropically, and therefore, it is the α-bromide that underwent the Williamson reaction. The fact that the α-bromide used its phenol group in the reaction is verified by the product's insolubility in base. Had this not been true, the hydroxyl group would lose its proton to form the anion and therefore dissolve. The product, coumarane, is therefore:

Q Write the principal reaction product or products, if any, of o-cresol with CO_2, K_2CO_3, 240°.

A The phenol o-cresol, in the presence of the base, K_2CO_3, ionizes to the phenoxide anion. This anion can be written as several resonance forms as a result of oxygen contributing a pair of electrons to the aromatic ring:

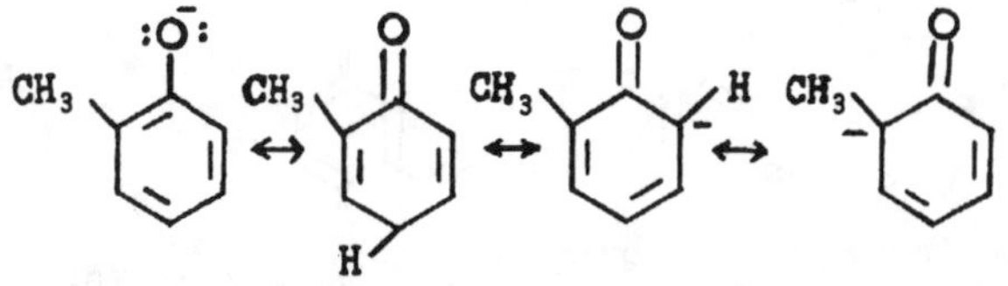

Note that the negative charge on the ring appears only ortho and para to the hydroxyl group. In the presence of CO_2, the phenoxide anion will attack the positively charged carbon $\begin{bmatrix} \delta- & \delta+ & \delta- \\ O & = C & = O \end{bmatrix}$ to give a salicylic acid after acidification (Kolbe reaction). While it is the ortho carbon of the phenol that attacks CO_2 at moderate temperatures, higher temperatures will promote attack by the para carbon.

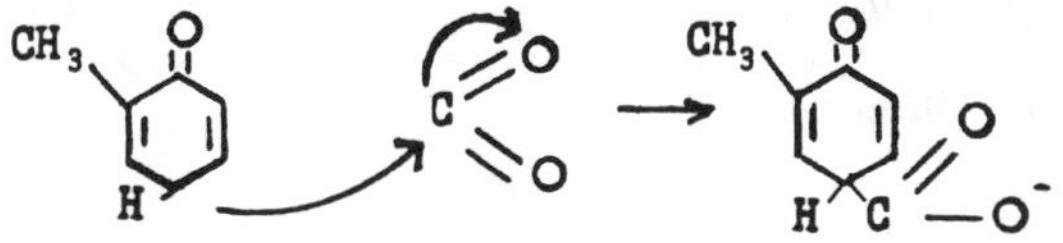

A tautomerization and acidification yields the final product.

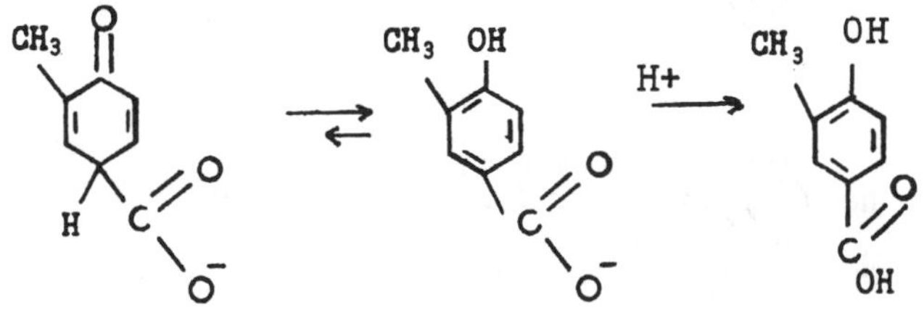

4-hydroxy-3-methylbenzoic acid

4.5   Quinones

Nomenclature

Quinones are cyclic, conjugated diketones named after the parent hydrocarbon.

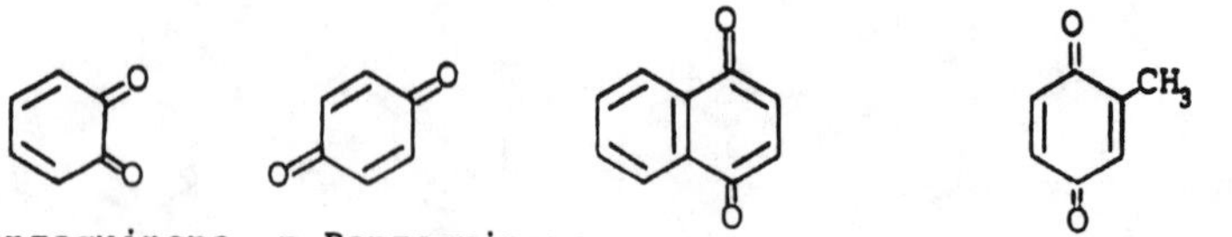

o-Benzoquinone p-Benzoquinone 1,4-Naphthoquinone Toluquinone (2-Methyl-1,4-benzoquinone)

Properties

Quinones are colored crystalline compounds. Since quinones are highly conjugated, they are closely balanced, energetically, against the corresponding hydroquinones. Quinones exhibit the properties of unsaturated cyclic ketone.

Preparation of Quinones

A)

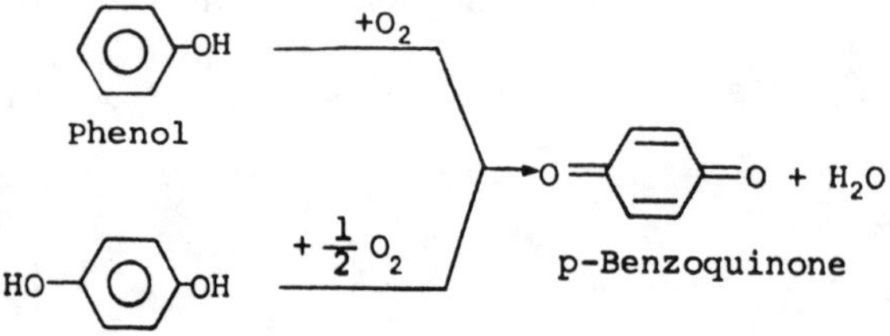

Hydroquinone

B)

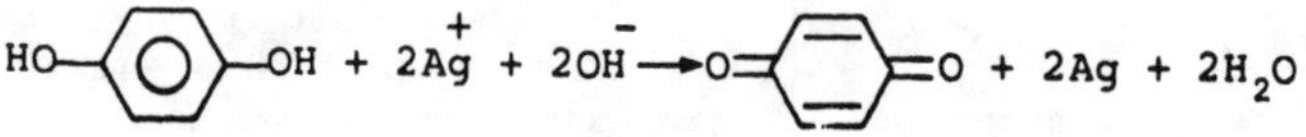

Hydroquinone p-Benzoquinone

C)

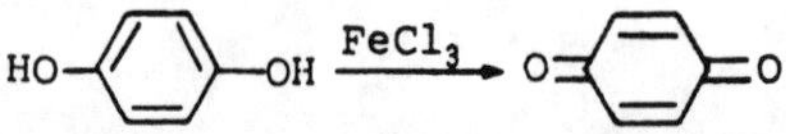

Hydroquinone p-Benzoquinone

D)

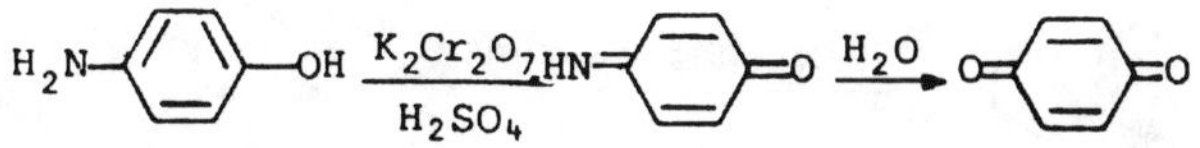

p-aminophenol   p-Benzoquinone

E)

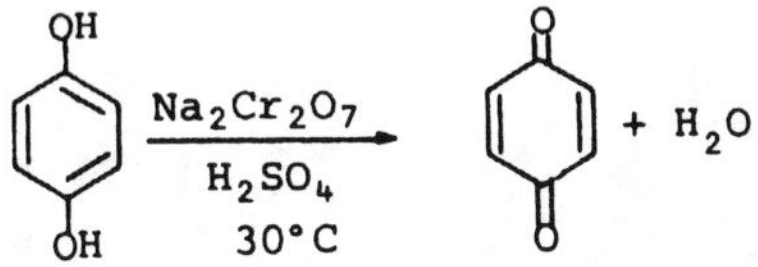

Hydroquinone   p-Benzoquinone

F)

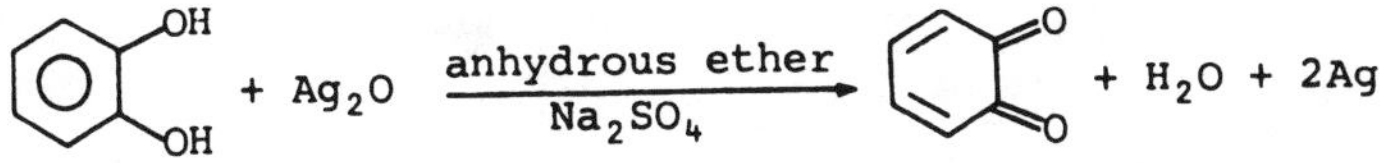

Catechol

G)

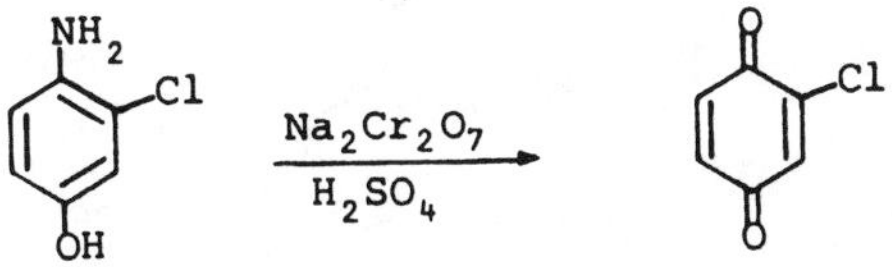

3-Chloro-4-aminophenol   2-Chloro-1,4-benzoquinone

H)

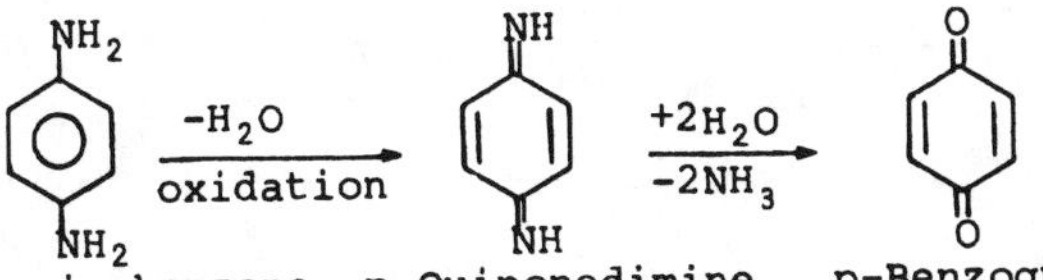

p-Diaminobenzene   p-Quinonedimine   p-Benzoquinone

I)

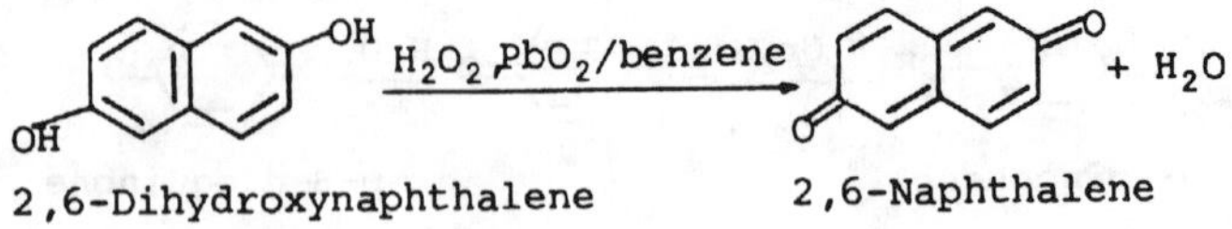
H_2O_2,PbO_2/benzene
+ H_2O
OH
OH
2,6-Dihydroxynaphthalene
2,6-Naphthalene

J)

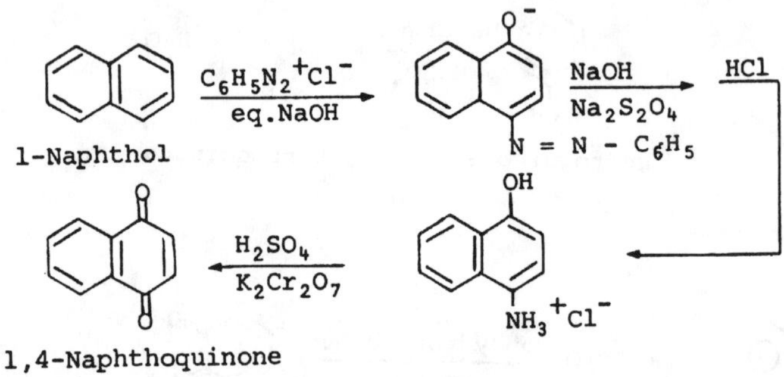
$C_6H_5N_2^+Cl^-$
eq.NaOH
1-Naphthol
O^-
$N = N - C_6H_5$
NaOH
$Na_2S_2O_4$
HCl
OH
$NH_3^+Cl^-$
H_2SO_4
$K_2Cr_2O_7$
1,4-Naphthoquinone

K)

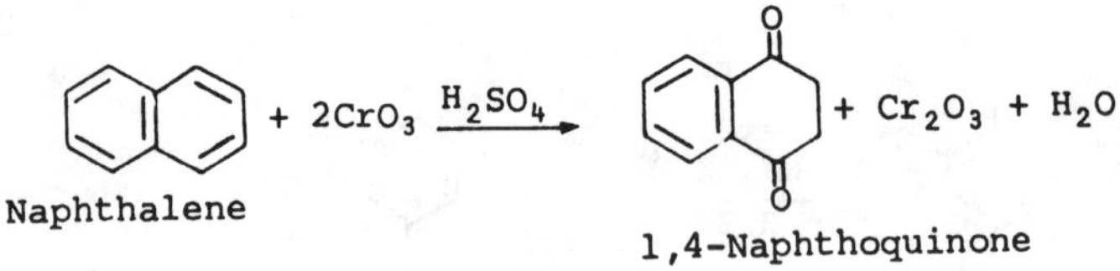
+ $2CrO_3$
H_2SO_4
+ Cr_2O_3 + H_2O
Naphthalene
1,4-Naphthoquinone

L)

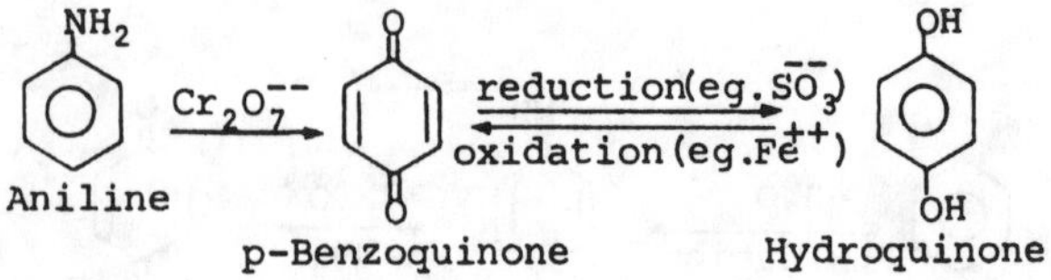
NH_2
Aniline
$Cr_2O_7^{--}$
p-Benzoquinone
reduction(eg.SO_3^{--})
oxidation (eg.Fe^{++})
OH
OH
Hydroquinone

M)

Friedel–Crafts Acylation

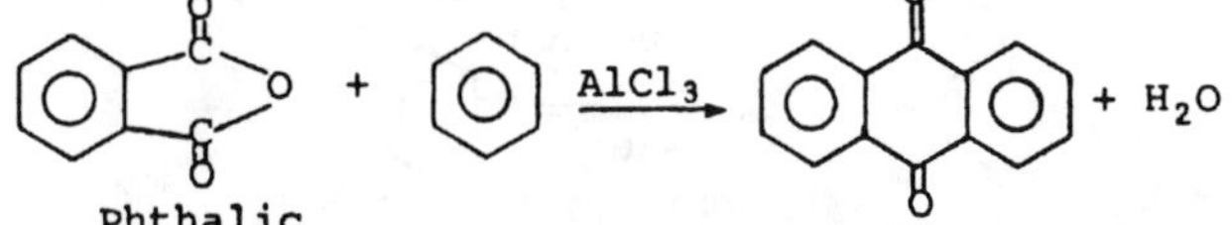

Phthalic anhydride

9,10 Anthraquinone

The preparation of quinones is achieved by oxidation of aromatic hydroxy and amino compounds and by Friedel-Crafts acylation.

Reactions of Quinones

A) Friedel-Crafts reactions

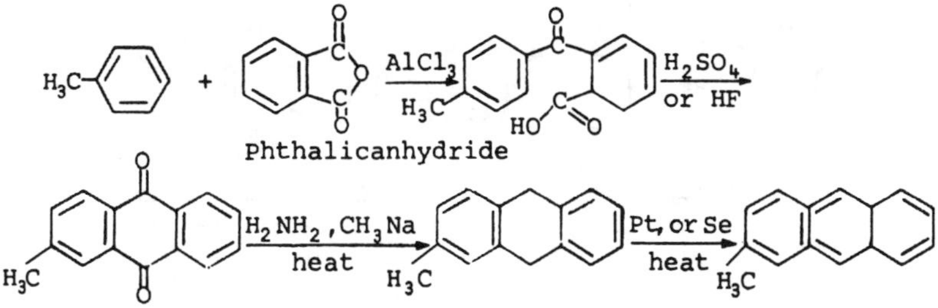

Phthalicanhydride

3-Methylanthracene

B) Diels-Alder reactions

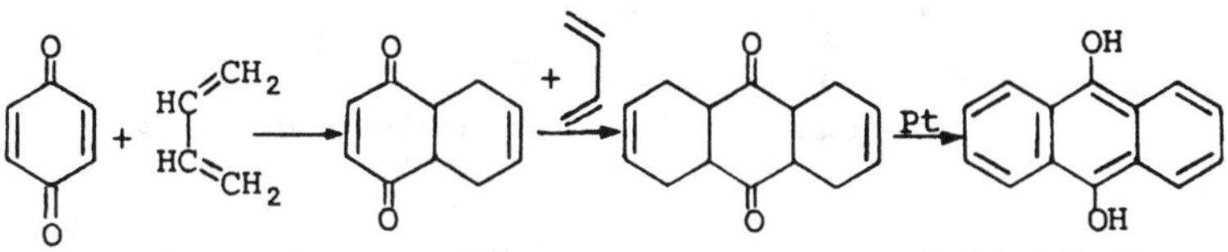

p-Benzoquinone   1,3 Butadiene

9,10 Dihydroxy-anthracene

C) Oxidation-reduction reactions

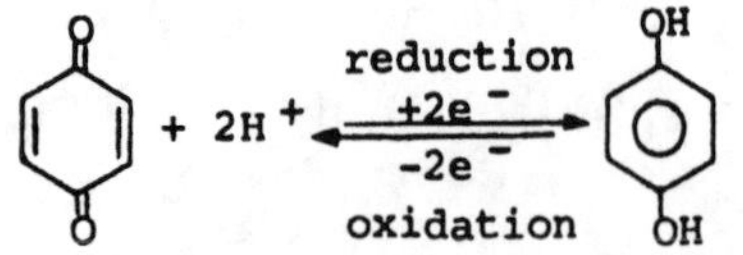

p-Benzoquinone   Hydroquinone

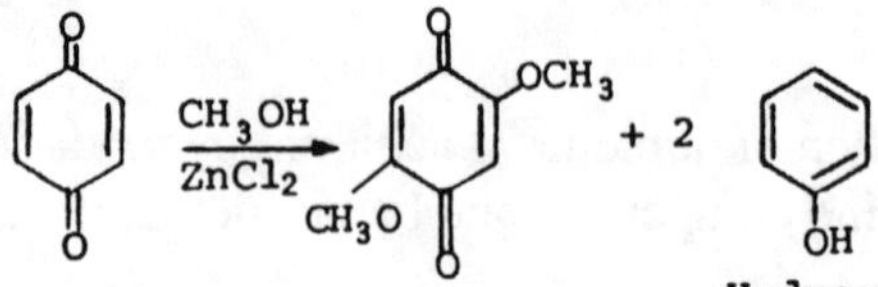

p-Benzoquinone 2,5 Dimethoxy-   Hydroquinone
                1,4-benzoquinone

D) Addition of halogen indirect substitution

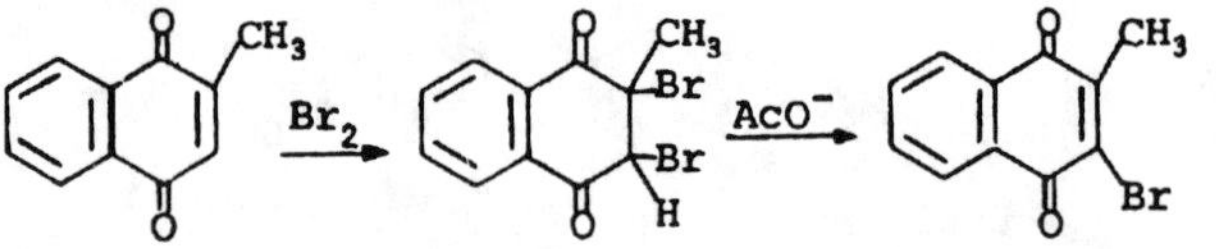

2-Methyl-1,4-naphthoquinone   2-Methyl-3-bromo-1,4-
                                  naphthoquinone

E) Addition of acetic acid

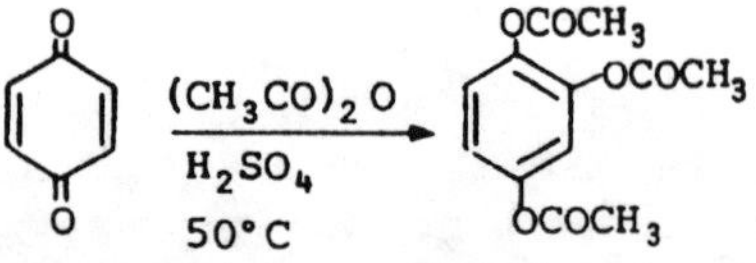

p-Benzoquinone   1,2,4-Triacetotoxybenzene

F) Reactions of keto group

Quinones undergo most reactions that are characteristic of ketones.

$$+ 2H_2N - OH \longrightarrow \quad + 2H_2O$$

Hydroxylamine

p-Benzoquinone   p-Benzoquinone dioxime

G) Addition reactions

Quinones undergo addition at the double bond by reagents such as bromine and dienes.

$$+Br_2 \longrightarrow$$

p-Benzoquinone   p-Benzoquinone tetrabromide

H) 1,4-additions

$$O= \hspace{-4pt} \bigcirc \hspace{-4pt} =O + HCl \xrightarrow{CHCl_3} \left[HO- \hspace{-4pt} \bigcirc \hspace{-4pt} =O \right] \longrightarrow HO- \hspace{-4pt} \bigcirc \hspace{-4pt} -OH$$

p-Benzoquinone   Chlorohydroquinone

Problem Solving Example:

(a) Hydroquinone is used in photographic developers to aid in the conversion of silver ion into free silver. What property of hydroquinone is being taken advantage of here?

(b) p-Benzoquinone can be used to convert iodide ion into iodine. What property of the quinone is being taken ad vantage of here?

(a) Parabenzoquinone and hydroquinone are in a rapid and reversible equilibrium as are all quinones and phenols. Hy-

droquinone, (OH/OH structure), has the ability to convert silver ion to

metallic silver by reducing it, i.e., by donating electrons to the cation. While Ag+ is reduced to Ag, the phenol is

oxidized to the quinone (O/O structure). The net reaction for this

commercially important process is:

$$2\ Ag^+ + \text{(hydroquinone)} \rightleftharpoons \text{(quinone)} \quad 2\ H^+ + 2\ Ag^\circ$$

Thus, it is the ability of hydroquinone to be oxidized that we are using here.

(b) As mentioned, the phenol-quinone equilibrium is reversible and parabenzoquinone may be easily reduced by a suitable electron source such as the iodide ion (I^-). The products of this process where we use p-benzoquinone's ability to be reduced are hydroquinone and iodine.

$$2\ H^+ + \text{(quinone)} + 2\ I^- \rightleftharpoons I_2 + \text{(hydroquinone)}$$

CHAPTER 5

Organometallic

Compounds

Organometallic compounds are defined as compounds that possess direct carbon-metal bonds. This excludes salts of organic acids, metal amines, and Lewis acid complexes on heteroatoms of organic molecules.

5.1 Nomenclature

Organometallic compounds are named by prefixing the name of the metal with the appropriate organic radical name. A rough classification of different types of organometallic compounds can be given by the periodic table of elements in Table 5.1

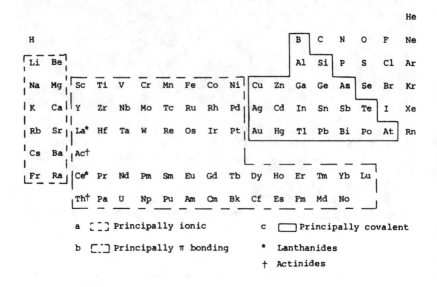

He

H												B	C	N	O	F	Ne
Li	Be											Al	Si	P	S	Cl	Ar
Na	Mg	Sc	Ti	V	Cr	Mn	Fe	Co	Ni	Cu	Zn	Ga	Ge	As	Se	Br	Kr
K	Ca	Y	Zr	Nb	Mo	Tc	Ru	Rh	Pd	Ag	Cd	In	Sn	Sb	Te	I	Xe
Rb	Sr	La*	Hf	Ta	W	Re	Os	Ir	Pt	Au	Hg	Tl	Pb	Bi	Po	At	Rn
Cs	Ba	Ac†															
Fr	Ra	Ce*	Pr	Nd	Pm	Sm	Eu	Gd	Tb	Dy	Ho	Er	Tm	Yb	Lu		
		Th†	Pa	U	Np	Pu	Am	Cm	Bk	Cf	Es	Fm	Md	No			

a ⌐ ┐ Principally ionic   c ☐ Principally covalent

b ⌐ ┐ Principally π bonding   * Lanthanides

† Actinides

Table 5.1 Characteristics of Carbon-metal Bonds

5.2 Properties of Organometallic Compounds

A) Many organometallic compounds react vigorously with water or other protic compounds and with oxygen.

B) Many organometallic compounds decompose in water, but they are soluble in various inert aprotic organic solvents.

C) Methylsodium and methylpotassium are colorless amorphous solids.

D) Dimethylzinc is a colorless, mobile, strongly refractive liquid.

E) Dimethylmercury, trimethylaluminum, tetramethyltin, and tetra-methyllead are volatile liquids that distill without decomposition.

F) Other characteristics are given in Table 5.2.

Organometallic compound	Bond character	Physical properties	Reactive
Carbon-alkali metals Li < Na < K < Rb < C	Ionic Very polar	Saltlike Nonvolatile	Highly reactive Inflammable in air
		Insoluble in nonpolar solvents	Fast hydrolysis
Carbon-earth alkali metals Be < Mg < Ca < Sr < Ba Carbon transition metals Cu > Ag > Au; Zn > Cd > Hg	Covalent Weakly polar	Volatile Soluble in non-polar solvents	Less reactive Stable in air Slow hydrolysis

Table 5.2 Properties of Organometallic Compounds

5.3   Preparation of Organometallic Compounds

By the action of an alkyl halide on zinc dust:

$$R - X + Zn, \text{ finely divided} \rightarrow R - Zn - X$$

By the action of alkyl halides on magnesium in dry ether to give organometallic halides:

$$R - X + Mg, \text{ dry ether} \rightarrow R - Mg - X \text{ (Grignard reagent)}$$

By the action of alkyl magnesium or alkyl zinc halides on metallic halides of less active metals:

$$R - MgX/\text{dry ether} + HgX_2 \rightarrow R - Hg - X + MgX_2$$

$$R - ZnX/\text{dry ether} + HgX_2 \rightarrow R - Hg - X + MgX_2$$

by the action of metallic sodium on dialkyl zinc or dialkyl mercury in dry benzene to give alkyl sodium:

Example

$$R_2 Zn + 2\,Na, \text{dry benzene} \rightarrow 2\,R^- Na^+ + Zn$$

$$R_2 Hg + 2\,Na, \text{dry benzene} \rightarrow 2\,R^- Na^+ + Hg$$

By the action of metallic sodium on diphenyl mercury in pure thiophene to given alkyl sodium:

$$(C_6H_5)_2 Hg + 2Na, \text{dry thiophene} \rightarrow 2C_6H_5 - Na + Hg$$

By reaction between hydrocarbons and free metals:

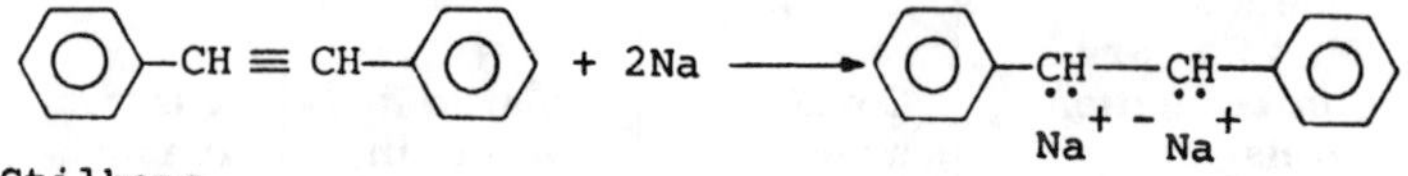

Stilbene

Brownish violet,
"ionic-like"

By reaction of alkali metal and hydrocarbons that have acidic hydrogens:

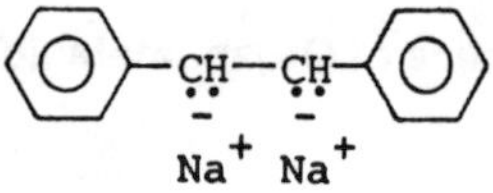

Example

$$2(C_6H_5)_3 C - H + 2K \rightarrow 2(C_6H_5)_3 C^- K^+ + H_2$$

$$H - C \equiv C - H \xrightarrow[-H]{+Na} H - C \equiv C^- Na^+ \xrightarrow[+Na]{-H} Na^{+-}C \equiv C^- Na^+$$

By the reaction between organic halides and metals:

$$\text{RX or } C_6H_5X + \text{powdered metal} \xrightarrow[400°C]{200°} \text{organometallic compound}$$

Powdered metals = Si, Al, Ge, Zn, Te, Sn.

Example

$$CH_3CH_2 - Br + 2Li \xrightarrow{\text{in ether}} CH_3CH_2Li + LiBr$$

Halogen-metal interconversion:

$$RX + R'\,Li \rightleftarrows RLi + R'\,X$$

RX = vinyl, alkyl, ethylarylbromides, or iodides

R' = alkyl

$$Cl-\!\!\left\langle\!\bigcirc\!\right\rangle\!\!-Br \; + \; CH_3CH_2CH_2CH_2Li \; \rightleftharpoons \; Cl-\!\!\left\langle\!\bigcirc\!\right\rangle\!\!-Li$$
$$+ \; CH_3(CH_2)_3Br$$

By the addition of a metal and hydrogen to alkenes:

Example

$$3(CH_3)2C = CH_2 + Al + \frac{3}{2}H_2 \rightarrow \left[(CH_3)_2CH - CH_2\right]_3 Al$$

By metal-metal exchange:

Example

$$(CH_3CH_2)_2 Hg + 2Li \xrightarrow[65°, \; 3 \; days]{Ligroin} 2CH_3CH_2Li + Hg$$

By the reaction of metallic halides of less reactive metals with alkylmagnesium halides:

$$4RMgX, \text{dry ether} + 2PbCl_2 \rightarrow R_4Pb + Pb + 2MgX_2 + 2MgCl_2$$

$$2RMgX, \text{dry ether} + HgCl_2 \rightarrow R_2Hg + MgX_2 + MgCl_2$$

By the action of metallic halides of less reactive metals with dialkyl zinc:

$$R_2Zn, \text{inert solvent} + HgCl_2 \rightarrow R_2Hg + ZnCl_2$$

By the action of intermetallic compounds with alkyl halides:

$$2RI + Na_2Hg \rightarrow R_2Hg + 2NaI$$

$$4RX + 4NaPb \rightarrow R_4Pb + 4NaX + 3Pb$$

Example

$$4CH_3CH_2Cl + 4Na - Pb \rightarrow (CH_3 - CH_2)4Pb + 4NaCl + 3Pb$$

$$2C_6H_5Br + 2Na - Hg, \text{dry toluene} \rightarrow (C_6H_5)_2Hg + Hg + 2NaBr$$

Problem Solving Examples:

 What would be the expected organic product from the reaction between water and the following organometallic compounds?

(a)

$$CH_3 - \underset{\underset{H}{|}}{\overset{\overset{CH_3}{|}}{C}} - \underset{\underset{MgBr}{|}}{\overset{\overset{H}{|}}{C}} - CH_3$$

(b)

 CH_2MgBr

(c)

$$CH_3 - \underset{\underset{MgBr}{|}}{\overset{\overset{CH_3}{|}}{C}} - (CH_2)_3 - \underset{\underset{MgBr}{|}}{\overset{\overset{H}{}}{CH}}$$

 Grignard reagents will react with water to form the corresponding alkane.

(a)

$$CH_3 - \underset{\underset{H}{|}}{\overset{\overset{CH_3}{|}}{C}} - \underset{\underset{MgBr}{|}}{\overset{\overset{H}{|}}{C}} - CH_3 \xrightarrow{H_2O} CH_3 - \underset{\underset{H}{|}}{\overset{\overset{CH_3}{|}}{C}} - \underset{\underset{H}{|}}{\overset{\overset{H}{|}}{C}} - CH_3$$

(b)

$CH_2MgBr \xrightarrow{H_2O}$ CH_3

(c)

$$CH_3 - \underset{\underset{MgBr}{|}}{\overset{\overset{CH_3}{|}}{C}} - (CH_2)_3 - \underset{\underset{MgBr}{|}}{\overset{\overset{H}{}}{CH}} \xrightarrow{H_2O} CH_3 - \underset{\underset{H}{|}}{\overset{\overset{CH_3}{|}}{C}} - (CH_2)_3 - CH_3$$

Write balanced equations for the preparation of each of the following organometallic compounds by two different reactions starting from suitable alkyl halides and inorganic reagents. Specify reaction conditions and solvents. In each case, indicate which method of preparation you would prefer from standpoints of yield, convenience, etc.

    (a) $(CH_3)_2Zn$ (c) $(CH_3CH_2)_4Pb$

    (b) CH_3MgCl (d) $(CH_3)_2CHLi$

A   Organometallic compounds can be prepared by a direct or an indirect method. The direct method of preparing alkyl organometallic compounds is to react an alkyl halide (RX) with the free metal.

$$R\text{---}X + 2\,Li \rightarrow Li^+X^- + R\text{---}Li$$

The indirect method of preparing organometallic alkanes involves first the formation of an organometallic alkane. Next, the free metal of the desired product is reacted with the aforementioned alkyl organometallic compound to give the desired product. For example, methyl sodium can be prepared by first forming methylmagnesium chloride. Subsequent reaction with sodium forms methyl sodium, sodium chloride, and magnesium:

$$CH_3Cl \xrightarrow[\text{ether}]{Mg} CH_3MgCl$$

$$CH_3MgCl \xrightarrow[\text{n - hexane}]{Na} CH_3Na + Mg + \overset{+}{Na}\,\overset{-}{Cl}$$

Generally, the direct method of preparing organometallic alkanes is preferred over the indirect method. This is because the direct method involves only one step and will hence give better yields than the indirect method.

    (a) A direct method of preparing the organozinc compound $(CH_3)_2Zn$ can be shown as:

$$2CH_3I + 2Zn \rightarrow (CH_3)_2Zn + ZnI_2$$

An indirect method is as follows:

$$2CH_3I + Hg(Na)_2 \rightarrow (CH_3)_2Hg + 2NaI$$

$$(CH_3)_2Hg + Zn \rightarrow (CH_3)_2Zn + Hg$$

The direct method (the first one) is preferred because it gives a better yield, and it does not proceed through any highly poisonous intermediates. ($(CH_3)_2Hg$ is a very poisonous compound.) A good solvent for the preparation and reaction of organozinc compounds is diethylether.

(b) The direct method of preparing methylmagnesium chloride is:

$$CH_3Cl + Mg \xrightarrow{\text{ether}} CH_3MgCl$$

Methylmagnesium chloride is an example of a Grignard reagent (organomagnesium compounds); ether is a good solvent for reactions of Grignard reagents. An indirect method of preparing CH_3MgCl is:

$$CH_3Cl + 2Li \rightarrow Li^+Cl^- + CH_3Li$$

$$CH_3Li + MgCl_2 \rightarrow CH_3MgCl + Li^+Cl^-$$

The direct method is preferred; it involves only one step and can be carried out at low temperatures due to methylchloride's low boiling point.

(c) Tetraethyl lead ($Pb(C_2H_5)_4$) is added to automobile fuel to prevent "knocking" in the engine. A direct method of preparing it is:

$$4CH_3CH_2Cl + 4Pb(Na) \ \text{Æ} \ (CH_3CH_2)_4Pb + 4Na^+Cl^- + 3Pb$$

An indirect method of preparation is:

$$2CH_3CH_2Cl + Hg(Na)_2 \rightarrow (CH_3CH_2)_2Hg + 2Na^+Cl^-$$

$$2(CH_3CH_2)_2Hg + Pb \rightarrow 2Hg + (CH_3CH_2)_4Pb$$

The direct method is preferred because it gives a better yield and it does not proceed through any highly poisonous intermediates.

(d) A direct method for preparing isopropyl lithium is:

$$(CH_3)_2CHBr + 2Li \rightarrow (CH_3)_2CHLi + \overset{+}{Li}\overset{-}{Br}$$

An indirect method is:

$$(CH_3)_2CHBr + Mg \xrightarrow{\text{ether}} (CH_3)_2CHMgBr$$
$$(CH_3)_2CHMgBr + 2Li \rightarrow (CH_3)_2CHLi + Mg + \overset{+}{Li}\overset{-}{Br}$$

The direct method is preferred and ether can be used as a solvent.

Q What products would you expect to be formed in an attempt to synthesize hexamethylethane from t-butyl chloride and sodium? Write equations for the reactions involved.

A Organosodium compounds are unique in that they are one of the most highly reactive organometallic compounds. Organosodium compounds react rapidly with ethers and alkyl halides. They react with alkyl halides to produce hydrocarbons by either S_N2 and/or E2 reactions. For example, ethylsodium ($CH_3CH_2^-: Na^+$) undergoes S_N2 and E2 reactions with ethyl bromide as shown:

```
E2 reaction:
```

$$CH_3CH_2^-: Na^+ + H-CH_2-CH_2-Br \rightarrow CH_3CH_3 + Na^+Br^- + CH_2 = CH_2$$

```
SN2 reaction (Wurtz coupling reaction):
```

$$CH_3CH_2^-: Na^+ + CH_3CH_2-Br \rightarrow CH_3CH_2CH_2CH_3 + Na^+Br^-$$

When t-butyl chloride is treated with metallic sodium, the organosodium compound (t-butyl sodium) is first formed; sodium chloride is a by-product of this reaction:

$$(CH_3)_3C-Cl + 2 Na \rightarrow (CH_3)_3C^-: Na^+ + Na^+Cl^-$$

As t-butylsodium is formed, it reacts with the unconsumed t-butyl chloride to form hydrocarbons. The reaction is strictly E2; there is no

Wurtz coupling occurring due to the steric hindrance of the t-butyl groups. Hence, the hydrocarbons formed are 2-methylpropane and isobutylene; no hexamethylethane is produced. The reaction can be shown as:

E2 mechanism

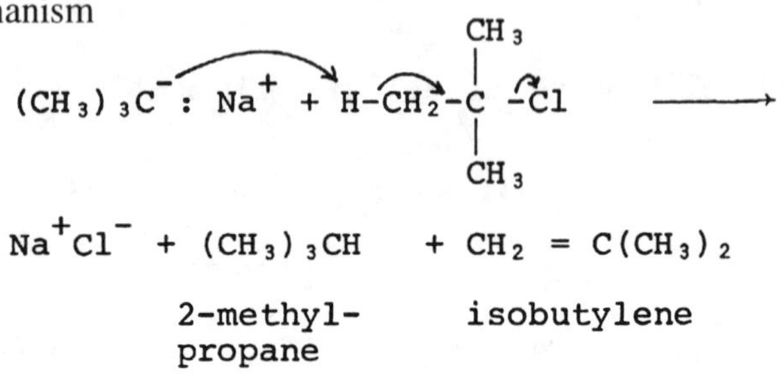

$$(CH_3)_3C^- : Na^+ + H-CH_2-C \overset{CH_3}{\underset{CH_3}{|}} Cl \longrightarrow$$

$$Na^+Cl^- + (CH_3)_3CH + CH_2 = C(CH_3)_2$$

2-methyl-   isobutylene
propane

5.4 Reactions of Organometallic Compounds

Reactions of Grignard Reagents

Grignard reagents react under proper conditions to give hydrocarbons when treated with reagents containing labile hydrogen atoms such as:

A) Acids

$$R' - MgX + R - CO - OH \rightarrow R' - H + R - CO - O - MgX$$

B) Alcohols

$$R' - MgX + R - OH \rightarrow R' - H + R - O - MgX$$

C) Thiols

$$R' - MgX + R - SH \rightarrow R' - H + R - S - MgX$$

D) Water

$$R' - MgX + H_2O \rightarrow R' - H + HO - MgX$$

E) Amines

$$R' - MgX + R - NH_2 \rightarrow R' - H + R - NH - MgX$$

F) Amides

$$R' - MgX + R - CO - NH_2 \rightarrow R' - H + R - CO - NHO - MgX$$

Grignard reagents react under proper conditions to give condensation products when treated with reagents containing labile halogen atoms such as:

A) Phosgene

$$2R' - MgX + Cl_2C = O \rightarrow R'_2C = O + MgX_2 + MgCl_2$$

B) Alkyl halides

$$R' - MgX + R - X \rightarrow R' - R - O - MgX_2$$

C) Acid halides

$$R' - MgX + R - \overset{\overset{\displaystyle O}{\|}}{C} - X \rightarrow R' - \overset{\overset{\displaystyle O}{\|}}{C} - R' + MgX_2$$

D) Aryl halides

$$R' - MgX + Ar - X \rightarrow R' - Ar + MgX_2$$

E) Halo derivatives

$$R' - MgX + XCH_2 - O - R \rightarrow MgX_2 + R' - CH_2 - O - R$$

Grignard reagents react under proper conditions to give double decomposition products with:

A) Organic anhydrides

$$R - MgX + R - \overset{\overset{\displaystyle O}{\|}}{C} - O - \overset{\overset{\displaystyle O}{\|}}{C} - R \rightarrow R - \overset{\overset{\displaystyle O}{\|}}{C} - R + R - \overset{\overset{\displaystyle O}{\|}}{C} - O - MgX$$

B) Sulfonic acid esters

$$R - MgX + ArSO_2OR \rightarrow 2R + Ar - SO_2 - O - MgX$$

C) Inorganic halides

$$(AsCl_3, BiCl_3, GeCl_4, HgCl_2, PCl_3, SbCl_3, SiCl_4)$$

$$3R - MgX + AsCl_3 \rightarrow 3MgX - Cl_3 + R_3As$$

Grignard reagents also react with the following:

A) With compounds possessing an active hydrogen.

$$R - C \equiv CH + R' \; MgX \xrightarrow{-R'H} R - C \equiv C - MgX \xrightarrow{H_2O} R - C \equiv C - H$$

$$R - CH_2OH + R' \; MgX \xrightarrow{-R'H} R - CH_2 - OMgX \xrightarrow{H_2O} R - CH_2OH$$

$$R_2NH + R' \; MgX \xrightarrow{-R'H} R_2N - MgX \xrightarrow{H_2O} R_2NH$$

B) With oxygen.

$$RMgX + O_2 (\text{in ether}) \rightarrow R - O - O - MgX \xrightarrow{H_2O} R - O - O - H$$
$$\text{hydroperoxides}$$

Problem Solving Examples:

 Convert the following aldehydes into alcohols using the Grignard synthesis method.

(a)
$$\underset{\text{Propionaldehyde}}{CH_3CH_2 \overset{\overset{H}{|}}{C}=O}$$

(b)
$$\underset{\text{Isobutyraldehyde}}{(CH_3)_2 CH\overset{\overset{H}{|}}{C}=O}$$

The Grignard synthesis of alcohols is done with one of the most powerful tools used by the organic chemist, which is the Grignard reagent. The Grignard reagent, as has been shown in previous chapters, has the formula RMgX, and is prepared by the reaction of metallic magnesium with the appropriate organic halide. This halide can be alkyl, allylic, benzylic, or aryl. The halogen may be –Cl, –Br, or I. An example of an aldehyde-Grignard reaction is:

$$H - \overset{\overset{H}{|}}{C}=O + RMgX \rightarrow H - \overset{\overset{H}{|}}{\underset{\underset{R}{|}}{C}} - \overset{-}{O}\overset{+}{M}gX \xrightarrow{H_2O} H - \overset{\overset{H}{|}}{\underset{\underset{R}{|}}{C}} - OH,$$

showing formaldehyde reacting with the Grignard reagent. This yields an intermediate that is an organic salt, and the addition of water converts the salt into an alcohol.

With the principles discussed, and using the example as a reference, and using the example as a reference, the aldehydes (a) and (b) can be converted to an alcohol by reacting with the Grignard reagent.

(a)

$$CH_3CH_2\overset{\overset{\displaystyle H}{|}}{C}{=}O + CH_3MgBr \rightarrow CH_3CH_2\overset{\overset{\displaystyle H}{|}}{\underset{\underset{\displaystyle CH_3}{|}}{\overset{-}{C}}}\overset{+}{O}MgBr \xrightarrow{H_2O} CH_3CH_2\overset{\overset{\displaystyle H}{|}}{\underset{\underset{\displaystyle CH_3}{|}}{C}}{-}OH$$

Propionaldehyde   2-butanol

(b)

$$(CH_3)_2CH\overset{\overset{\displaystyle H}{|}}{C}{=}O + CH_3CH_2MgBr \rightarrow (CH_3)_2CH\overset{\overset{\displaystyle H}{|}}{\underset{\underset{\underset{\displaystyle CH_3}{|}}{CH_2}}{\overset{-}{C}}}\overset{+}{O}MgBr$$

Isobutyraldehyde

$$\xrightarrow{H_2O} \quad \overset{CH_3}{\underset{CH_3}{\diagdown\diagup}}CH\overset{}{\underset{\underset{\underset{\displaystyle CH_3}{|}}{CH_2}}{\overset{|}{C}}}HOH$$

2-methyl-3-pentanol

Show the product of the following ketones after addition of the Grignard reagent.

(a) $CH_3\overset{\overset{\displaystyle O}{\|}}{C}CH_3$ (b) $CH_3CH_2CH_2CH_2\overset{\overset{\displaystyle O}{\|}}{C}CH_3$

Acetone   2-Hexanone

A The Grignard reagent has been given as an organic magnesium halide, with the structure generally as RMgX. Of course as shown before, the organic group does not necessarily have to be alkyl; it may be allyl, aryl, or arenyl. The halogen (X) may be –Cl, –Br, or –I, and the entire organic magnesium halide reacts with ketones to yield an alcohol.

The general equation for the reaction is

$$-\underset{\overset{\|}{O}}{C}- \ + \ RMgX \ \longrightarrow \ -\underset{\overset{|}{O\overset{+}{M}gX}}{\overset{|}{C}}-R \ \xrightarrow{\ H_2O\ } \ -\underset{\overset{|}{OH}}{\overset{|}{C}}-R$$

When the Grignard reagent is added to the carbonyl group, the double bond is broken and the intermediate is the magnesium salt of the weakly acidic alcohol and is easily converted into the alcohol itself by the addition of the stronger acid, water. Sometimes a diluted solution of H_2SO_4 or HCl is used instead of water.

With these concepts and by use of the general equation, problems (a) and (b) can be solved.

(a)

$$\underset{\text{acetone}}{\overset{O}{\overset{\|}{CH_3CCH_3}}} \ + \ CH_3MgBr \longrightarrow \underset{\overset{|}{CH_3}}{\overset{\overset{-}{O}\overset{+}{M}gBr}{\overset{|}{CH_3\overset{|}{C}-CH_3}}} \ \xrightarrow{\ H_2O\ } \ \underset{\overset{|}{CH_3}}{\overset{OH}{\overset{|}{CH_3\overset{|}{C}H-CH_3}}}$$

2-methyl-2-propanol

(b)

$$\underset{\text{2-hexanone}}{\overset{O}{\overset{\|}{CH_3CH_2CH_2CH_2CCH_3}}} \ + CH_3MgBr \ \rightarrow \ \underset{\overset{|}{CH_3}}{\overset{\overset{-}{O}\overset{+}{M}gBr}{\overset{|}{CH_3CH_2CH_2CH_2\overset{|}{C}-CH_3}}}$$

$$\underset{\overset{|}{CH_3}}{\overset{OH}{\overset{|}{CH_3CH_2CH_2CH_2\overset{|}{C}CH_3}}} \ \xleftarrow{\ H_2O\ }$$

2-methyl-2-hexanol

Reactions of Other Organometallic Compounds

Organometallic compounds in which the metal has an electronegativity value of 1.7 or less react with water to give a hydrocarbon and a metal hydroxide.

Example

$$CH_3Li + H_2O \rightarrow CH_4 + LiOH$$

$$CH_3CH_2MgBr + H_2O \rightarrow CH_3CH_3 + HOMgBr$$

$$(CH_3)_3Al + 3H_2O \rightarrow 3CH_4 + Al(OH)_3$$

Reactions with halogens:

$$RM + Cl_2 \rightarrow RCl + M^+Cl^- \text{ (vigorous)}$$

Reactions with other organometallic compounds:

Alkyl copper compounds react with alkyl lithium compounds to give cuprates.

Example

$$CH_3Li + (CH_3)Cu \rightarrow (CH_3)_2CuLi$$

Problem Solving Example:

Explain why the following reactions all yield 2-methyl butane as the product.

(a)

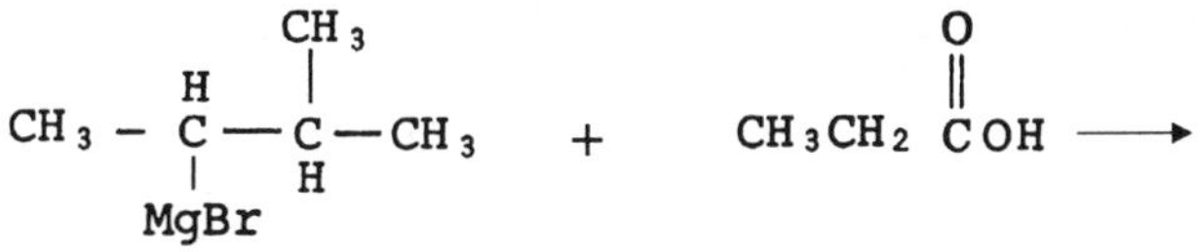

$$CH_3 - \underset{\underset{MgBr}{|}}{\overset{\overset{H}{|}}{C}} - \underset{\underset{H}{|}}{\overset{\overset{CH_3}{|}}{C}} - CH_3 \quad + \quad CH_3CH_2\overset{\overset{O}{\|}}{C}OH \longrightarrow$$

(b)

$$CH_3 - \underset{\underset{MgBr}{|}}{\overset{\overset{H}{|}}{C}} - \underset{\underset{H}{|}}{\overset{\overset{CH_3}{|}}{C}} - CH_3 \quad + \quad CH_3CH_2OH \longrightarrow$$

(c)

$$CH_3 - \underset{\underset{MgBr}{|}}{\overset{\overset{H}{|}}{C}} - \underset{\underset{H}{|}}{\overset{\overset{CH_3}{|}}{C}} - CH_3 \quad + \quad CH_3CH_2SH \longrightarrow$$

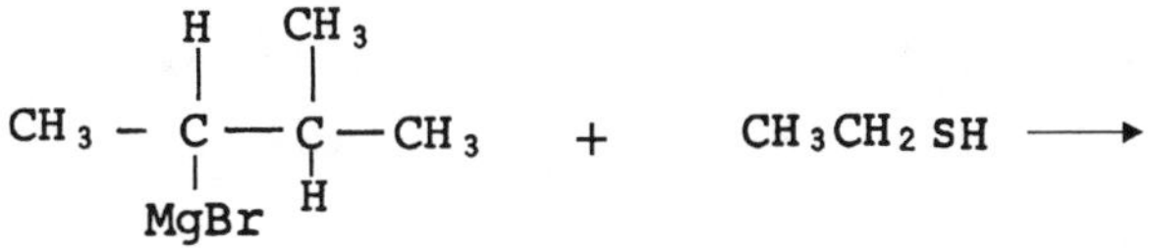

(d)

$$CH_3 - \underset{\underset{MgBr}{|}}{\overset{\overset{H}{|}}{C}} - \underset{\underset{H}{|}}{\overset{\overset{CH_3}{|}}{C}} - CH_3 \quad + \quad CH_3CH_2NH_2 \longrightarrow$$

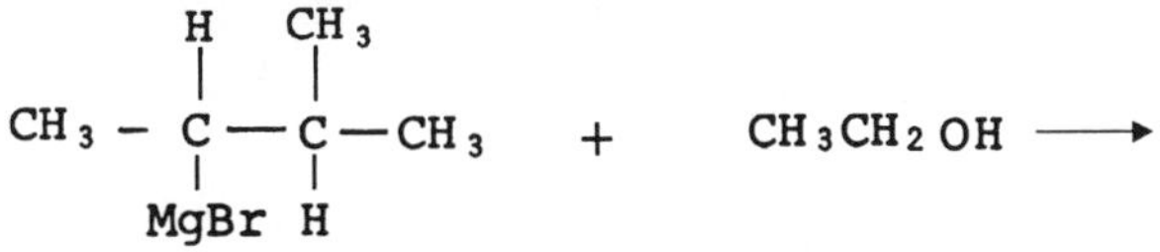

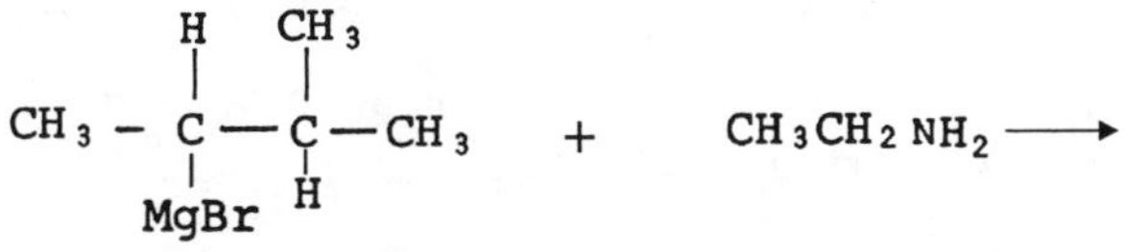

A   Grignard reagents will react with compounds containing labile hydrogen atoms to give hydrocarbons. The reactions occur as follows:

(a) $CH_3 - \overset{\overset{\displaystyle H}{|}}{\underset{\underset{\displaystyle MgBr}{|}}{C}} - \overset{\overset{\displaystyle CH_3}{|}}{\underset{\underset{\displaystyle H}{|}}{C}} - CH_3$ + $CH_3CH_2\overset{\overset{\displaystyle O}{\|}}{C}OH$ ⟶

$CH_3 - \overset{\overset{\displaystyle H}{|}}{\underset{\underset{\displaystyle H}{|}}{C}} - \overset{\overset{\displaystyle CH_3}{|}}{\underset{\underset{\displaystyle H}{|}}{C}} - CH_3$ + $CH_3CH_2\overset{\overset{\displaystyle O}{\|}}{C}O\,MgBr$

(b) $CH_3 - \overset{\overset{\displaystyle H}{|}}{\underset{\underset{\displaystyle MgBr}{|}}{C}} - \overset{\overset{\displaystyle CH_3}{|}}{\underset{\underset{\displaystyle H}{|}}{C}} - CH_3$ + CH_3CH_2OH ⟶

$CH_3 - \overset{\overset{\displaystyle H}{|}}{\underset{\underset{\displaystyle H}{|}}{C}} - \overset{\overset{\displaystyle CH_3}{|}}{\underset{\underset{\displaystyle H}{|}}{C}} - CH_3$ + CH_3CH_2OMgBr

(c) $CH_3 - \overset{\overset{\displaystyle H}{|}}{\underset{\underset{\displaystyle MgBr}{|}}{C}} - \overset{\overset{\displaystyle CH_3}{|}}{\underset{\underset{\displaystyle H}{|}}{C}} - CH_3$ + CH_3CH_2SH ⟶

$CH_3 - \overset{\overset{\displaystyle H}{|}}{\underset{\underset{\displaystyle H}{|}}{C}} - \overset{\overset{\displaystyle CH_3}{|}}{\underset{\underset{\displaystyle H}{|}}{C}} - CH_3$ + $CH_3CH_2S{-}MgBr$

(d) $CH_3 - \overset{\overset{\displaystyle H}{|}}{\underset{\underset{\displaystyle MgBr}{|}}{C}} - \overset{\overset{\displaystyle CH_3}{|}}{\underset{\underset{\displaystyle H}{|}}{C}} - CH_3$ + $CH_3CH_2NH_2$ ⟶

$CH_3 - \overset{\overset{\displaystyle H}{|}}{\underset{\underset{\displaystyle H}{|}}{C}} - \overset{\overset{\displaystyle CH_3}{|}}{\underset{\underset{\displaystyle H}{|}}{C}} - CH_3$ + CH_3CH_2NMgBr

5.5   Uses of Organometallic Compounds

Organometallic compounds are an important reagent for the synthesis of organic compounds (e.g., Grignard reagents).

Alkylboranes are important, and hydroboration is a method for preparing alcohols.

$$6RCH = CH_2 + (BH_3)_2 \rightarrow 2(RCH_2CH_2)_3B \rightarrow \frac{H_2O_2}{6OH^-}$$
$$\text{diborane} \quad 6RCH_2CH_2OH + 2B(OH)_3$$

Alkylmercury compounds are useful as intermediates in the formation of alcohols and ethers.

$$\overset{\diagdown}{\diagup}C=C\overset{\diagup}{\diagdown} + Hg(OAc)_2 + H_2O \longrightarrow -\overset{|}{\underset{OH}{C}}-\overset{|}{\underset{HgOAc}{C}}- \xrightarrow{NaBH_4} -\overset{|}{\underset{OH}{C}}-\overset{|}{C}-$$

Organic compounds of cadmium and cuprates are often used for the preparation of ketones from acyl halides.

$$R_2'Cd + R\overset{O}{\overset{||}{C}}Cl \longrightarrow R'\overset{O}{\overset{||}{C}}R$$

Mercuric compounds are toxic toward plant life and are used as bactericides, algicides, fungicides, and herbicides.

Mercurochrome and methiolate are used medically.

Problem Solving Examples:

Predict the products of hydroboration on the following compounds:

(a) $CH_3 \overset{\overset{\textstyle CH_3}{\textstyle |}}{\underset{\underset{\textstyle H}{\textstyle |}}{C}} — CH = CH_2$

(b) $CH_3 - \overset{\overset{\textstyle H}{\textstyle ||}}{\underset{\underset{\textstyle CH_3}{\textstyle |}}{C}} = C - CH_3$

Hydroboration converts alkenes into alcohols. Rearrangement does not occur because carbonium ions are not intermediates.

(a)

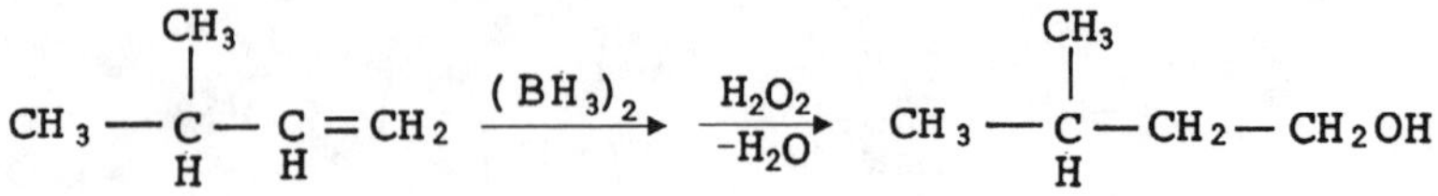

$$CH_3 - \underset{\underset{H}{|}}{\overset{\overset{CH_3}{|}}{C}} - CH = CH_2 \xrightarrow{(BH_3)_2} \xrightarrow[-H_2O]{H_2O_2} CH_3 - \underset{\underset{H}{|}}{\overset{\overset{CH_3}{|}}{C}} - CH_2 - CH_2OH$$

3-methyl 1-butanol

(b)

$$CH_3 - \underset{\underset{CH_3}{|}}{C} = \overset{\overset{H}{|}}{C} - CH_3 \xrightarrow{(BH_3)_2} \xrightarrow[-H_2O]{H_2O_2} CH_3 - \underset{\underset{H}{|}}{\overset{\overset{CH_3}{|}}{C}} - \underset{\underset{OH}{|}}{\overset{\overset{H}{|}}{C}} - CH_3$$

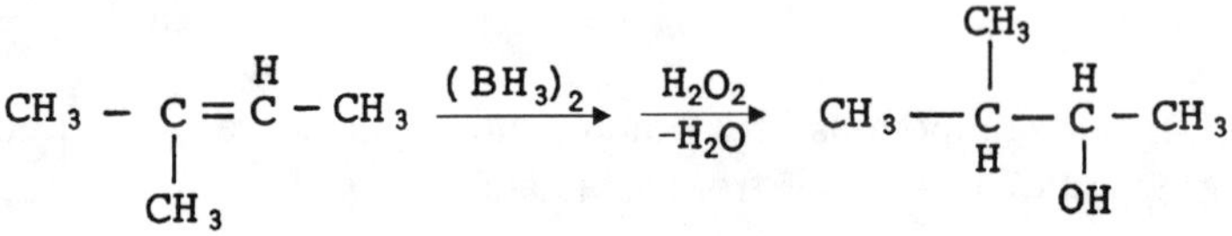

3-methyl 2-butanol

Q Predict the products of the reactions between the following alkenes and mercuric acetate in the presence of water upon reduction with sodium borohydride.

(a) 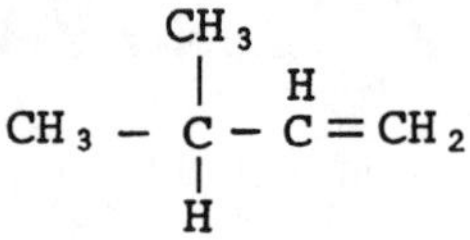 $$CH_3 - \underset{\underset{H}{|}}{\overset{\overset{CH_3}{|}}{C}} - \overset{\overset{H}{|}}{C} = CH_2$$

(b) 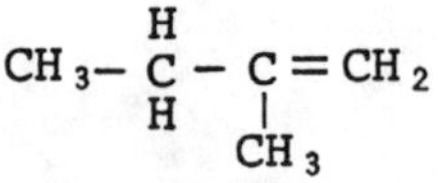 $$CH_3 - \underset{\underset{H}{|}}{\overset{\overset{H}{|}}{C}} - \underset{\underset{CH_3}{|}}{C} = CH_2$$

A The first step in this reaction is the addition of the hydroxide ion from water and HgOAc to the carbon-carbon double bond. In the reduction step with sodium borohydride, the –HgOAc is replaced with hydrogen. This reaction results in alcohols corresponding to Markovnikov addition of water to the double bond.

(a)

$$CH_3 - \underset{\underset{H}{|}}{\overset{\overset{CH_3}{|}}{C}} - \overset{H}{C} = CH_2 \xrightarrow[H_2O]{Hg\,(OAc)_2} \xrightarrow{Na\,BH_4} CH_3 - \underset{\underset{H}{|}}{\overset{\overset{CH_3}{|}}{C}} - \underset{\underset{OH}{|}}{\overset{H}{C}} - CH_3$$

3-methvl 2-butanol

(b)

$$CH_3 - \underset{\underset{H}{|}}{\overset{\overset{H}{|}}{C}} - \underset{\underset{CH_3}{|}}{C} = CH_2 \xrightarrow[H_2O]{Hg\,(OAc)_2} \xrightarrow{Na\,BH_4} CH_3 - CH_2 - \underset{\underset{CH_3}{|}}{\overset{\overset{OH}{|}}{C}} - CH_3$$

2-methyl 2-butanol

   Predict the products of the reactions of the following acid chlorides with diethyl cadmium.

(a) $$CH_3 - \underset{\underset{CH_3}{|}}{\overset{H}{C}} - \overset{\overset{O}{\|}}{C} - Cl$$

(b) $$CH_3 \overset{\overset{O}{\|}}{C} - CH_2 CH_2 \overset{\overset{O}{\|}}{C} - Cl$$

A   Organocadmium compounds react with acid chlorides in a nucleophilic substitution reaction to yield ketones. Organocadmium compounds containing aryl or primary alkyl groups are stable. They do not react with many of the functional groups with which the Grignard reagent will react.

(a)

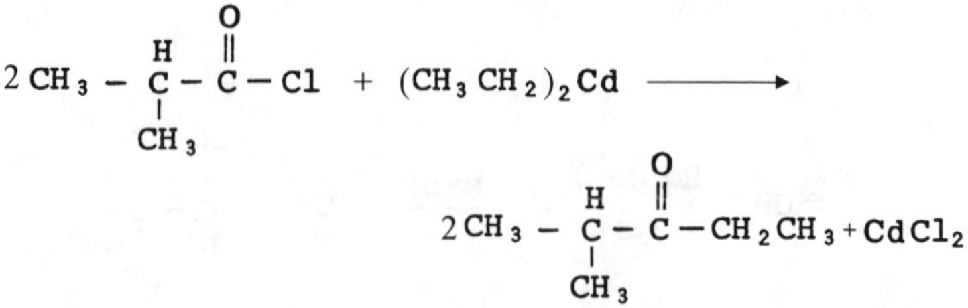

$$2\ CH_3 - \underset{\underset{CH_3}{|}}{\overset{H}{\underset{|}{C}}} - \overset{\overset{O}{\|}}{C} - Cl\ +\ (CH_3\ CH_2)_2\ Cd\ \longrightarrow$$

$$2\ CH_3 - \underset{\underset{CH_3}{|}}{\overset{H}{\underset{|}{C}}} - \overset{\overset{O}{\|}}{C} - CH_2\ CH_3 + CdCl_2$$

(b)

$$2\ CH_3\ \overset{\overset{O}{\|}}{C} - CH_2\ CH_2\ \overset{\overset{O}{\|}}{C}\ Cl\ +\ (CH_3\ CH_2)_2\ Cd\ \longrightarrow$$

$$CH_3\ \overset{\overset{O}{\|}}{C}\ CH_2\ CH_2\ \overset{\overset{O}{\|}}{C}\ CH_2\ CH_3 + CdCl_2$$

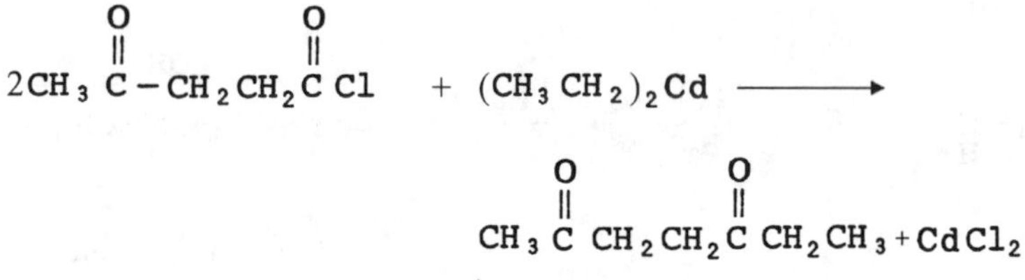

Quiz: Phenols, Quinones, and Organometallic Compounds

1. Which one of the following reactions below represents an aldol condensation?

   (A)

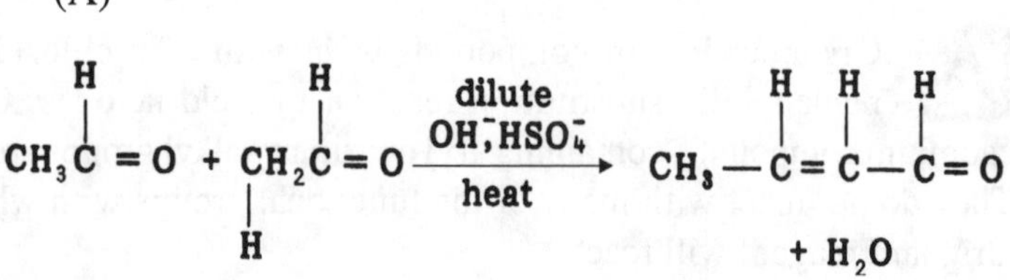

   $$CH_3 \overset{H}{\underset{|}{C}} = O\ +\ CH_2 \overset{H}{\underset{\underset{H}{|}}{\overset{|}{C}}} = O \xrightarrow[\text{heat}]{\substack{\text{dilute} \\ OH^-, HSO_4^-}} CH_3 - \overset{H}{\underset{|}{C}} = \overset{H}{\underset{|}{C}} - \overset{H}{\underset{|}{C}} = O$$
   $$+\ H_2O$$

   (B)

   cyclohexanone $+ Br_2 \xrightarrow{H^+}$ 2-bromocyclohexanone $+ HBr$

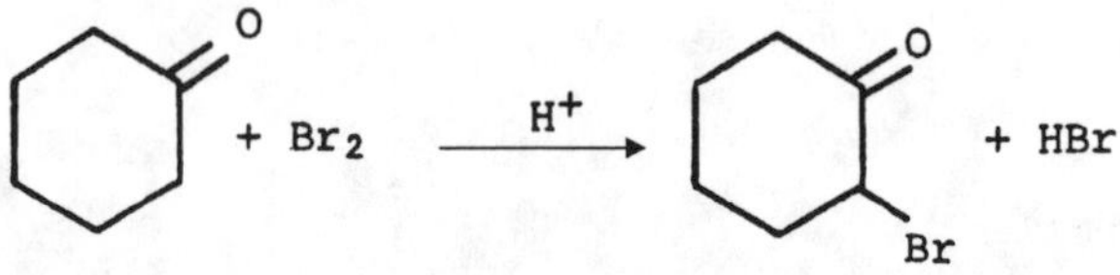

(C)

(D)

(E)

2. Which of the compounds is the strongest acid?

(A)

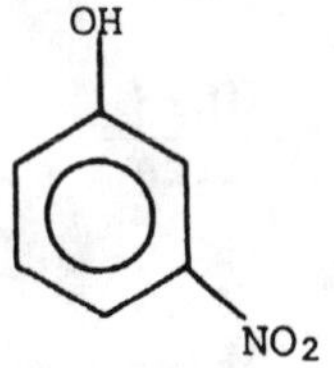

(D)

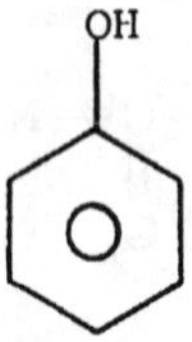

(B)

(E)

(C)

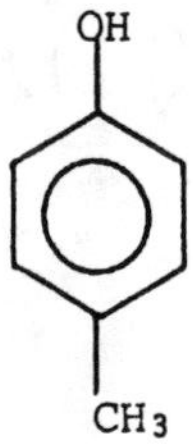

3. Catechol is treated with four reagents, as follows:

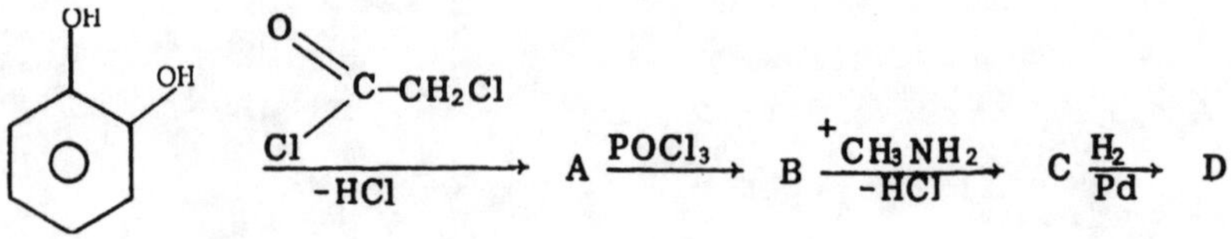

What is the structure of D?

(A)

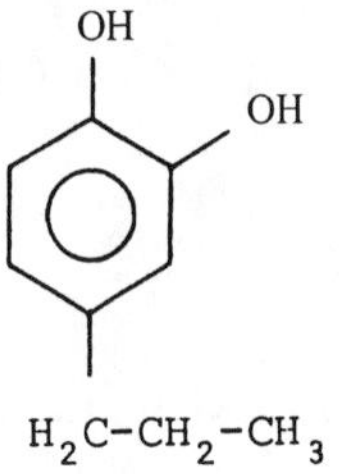

OH
OH

$H_2C-CH_2-CH_3$

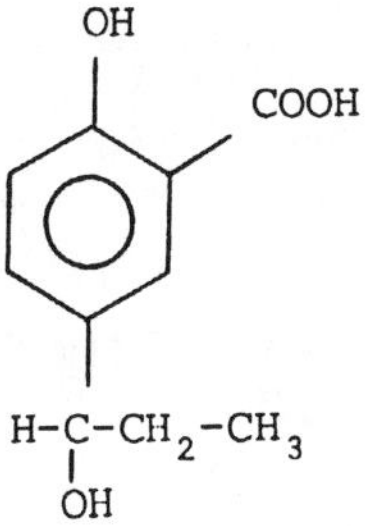

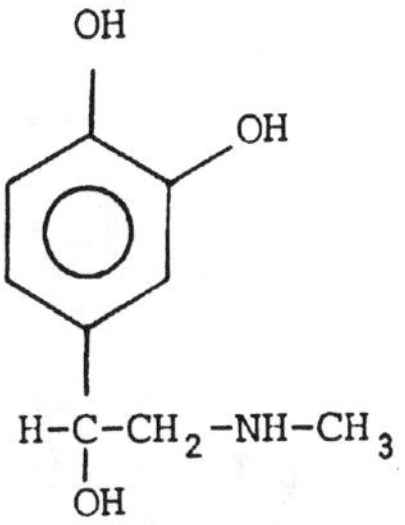

(D)

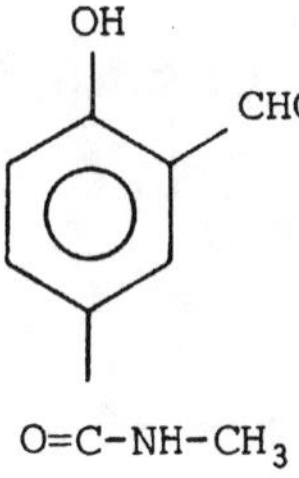

OH
CHO

$O=C-NH-CH_3$

(B)

OH
COOH

$H-C-CH_2-CH_3$
|
OH

(E)

OH
COOH

$H_2C-CHOH-NH_2$

(C)

OH
OH

$H-C-CH_2-NH-CH_3$
|
OH

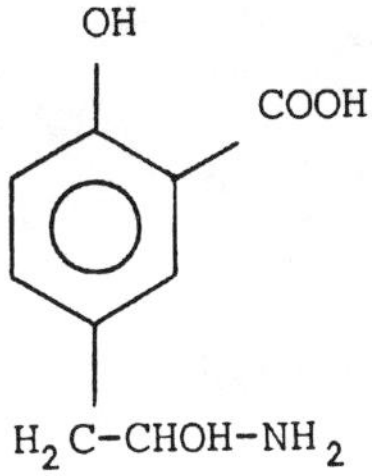

4. Transmetallation of an aryl bromide (or iodide) with an alkyllithium results in the production of a reagent that is capable of undergoing the same reactions as the Grignard reagents; this reagent is

(A) arylmetal.   (D) arylhalide.

(B) aryllithium.   (E) alkyllithium-bromide.

(C) alkylbromide (or idodide).

5. Compound Y in the following reaction is an:

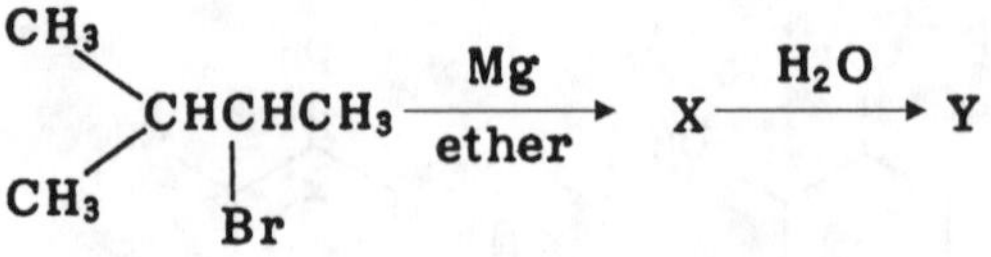

$$\begin{array}{c} CH_3 \\ \\ CH_3 \end{array} CHCHCH_3 \xrightarrow[\text{ether}]{Mg} X \xrightarrow{H_2O} Y$$
$$\qquad\quad | \\ \qquad\quad Br$$

(A) alkane.

(D) alkylhalide.

(B) alkene.

(E) organometallic haloalcohol.

(C) alcohol.

6. Product X is

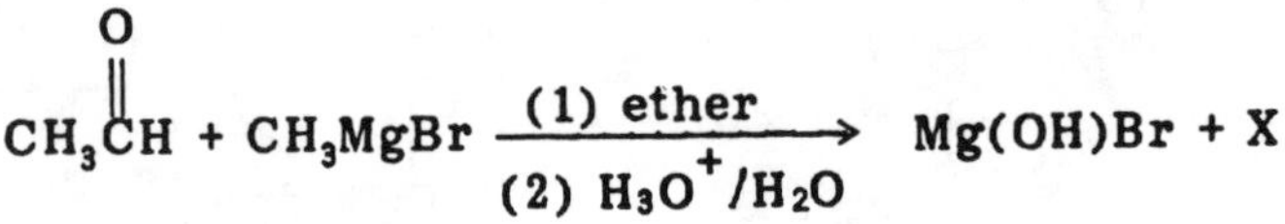

$$CH_3\overset{\displaystyle O}{\overset{\|}{C}}H + CH_3MgBr \xrightarrow[\text{(2) } H_3O^+/H_2O]{\text{(1) ether}} Mg(OH)Br + X$$

(A) 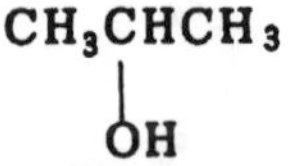 $\begin{array}{c} CH_3CHCH_3 \\ | \\ OH \end{array}$

(D) 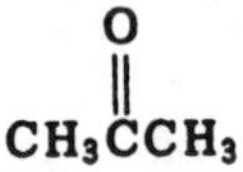 $CH_3\overset{\displaystyle O}{\overset{\|}{C}}CH_3$

(B) $CH_3CH_2CH_2OH$

(E) $CH_3CH = CH_2$

(C) CH_3CH_2CHO

7. In the preparation of phenol from cumene, which one of the following is a reactant?

(A) NaOH

(D) HCl

(B) O_2

(E) $Pb(OAc)_4$

(C) ArN_2^+

8. Which one is the product of the following reaction?

    hydroquinone $\xrightarrow{\text{FeCl}_3}$

    (A)  HO—⟨ ⟩—OH   (D) 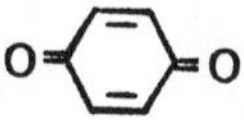 O=⟨ ⟩=O

    (B) 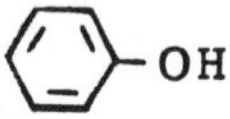 ⟨ ⟩—OH   (E) 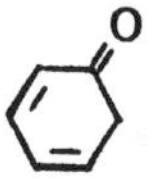 ⟨ ⟩=O

    (C) 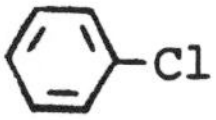 ⟨ ⟩—Cl

9. Organometallic compounds containing potassium, K, have which one of the following properties?

    (A) Ionic   (D) Stable in air

    (B) Volatile   (E) Weakly polar

    (C) Soluble in nonpolar solvents

10. The reaction of a Grignard reagent with nitriles yields a(n)

    (A) ether.   (D) peroxide.

    (B) primary alcohol.   (E) ketone.

    (C) tertiary alcohol.

ANSWER KEY

1. (A)
2. (E)
3. (C)
4. (B)
5. (A)

6. (A)
7. (B)
8. (D)
9. (A)
10. (E)

CHAPTER 6

Heterocyclic Compounds

6.1 Structure

Heterocyclics are cyclic compounds in which one or more of the ring atoms are not carbon. Heterocyclics include saturated and unsaturated cyclic ethers, thioethers, and amines.

Nomenclature

Number 1 is given to the heteroatom and numbering proceeds clockwise or counterclockwise so that the other substituents or other heteroatoms get the lowest numbers:

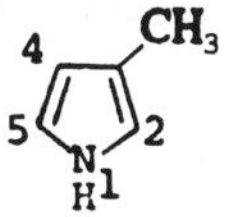

3-Methylpyrrole

When two or more different heteroatoms exist, oxygen takes preference over sulfur and nitrogen.

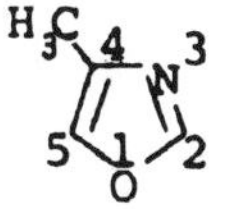

4-Methyloxazole

Problem Solving Examples:

  Name the following compounds by an accepted system.

(a) H₃C— S —CH₃

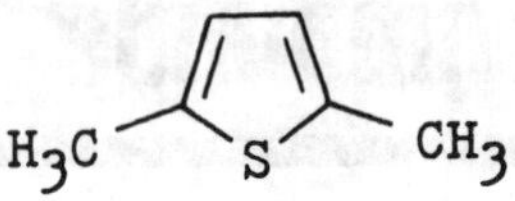

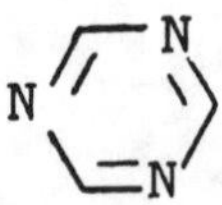

(e) CH₃Br (oxazolium with N⁺ and O)

(b) N-CH₃ ring with CH₃

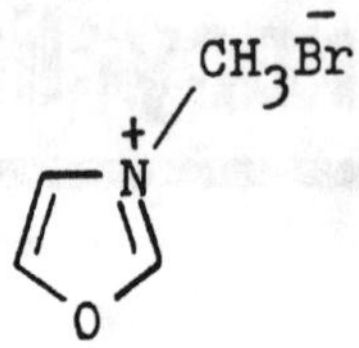

(f) pyridine fused to pyrrolidine N-CH₃

(c) S—NH ring

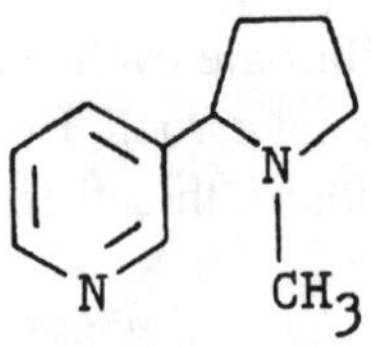

(g) OH, H₃C, N, N, OH pyrimidine

(d) N, N, N triazine ring

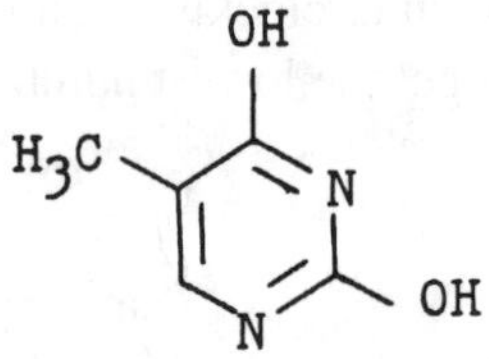

(h) H₃C— N(H) —CH₃ pyrrole ring

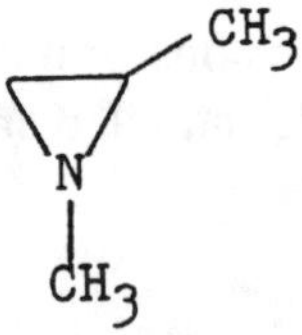

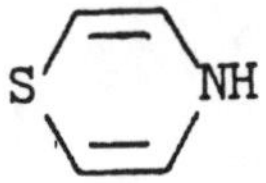

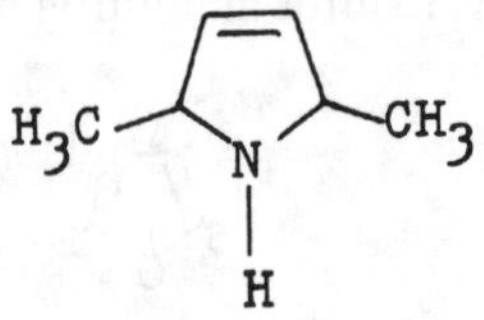

A Many heterocyclic compounds have acquired common names that are unlikely to be replaced by more systematic names. However, some rules have been adopted by IUPAC for the naming of monocyclic and polycyclic ring systems possessing one or more heteroatoms. We will briefly discuss the nomenclature for monocyclic rings (nomenclature for polycyclic rings is discussed in the next problem), and then we will name compounds (a) – (h). The method is to add the appropriate suffix and prefix to a given stem according to the following rules:

(1) The ring size is denoted by the stem (see Table).

(2) The heteroatom is denoted by the prefix: oxa and oxygen; thia and sulfur; aza and nitrogen. The presence of two oxygens, sulfurs, or nitrogens is indicated by dioxa, dithia, or diaza, respectively. If there are two or more different heteroatoms in the ring, they are written in order of preference: oxygen before sulfur before nitrogen.

(3) The degree of unsaturation is usually indicated by the suffix, which differs for nitrogen and non-nitrogen compounds (see Table)

(4) The numbering system starts with the heteroatom and proceeds around the ring so as to give substituents (or other heteroatoms) the lowest numbered positions. If there are two or more different heteroatoms, oxygen has priority over sulfur, and sulfur over nitrogen for the number one position.

(5) Partially reduced ring compounds are either referred to as dihydro or tetrahydro derivatives of the parent compound, or else they are indicated by attaching H with the number denoting the position of saturation to the name of the parent compound.

Stem + suffix

Ring size	Stem	Ring contains nitrogen		Ring contains no nitrogen	
		Unsaturated[a]	Saturated	Unsaturated[b]	Saturated
3	-ir-	-irine	-iridine	-irene	-irane
4	-et-	-ete	-etidine	-ete	-etane
5	-ol-	-ole	-olidine	-ole	-olane
6	-in-	-ine	b	-in	-ane
7	-ep-	-epine	b	-epin	-epane
8	-oc-	-ocine	b	-ocin	-ocane
9	-on-	-onine	b	-onin	-onane
10	-ec-	-ecine	b	-ecin	-ecane

[a] Corresponding to maximum number of double bonds, excluding cumulative double bonds.

[b] The prefix "perhydro" is attached to the stem and suffix of the parent unsaturated compound.

Table: Stems, Suffix, and Ring Size of Heterocyclic Compounds

Note that the systematic names are usually not used with the common heterocyclic compounds. Examples of these are (systematic names are in parentheses):

pyrrole
(azole)

furan
(oxole)

thiophene
(thiole)

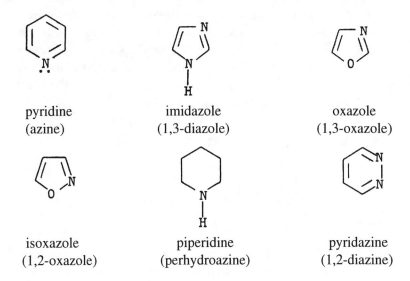

pyridine
(azine)

imidazole
(1,3-diazole)

oxazole
(1,3-oxazole)

isoxazole
(1,2-oxazole)

piperidine
(perhydroazine)

pyridazine
(1,2-diazine)

With this in mind, let us name compounds (a) – (h):

(a) The parent compound is a thiopene, with two methyl substituents at positions 2 and 5; therefore, its name is 2,5-dimethylthiophene.

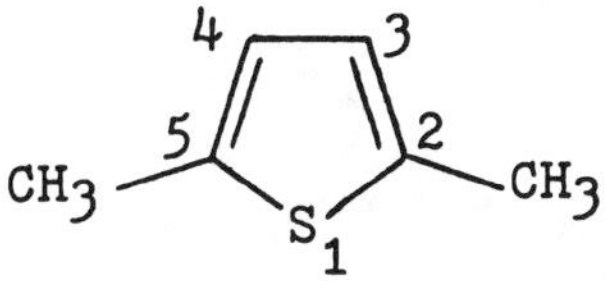

(b) The parent compound is aziridine; *aza* (the "a" drops out when it is followed by another vowel) because it is a nitrogen compound, and iridine because it is a three-membered saturated ring that contains nitrogen. The two methyl substituents are at positions 1 and 2; therefore, its name is 1,2-dimethylaziridine.

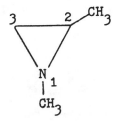

(c) The parent compound is 1,4-thiazine. The "1" indicates the position of the sulfur, the "4" indicates the position of nitrogen, and ine indicates an unsaturated six-membered ring that contains nitrogen. Since the nitrogen is hydrogenated, we must indicate this by using the prefix 4H, where the "4" indicates that hydrogenation is at the "4" position:

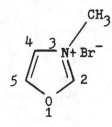

4H-1,4-thiazine

(d) This heterocyclic compound has three nitrogens at positions 1, 3, and 5 of an unsaturated six-membered ring that contains nitrogen; therefore, its name is 1,3,5-triazine.

(e) The parent structure is 1,3-oxazole. The methyl group is at position 3. Since this oxazole has a positive charge, it is called an oxazolium. The associated anion is bromide; therefore, its name is 3-methyl-1,3-oxazolium bromide:

(f) The parent structure is pyridine, with a substituent at position 3. The substituent itself is a heterocylic compound whose parent structure is azolidine (nitrogen/aza, olidine/five-membered saturated ring that contains nitrogen). The azolidino substituent has a methyl group at the 1 position; therefore, the substituent is called 1-methylazolidine and the compound is 3-(1-methylazolidino)-pyridine. Its common name is nicotine.

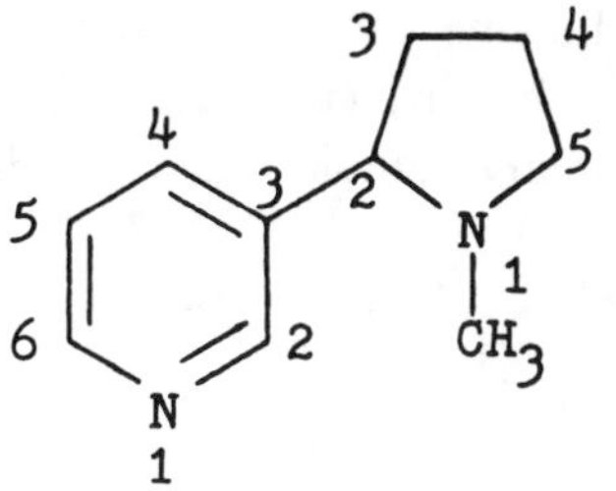

(g) The parent structure is pyrimidine with hydroxy groups at positions 2 and 4 and a methyl group at position 5; therefore, its name is 2,4-dihydroxy-5-methylpyrimidine. Its common name is thymine, one of the four bases in DNA.

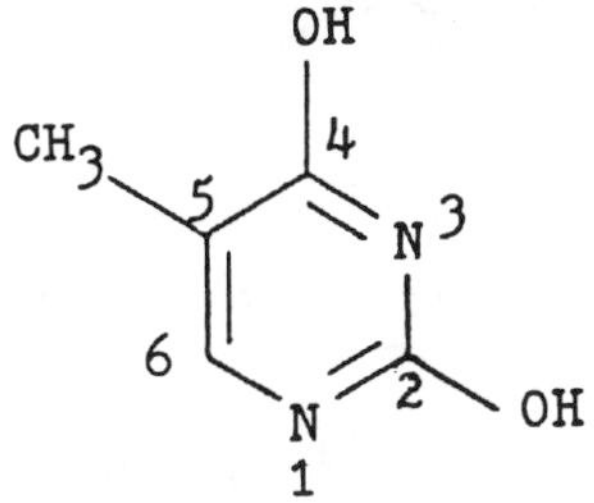

(h) Its parent structure is pyrrole. It is partially reduced with hydrogens at positions 2 and 5, and methyl groups at positions 2 and 5. Its name is 2,5-dihydro-2,5-dimethylpyrrole.

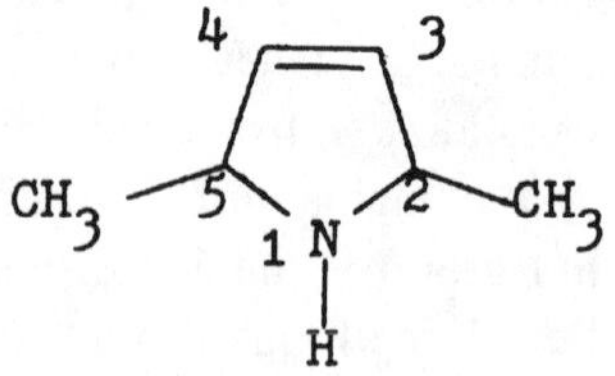

The structure shows a pyrrole ring with positions labeled 4, 3 at top, 5, 2, 1 N at bottom, with CH₃ groups attached at positions 5 and 2, and H on the N at position 1.

Name the following structures by an accepted system. Note that the numbering given in part (b) constitutes an exception to the rules; here the system resembles that for anthracene.

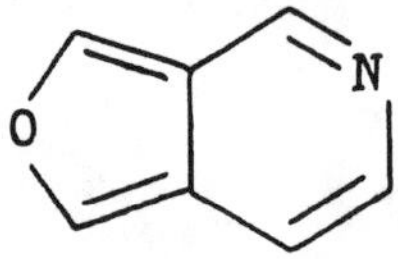

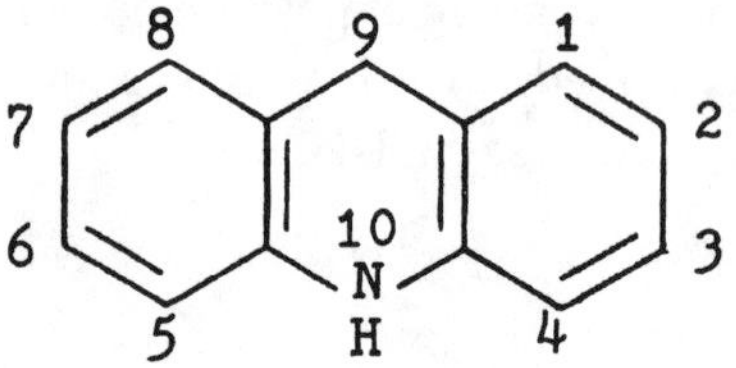

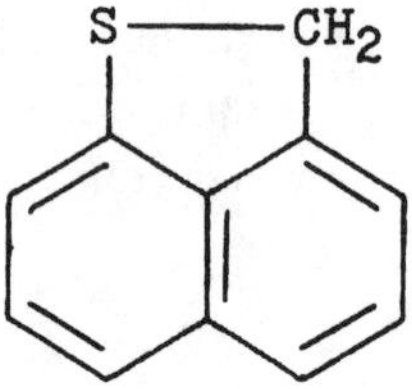

(a) O_2N structure — a quinoline ring system with N at lower right and O_2N substituent.

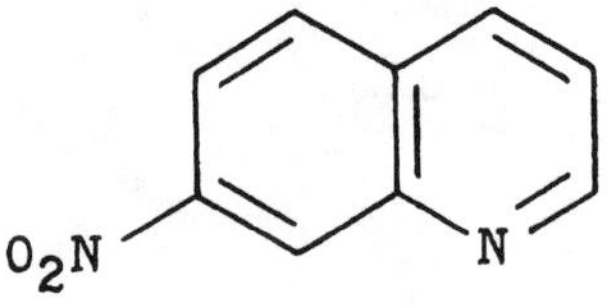

(c) structure with O and N — an isobenzofuran fused with pyridine.

(b) structure with positions labeled 8, 9, 1 across the top, 7, 2 on the sides, 6, 10 N, 3, 5, H, 4 — an acridine-type ring system.

(d) structure with S——CH_2 at top.

Polycyclic rings with heteroatoms are named according to the following rules:

(1) The name of the hetero ring is used as the parent compound, and the name of the fused ring is attached as a prefix.

(2) In the selection of the parent ring, when two or more hetero rings are present, the ring containing nitrogen has priority over the ring containing oxygen (and oxygen over sulfur). However, in the numbering system for the polycyclic compound, oxygen has priority over sulfur, and sulfur over nitrogen for

the number one position (see Rule 5 for numbering system).

(3) Preference is given to the largest hetero ring system which has a common name.

(4) To indicate the position of the ring junction, the sides of the parent ring are lettered a, b, c, etc., starting clockwise with the 1,2-bond; substituent fused rings that are not symmetrical are indicated by numbers. For example:

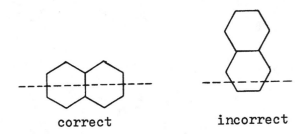

difura [2,3-b:2',3'-e] pyrazine   difura [2,3-b:3',2'-e] pyrazine

(5) In numbering the periphery of a polycyclic compound, the structure must be oriented according to the following rules:

(a) The greatest number of rings must lie along a horizontal axis:

correct   incorrect

(b) Of the other rings, a maximum must lie uppermost to the right above the horizontal axis (the first quadrant):

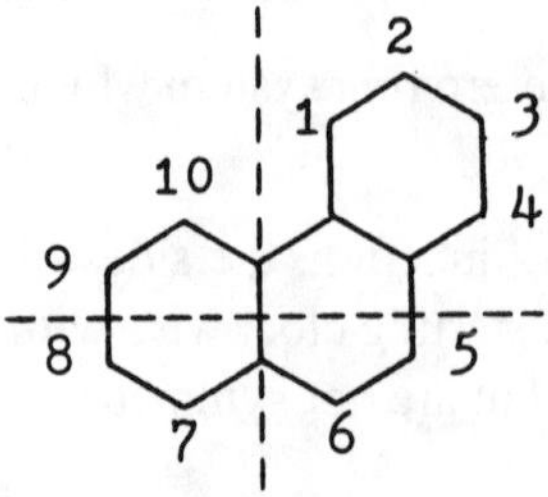

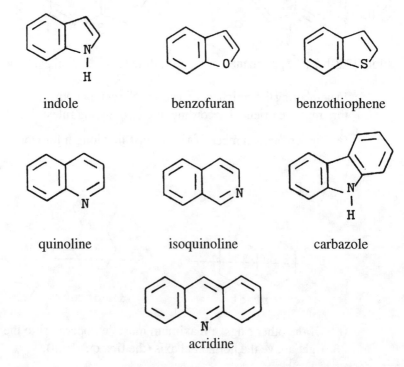

Numbering starts with the uppermost ring farthest to the right and proceeds in a clockwise direction, omitting ring junctions. Other things being equal, the orientation of the rings should be such as to lead to the lowest possible number for the heteroatoms.

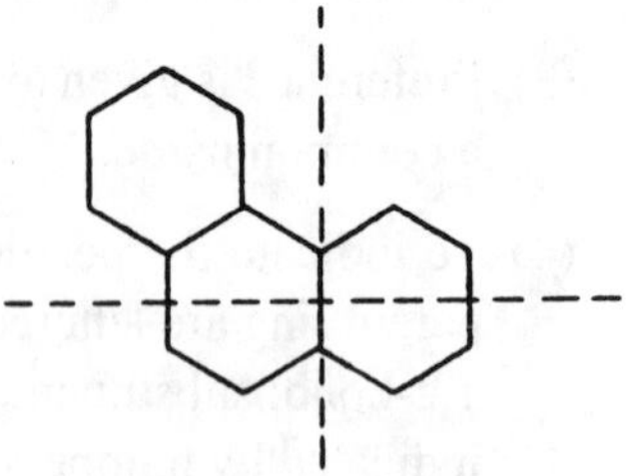

Common names for some polycyclic compounds are:

indole

benzofuran

benzothiophene

quinoline

isoquinoline

carbazole

acridine

With this in mind, let us name compounds (a) – (d):

(a) The parent structure is quinoline. There is a nitro substituent at position 7; therefore, its name is 7-nitro-quinoline.

(b) The parent structure is acridine. It is partially reduced, with hydrogens at positions 9 and 10. Therefore, its name is 9,10-dihydroacridine.

(c) The parent structure is pyridine. It has a furan ring fused to it. The position of this fusion is on side c of pyridine and between C_3 and C_4 of furan; therefore, its name is furo[3,4-c]pyridine.

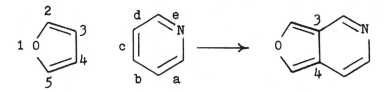

(d) The parent structure is thiophene. The ring fusion is on sides b and c of thiophene, and C_1 and C_8 of naphthalene. The thiophene is partially reduced, with the hydrogen on C_2, of the polycyclic compound. Therefore, its name is 2H-naphtho[1,8-*bc*]thiophene.

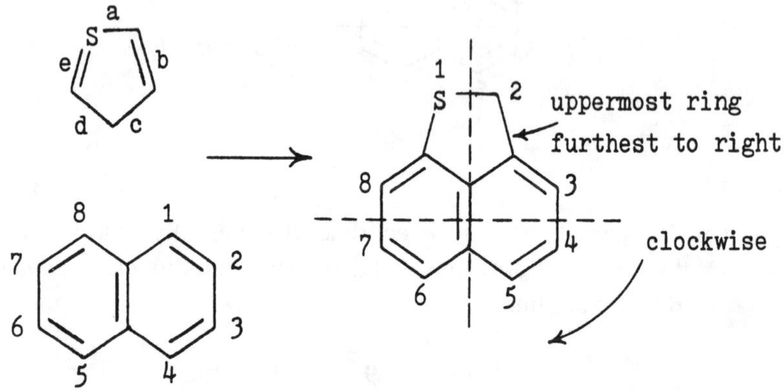

6.2 Properties of Furan, Pyrrole, and Thiophene

Pyrrole   Furan   Thiophene

Furan, pyrrole, and thiophene have the properties of unsaturated secondary amines, ethers, thioethers, and conjugated dienes.

Furan undergoes Diels-Alder cycloaddition with maleic anhydride.

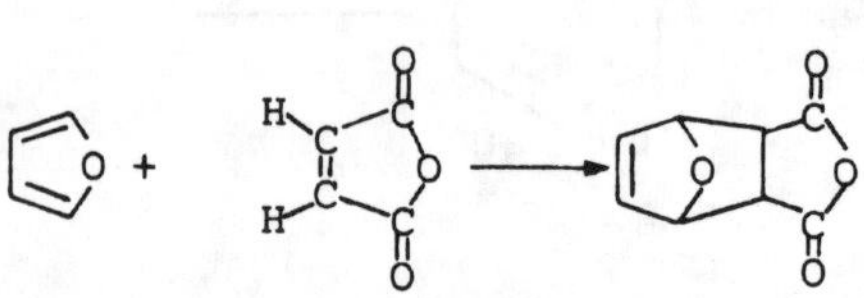

Furan, thiophene, and pyrrole undergo electrophilic substitution in preference to addition reactions.

The three have been shown by microwave spectra to be planar molecules. All three have aromatic characteristics.

Problem Solving Example:

Explain why furan, pyrrole, and thiophene and their derivatives commonly undergo electrophilic substitution reactions.

Looking at the structures of furan, pyrrole, and thiophene, we would expect each of these compounds to have the properties of a conjugated diene and an amine, ether, or thio ether. These compounds though do not behave as would be expected. There is overlap of the p orbitals, which gives rise to π clouds, one above and one below the plane of the rings. The π clouds contain six electrons that make up an aromatic sextet. The π electrons are delocalized causing the rings to be stable. These heterocycles tend to undergo electrophilic substitution reactions in order to retain this stabilization.

6.3 Synthesis of Furan, Pyrrole, and Thiophene

General Method: The Paal-Knorr Synthesis

Reaction occurs by heating 1,4-diketones in the presence of a dehydrating agent, ammonia, or a sulfide to obtain furan, pyrrole, or thiophene, respectively. Reaction proceeds by way of dienol.

$$\begin{array}{ccc}
\underset{\text{1,4-Diketone}}{R-\overset{\overset{\displaystyle H\;\;\;H}{\underset{|}{CH-CH}}}{\underset{\underset{O\;\;\;O}{}}{\Big\diagdown\Big/}}-R} & \rightleftharpoons \left[\underset{\text{Dienol}}{R-\overset{\overset{\displaystyle H\;\;\;H}{C=C}}{\underset{\underset{OH\;HO}{}}{\diagup\;\diagdown}}-R}\right] \xrightarrow[P_2O_5]{-H_2O} & \underset{\text{2,5-Dialkylfuran}}{R-\underset{O}{\Big\langle\!\!\Big\rangle}-R}
\end{array}$$

$$\xrightarrow[(NH_4)_2CO_3]{+NH_3,\,-2H_2O}\;\;\underset{\text{2,5-Dialkylpyrrole}}{R-\underset{N}{\Big\langle\!\!\Big\rangle}-R}\;\;\xrightarrow[P_2S_5]{+H_2S,\,2H_2O}\;\;\underset{\text{2,5-Dialkylthiophene}}{R-\underset{S}{\Big\langle\!\!\Big\rangle}-R}$$

Knorr Pyrrole Synthesis

An α-amino ketone and a ketone, or a β-ketoester react to give a substituted pyrrole.

$$\underset{\underset{H_5 C_2 OOC}{\overset{H_3 C}{\diagdown}} CH {\diagdown} NH_2}{\overset{C=O}{\diagup}} + \underset{\overset{H_2 C}{\underset{O}{\diagdown}} \overset{COOC_2 H_5}{\diagup} CH_3}{\overset{}{}} \xrightarrow{-2H_2 O} \underset{}{}$$

2,4-Dimethyl-3,5-dicarbethoxy-pyrrole

Thiophene is prepared as follows:

$$n - C_4 H_{10} + S \xrightarrow{600°C} \underset{S}{\boxed{}} + H_2 S$$

Problem Solving Example:

Q Using the Knorr Pyrrole synthesis, draw the synthesis of 2-carboxy-3-methyl pyrrole.

A Substituted pyrroles can be prepared via the Knorr Pyrrole synthesis, which utilizes the reaction between an α-amino ketone and a ketone or a β-ketoester.

$$\underset{\underset{\underset{Cl}{CH_2 NH_3}}{\overset{+}{}}}{\overset{CH_3}{\diagdown} C=O} + \underset{\overset{CH_2}{\underset{O}{\overset{C}{\diagdown}} CO_2 Mc}}{\overset{CO_2 Et}{\diagup}} \xrightarrow[\text{r.t.}]{\text{aq KOH}} \xrightarrow{\text{aq KOH}}$$

$$\xrightarrow[\triangle]{\text{aq KOH}}$$

2-carboxy-3-methyl pyrrole

6.4 Reactions

Electrophilic substitution in furan is at the α position.

a)

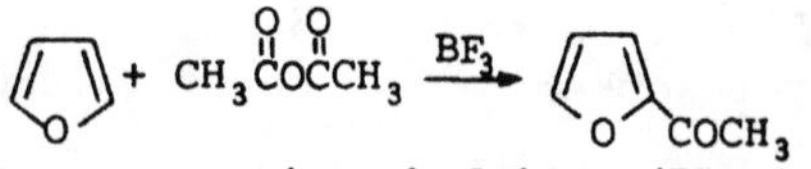

Furan   Acetic anhydride   (75-92%)
                              2-Acetylfuran

b)

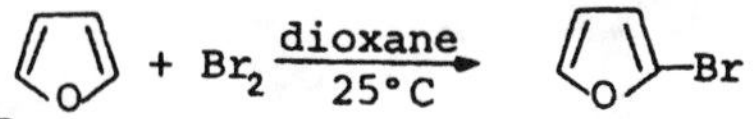

Furan   (75%)Bromofuran

c)

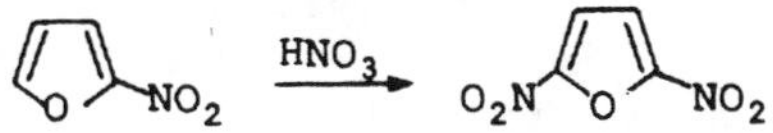

2-Nitrofuran   2,5-Dinitrofuran

Electrophilic substitution in thiophene is at the α position.

a)

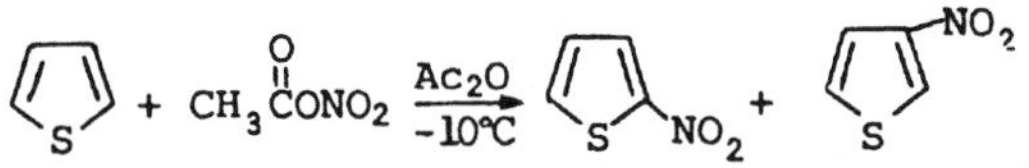

Thiophene   Acetyl-   2-Nitrothi-   3-Nitrothi-
             nitrate   ophene(70%)   ophene(5%)

b)

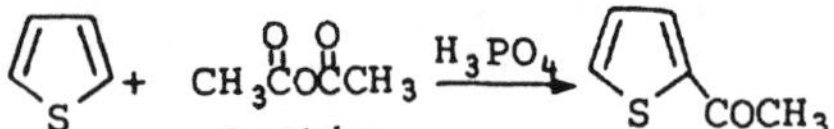

Thiophene   Acetic   (94%)
             anhydride   2-Acetylthiophene

c)

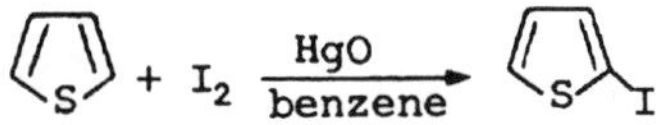

Thiophene   (75%) 2-Iodothiophene

d)

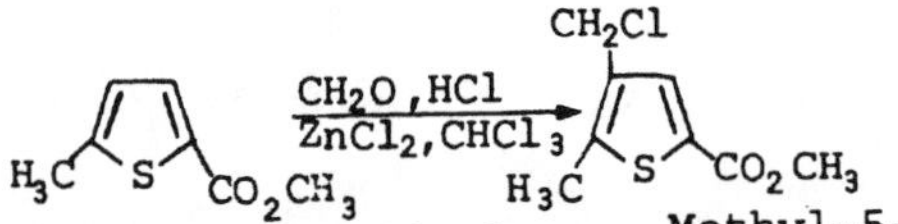

Methyl 5-methylthiophene   Methyl-5-Methyl-4-chloro   (93%)
2-Carboxylate   methyl-2-carboxylate

e)

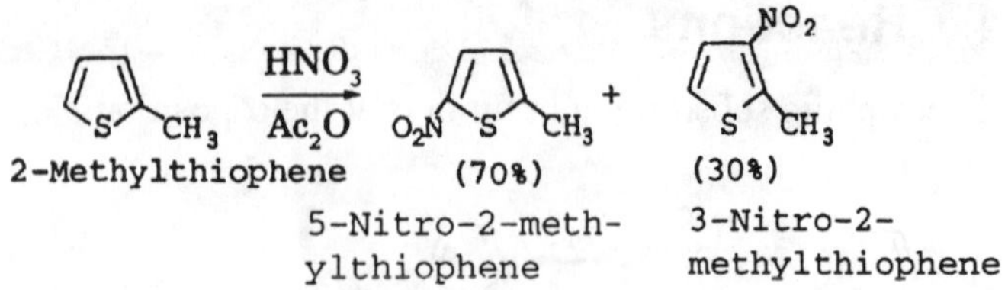

2-Methylthiophene $\xrightarrow[\text{Ac}_2\text{O}]{\text{HNO}_3}$ 5-Nitro-2-meth-
ylthiophene (70%)   +   3-Nitro-2-
methylthiophene (30%)

Electrophilic substitution in pyrroles is at the a position.

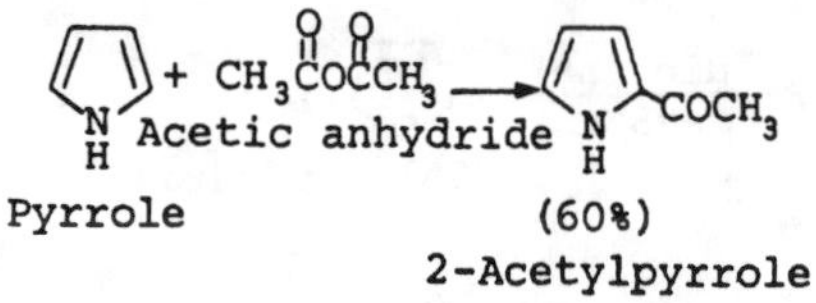

Pyrrole + Acetic anhydride $CH_3\overset{O}{\underset{}{C}}O\overset{O}{\underset{}{C}}CH_3$ ⟶ 2-Acetylpyrrole (60%)

Polymerization of pyrroles by dilute acids:

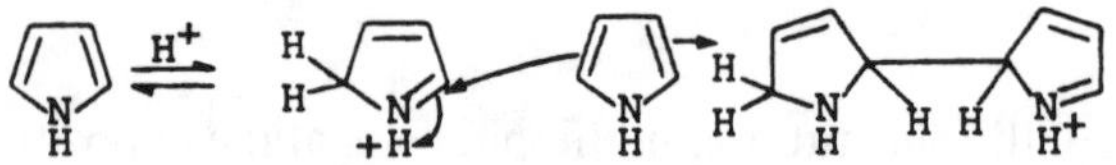

Electrophilic substitution at a position:

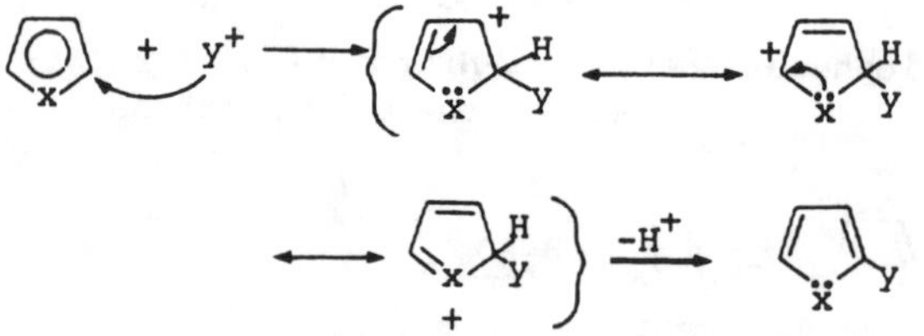

Acidic character of pyrroles:

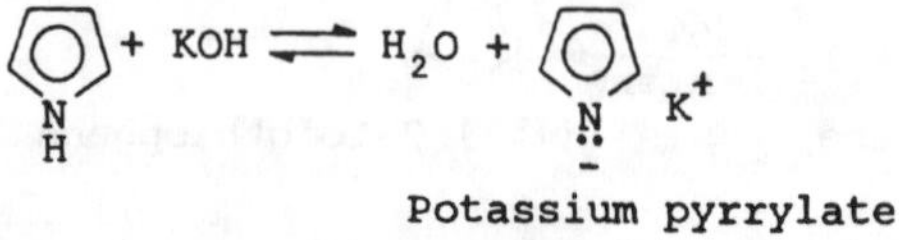

Potassium pyrrylate

Pyrrole is a weak acid that yields alkali-metal salts upon treatment with alkaline hydroxides.

Problem Solving Example:

Q Suggest a feasible synthesis of each of the following compounds from the indicated starting material:

(a) 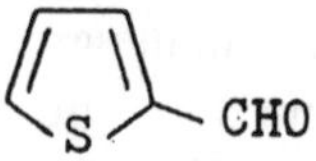 — CHO

from thiophene

(d) 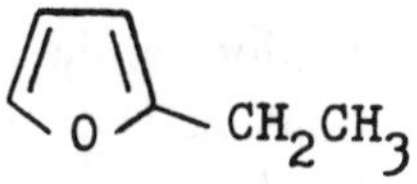 — CH₂CH₃

from furan

(b)

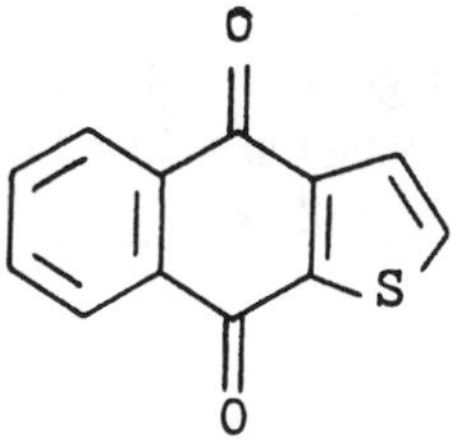

from thiophene

(e) 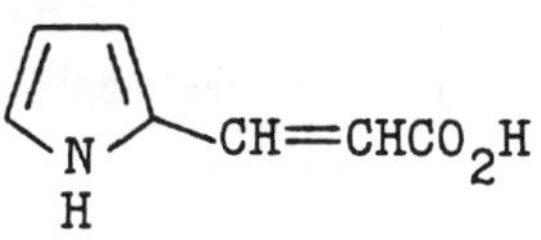 — CH=CHCO₂H

from pyrrole

(c)

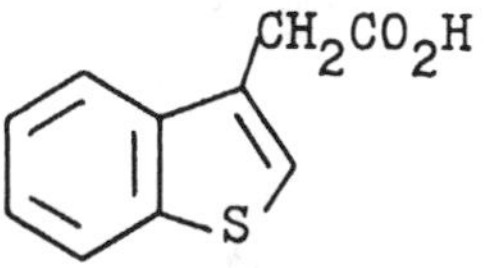

from benzothiophene

A (a) Before we answer this question, let us review some of the chemical properties of thiophene. Thiophene is an aromatic compound with a resonance energy of about 29–31 kcal/mole, as calculated from heats of combustion data. Although this is less than that of benzene (36.0 kcal/mole), it is almost the same per ring atom. Therefore, thiophene resembles benzene in many of its reactions, except it is more reactive and less stable. The resonance structures for thiophene are:

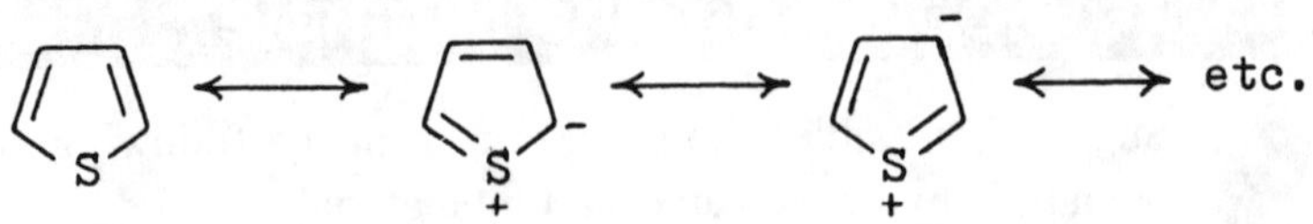

Calculations of the π-electron distribution in the ground state of thiophene show that electron distribution is greater at C_2 and C_5; therefore, electrophilic substitution should occur at C_2 and C_5, as it is in actuality.

The given compound has a $-\overset{\overset{O}{\|}}{C}-H$ group attached to thiophene at position 2. One method of achieving this is by chloromethylation via electrophilic substitution, followed by the Sommelet reaction:

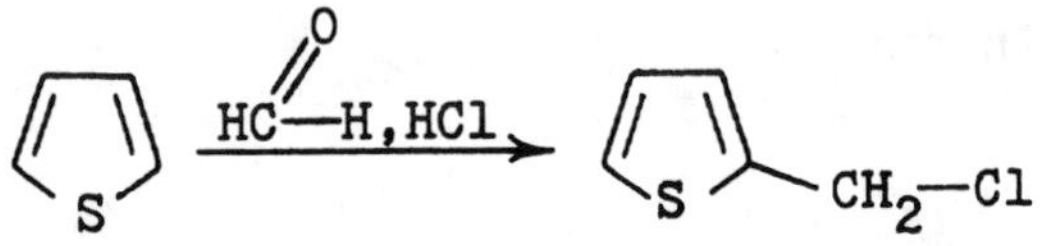

thiophene

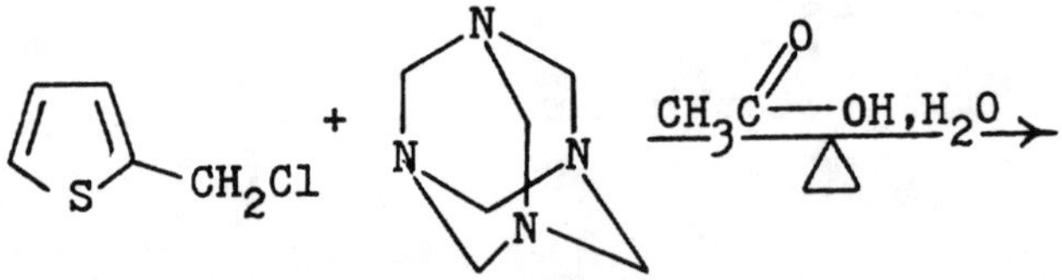

hexamethylene-
tetramine

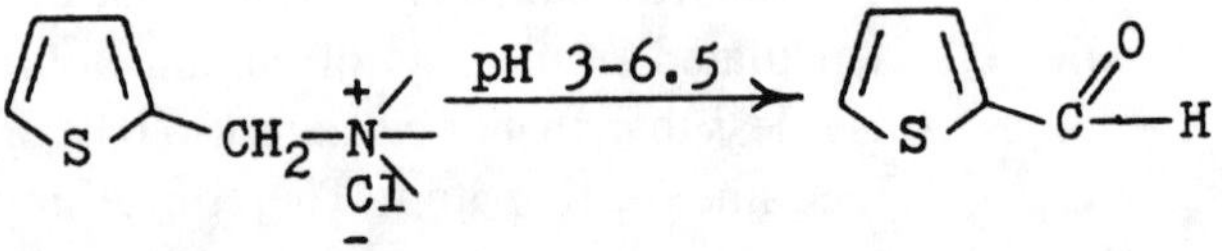

(b) This compound has a phthalic group attached to side "b" of the thiophene. One method of achieving this is by reacting thiophene with phthalic anhydride in $AlCl_3$ in an electrophilic substitution, followed by ring closure, this time via intramolecular electrophilic substitution.

phthalic
anhydride

thiophene

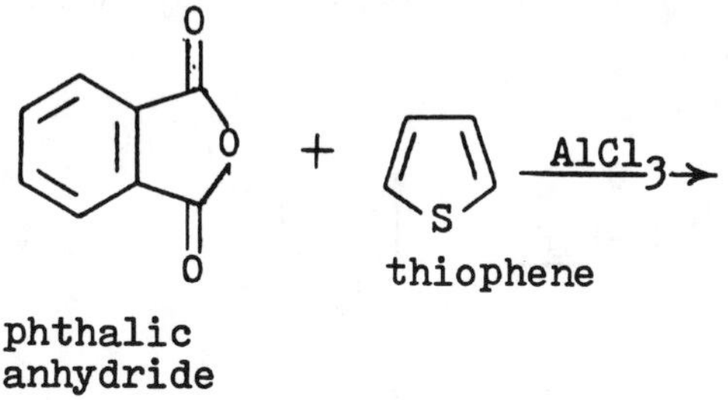

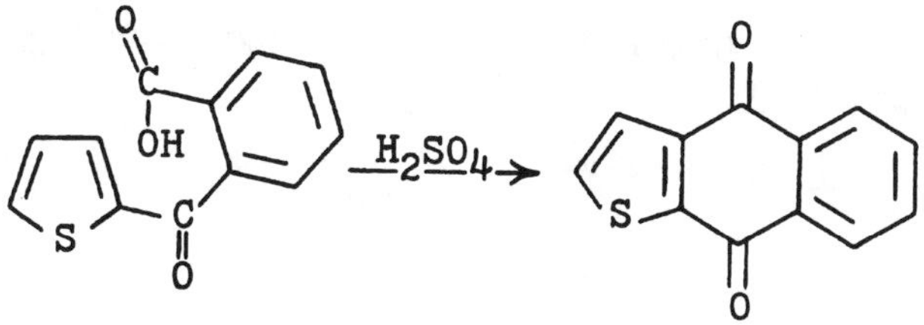

(c) Benzothiophene (specifically benzo[b]thiophene) is more stable and less reactive than thiophene. Benzothiophene is best represented as a resonance hybrid of structures I, II, and III in order of decreasing importance.

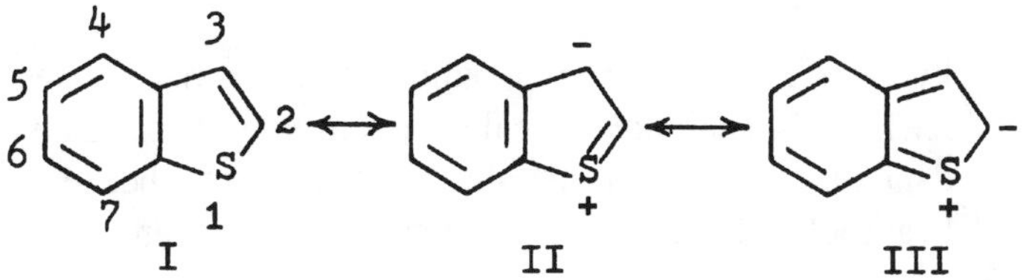

Since there is a greater electron density at C_3, electrophilic substitution occurs mainly at C_3.

The given compound has an acetic acid group attached to

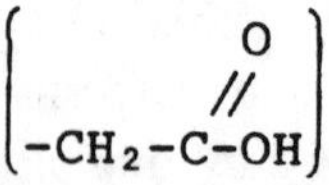

$$\left(\begin{array}{c} O \\ \parallel \\ -CH_2-C-OH \end{array} \right)$$

benzothiophene at C_3. This can be achieved by chloro-methylation of benzothiophene via electrophilic substitution, followed by a Grignard reaction, and subsequent carbonation:

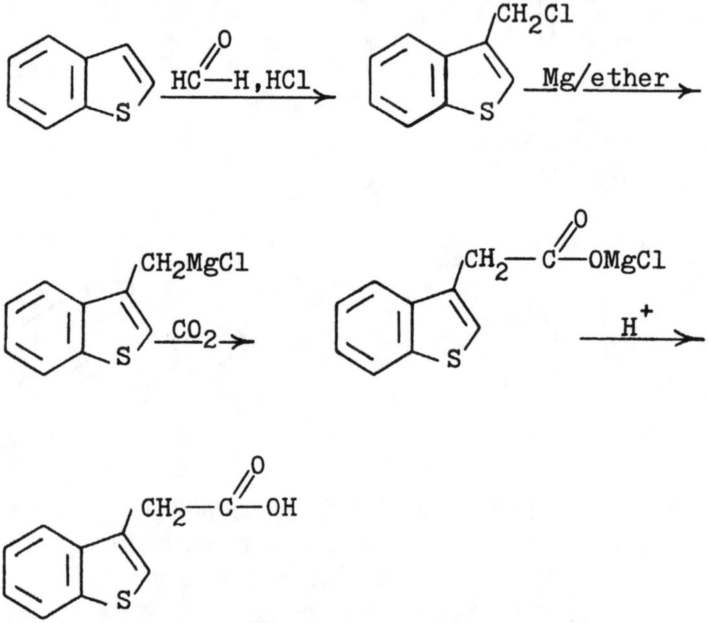

(d) Furan is an aromatic compound with a resonance energy of 23.7–26.54 kcal/mole, as calculated from heats of combustion data. This value is considerably less than that for benzene (36.0 kcal/mole). The resonance structures for furan are:

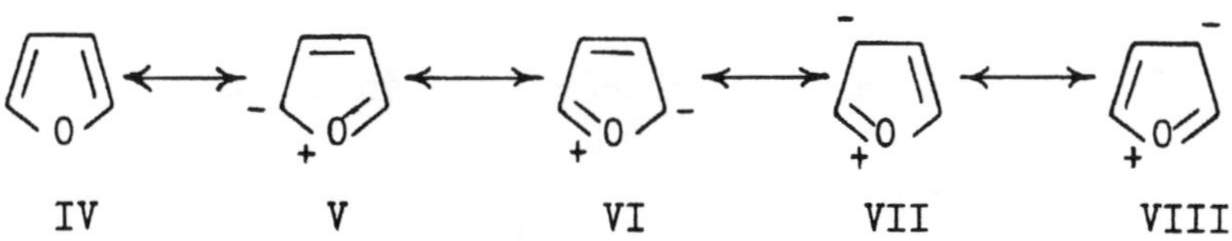

IV   V   VI   VII   VIII

The bond lengths data suggest that structure IV is the major contributor to the resonance hybrid. Structures V and VI are also of importance. The energy required to separate the charges is lower in these structures than in VII and VIII, as the distance involved is less. Therefore, electrophilic attack greatly favors substitution at positions 2 and 5. Since oxygen is more electronegative than sulfur or nitrogen, it shares its electrons less readily; therefore, furan is less aromatic than pyrrole or thiophene.

This compound has an ethyl group attached to furan at position 2. The best method for attaching an alkyl group to furan is by acylation followed by reduction:

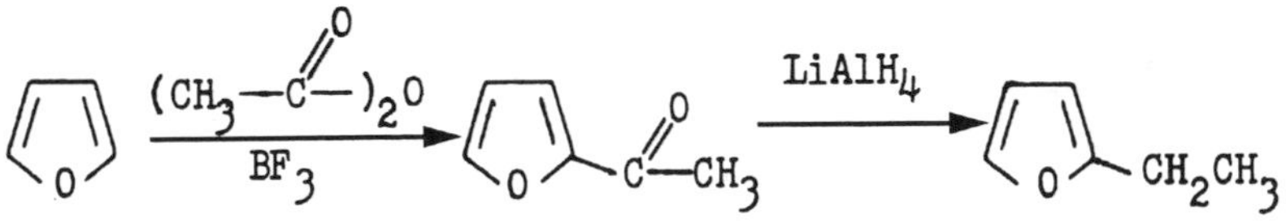

(e) Pyrrole is an aromatic compound with a resonance energy of 21–24 kcal/mole. The resonance structures for pyrrole are:

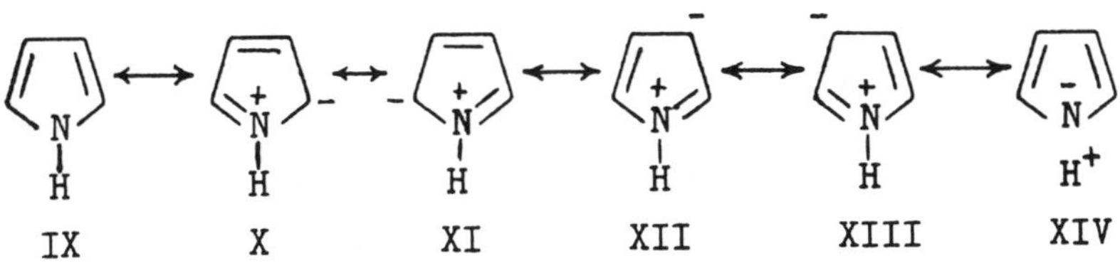

IX   X   XI   XII   XIII   XIV

Electrophilic substitution of pyrrole occurs preferentially at positions 2 and 5, as in the case of thiophene and furan.

This compound has a $-CH-CH\overset{O}{\overset{\|}{C}}-OH$ group attached to pyrrole at position 2. The best method of achieving this is by the Vilsmeier reaction to form pyrrole-2-carbaldehyde, followed by the Reformatsky reaction to form the given product:

Vilsmeier reaction:

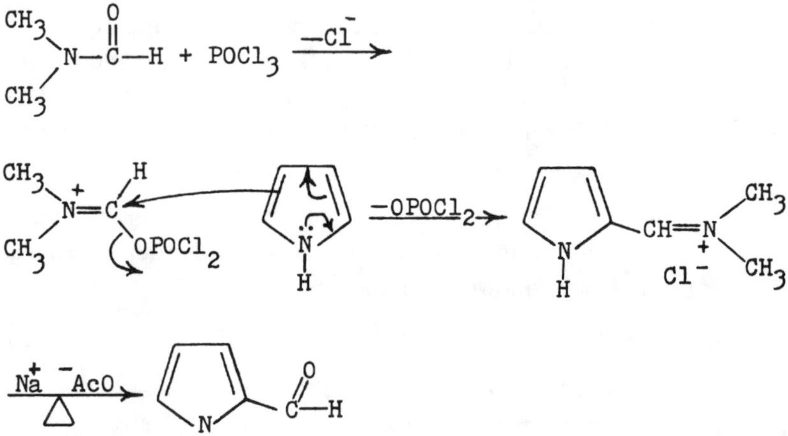

This reaction is basically an electrophilic substitution, followed by hydrolysis. The Reformatsky reaction, which follows, is an organometallic reaction that yields a β-hydroxy ester or acid. Upon heating, we get the corresponding α,β-unsaturated product:

Reformatsky Reaction

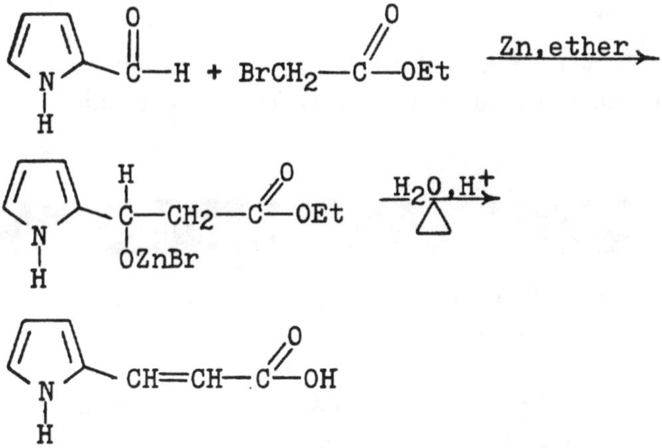

It is the carbanion character of the organozinc compound that allows this reaction to proceed. The formation and reaction of the organozinc compound is similar to that of a Grignard reagent. Zinc is used in place of magnesium because organozinc compounds do not react with the ester function, but only with the aldehyde or ketone. Hydrolysis of the ester yields the corresponding acid.

6.5 Pyridine, Quinoline, and Isoquinoline

Structure

The replacement of one CH group in benzene by nitrogen yields the pyridine molecule. It is a hybrid of the two Kekulé structures and the three polar structures shown below.

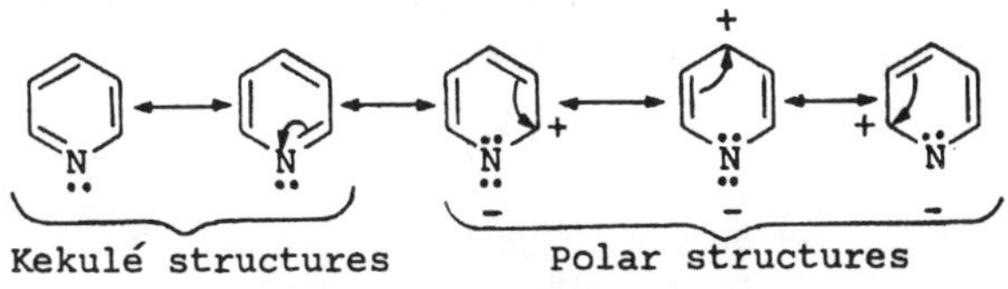

Kekulé structures   Polar structures

6.6 Structural Properties of Pyridines

Pyridine undergoes both nucleophilic substitution and electrophilic substitution.

The basic properties of pyridine (which are similar to those of the tertiary amines) are not influenced by the cyclic delocalization of the π electrons.

Problem Solving Example:

Explain how the nitrogen of pyridine is different from the nitrogen of pyrrole.

The nitrogen atom in pyridine, like each carbon atom, is bonded to other members of the ring using sp^2 orbitals. The nitrogen atom in pyridine provides one electron for the π cloud. The third sp^2 orbital of each carbon atom is used to form a bond with hydrogen. The third sp^2 orbital of nitrogen contains a pair of electrons. This electronic configuration makes pyridine a stronger base than pyrrole and also affects the reactivity of the ring.

6.7 Synthesis of Pyridine, Quinoline, and Isoquinoline

1,4 cycloaddition of nitriles to 1,3 dienes in the presence of an oxidizing agent produces 2-alkyl pyridine.

$$\underset{N}{\overset{\displaystyle\parallel}{}}\!\!\!C_{\diagdown R} \longrightarrow \underset{N}{\bigcirc}\!\!R \xrightarrow{\ K_3[FE(CN)_6]\ } \underset{N}{\bigcirc}\!\!R$$

The oxidation of 1,2-dihydroquinoline by nitrobenzene yields quinoline (Skraup quinoline synthesis).

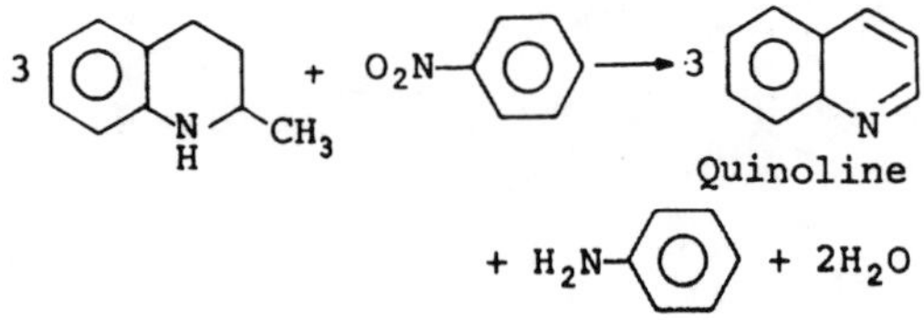

The cyclization of N-acyl-β-phenylethyl amines in the presence of phosphorus pentoxide as a dehydration catalyst yields isoquinolines (Bischler-Napieralski synthesis).

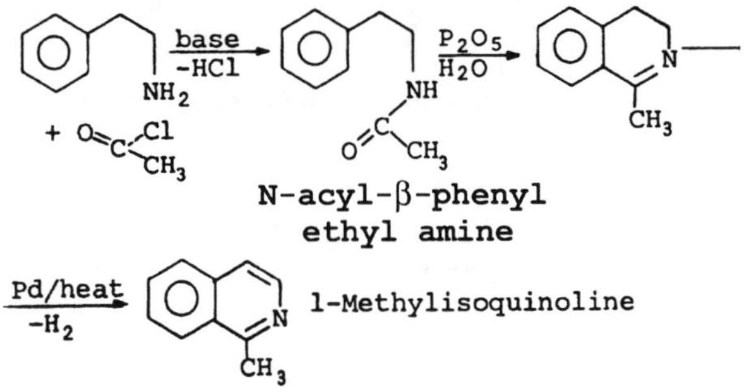

N-acyl-β-phenyl
ethyl amine

1-Methylisoquinoline

The condensation of o-aminobenzaldehyde with a ketone (Friedländer synthesis):

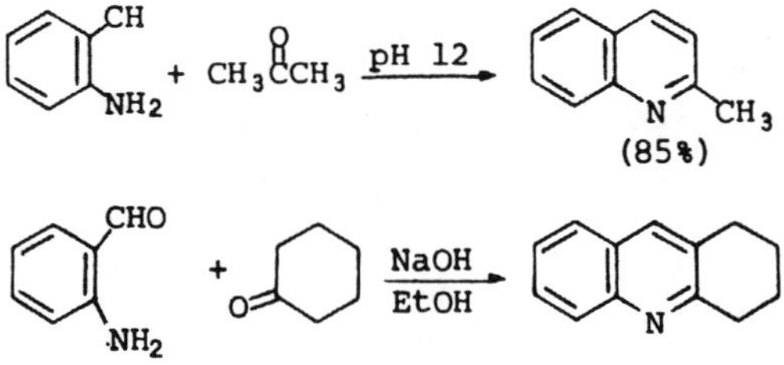

(85%)

Substitution of α,β-unsaturated ketone or aldehyde for glycerol.

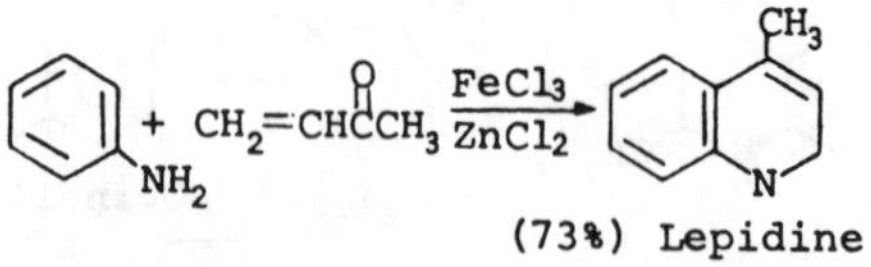

$$\text{aniline} + CH_2=CHCCH_3 \xrightarrow[ZnCl_2]{FeCl_3} \text{quinoline}$$

(73%) Lepidine

Heating a mixture of aniline, glycerol, nitrobenzene, and sulfuric acid (Skraup synthesis) to obtain quinoline:

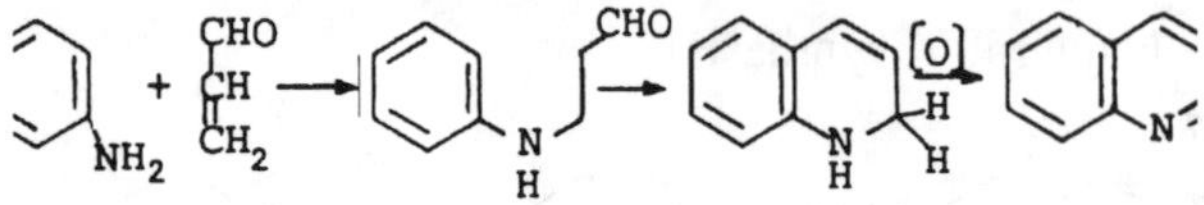

By heating the oxime of cinnamic aldehyde with phosphorus pentoxide:

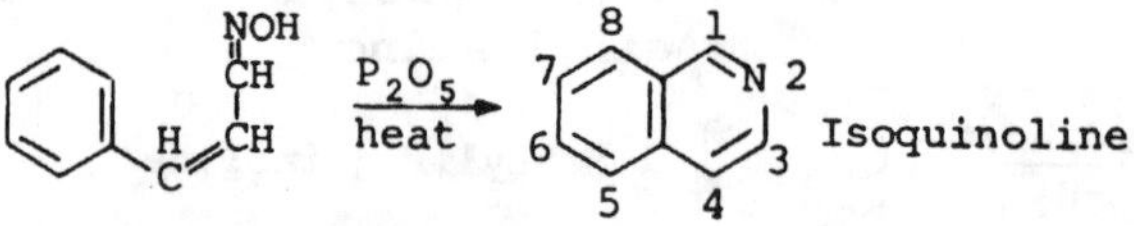

$$\xrightarrow[\text{heat}]{P_2O_5}$$

Isoquinoline

Problem Solving Example:

Two mechanisms for the Skraup synthesis are:

1.

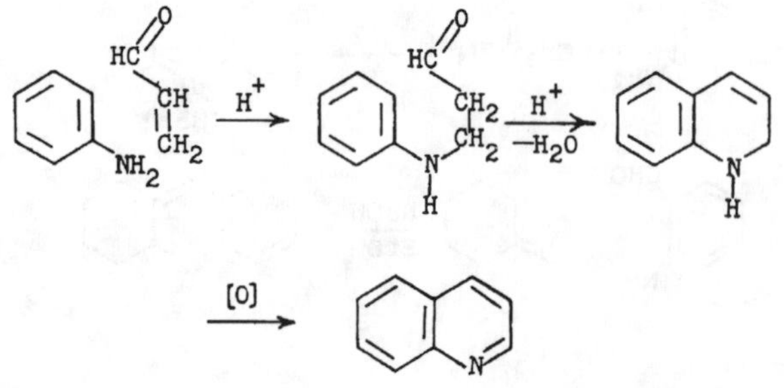

$$\xrightarrow{H^+} \qquad \xrightarrow[-H_2O]{H^+}$$

$$\xrightarrow{[O]}$$

2.

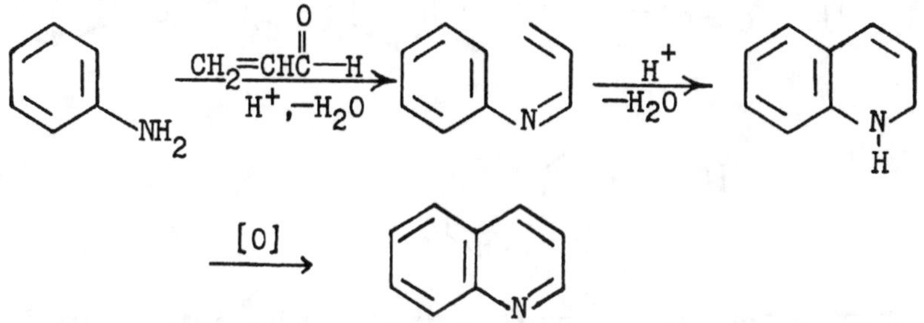

Devise an experiment whereby we can tell which is the actual mechanism.

A The difference between the two mechanisms is the place of attack by the nucleophilic nitrogen atom. The first mechanism involves attack on the β-carbon of the α,β-unsaturated aldehyde. The second mechanism involves attack at the aldehydic carbon. By using ^{14}C labeled acrolein, we can deduce the actual mechanism by the position of the label in the product:

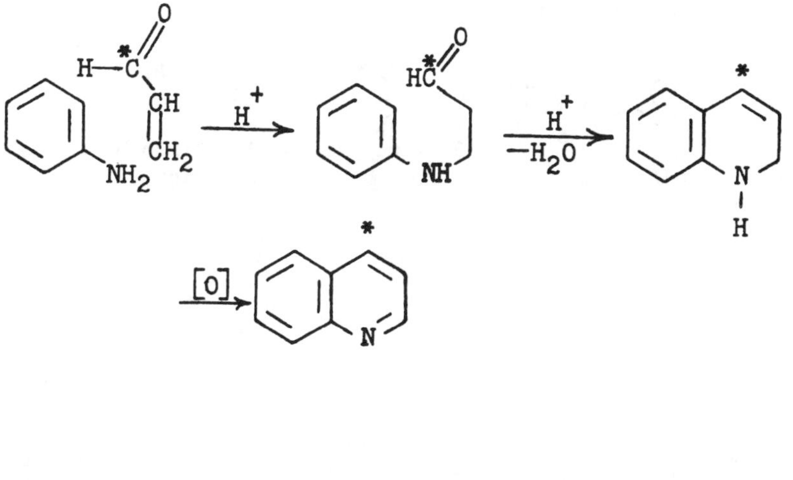

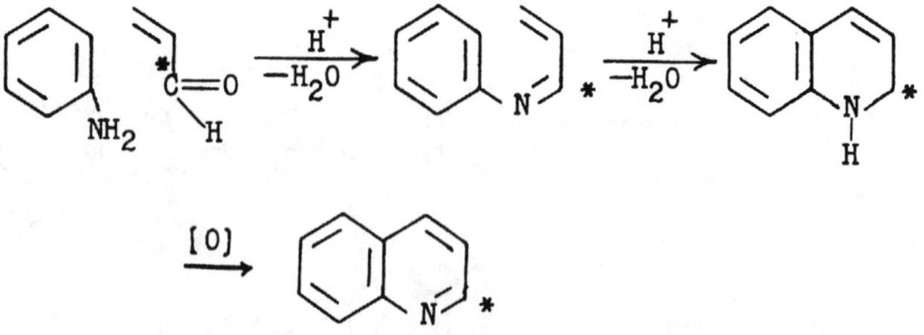

6.8 Reactions of Pyridine, Quinoline, and Isoquinoline

Methylation of Pyridine.

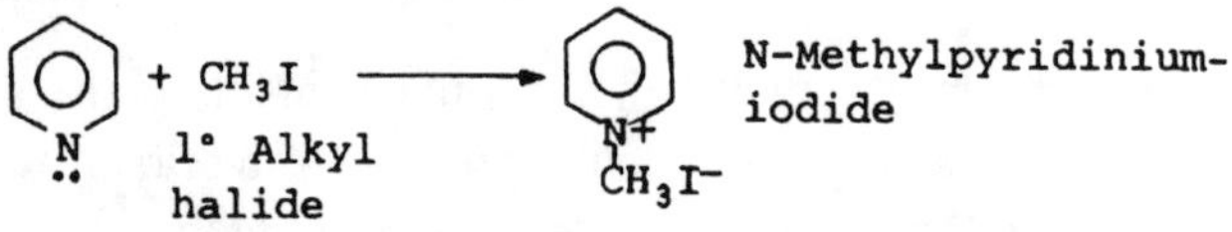

Sulfonation of Pyridine with Sulfur Trioxide.

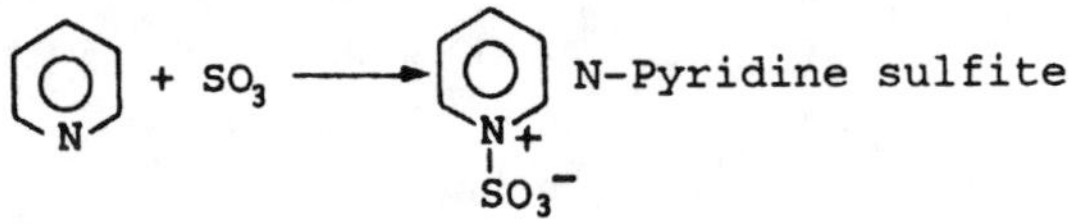

Bromination reaction.

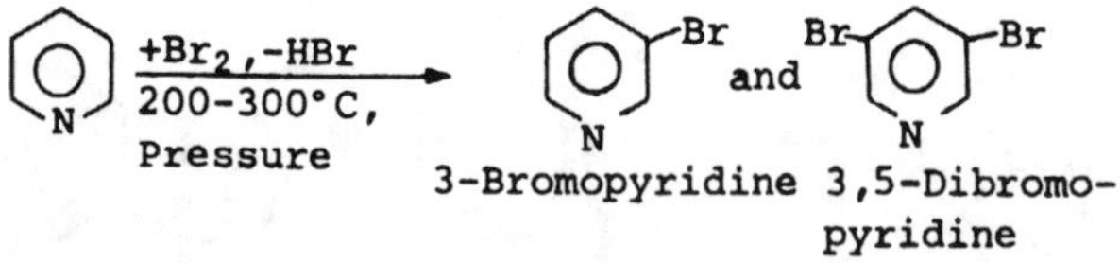

3-Bromopyridine   3,5-Dibromo-
                   pyridine

Oxidation of Pyridine to Yield Pyridine N-oxide.

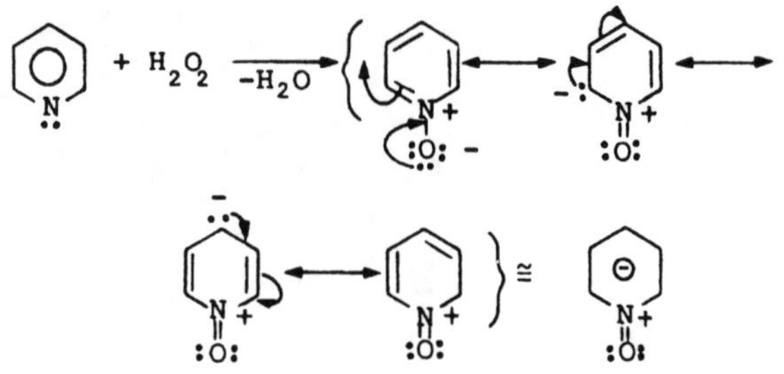

Amination of Pyridines to Give 2-Aminopyridine.

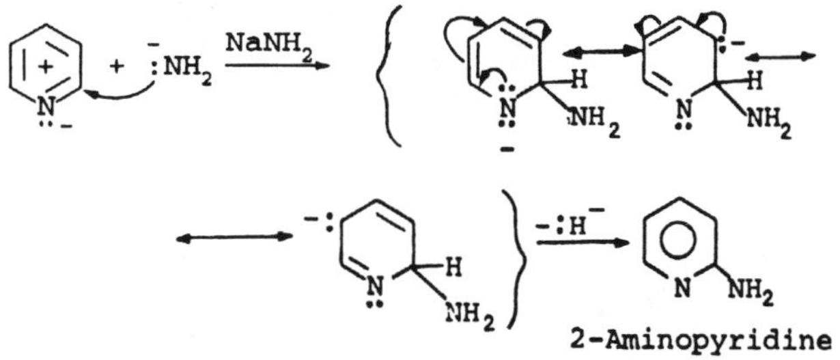

Nitration of Pyridine-N-Oxide to Yield 4-Nitropyridine-N-Oxide.

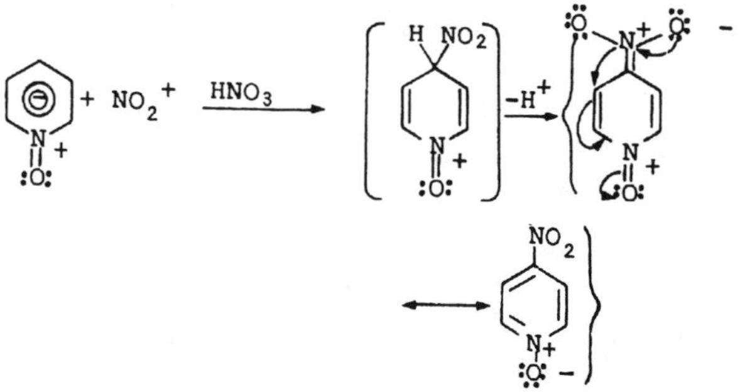

Arylation and alkylation of pyridines with lithium aryls and alkyls, respectively, yield the corresponding 2-aryl and 2-alkylpyridines.

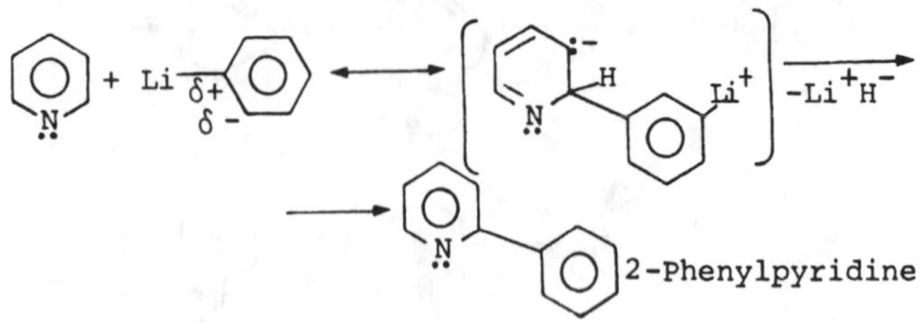
-Li⁺H⁻
2-Phenylpyridine

Electrophilic Substitution Reactions.

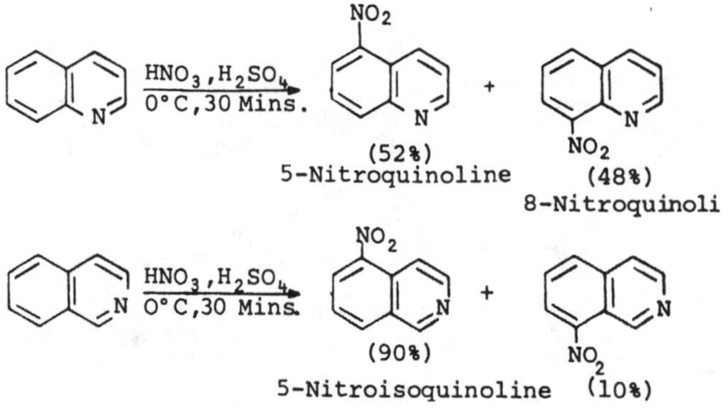
HNO3,H2SO4
0°C,30 Mins.
(52%)
5-Nitroquinoline
(48%)
8-Nitroquinoli
(90%)
5-Nitroisoquinoline
(10%)

Nucleophilic Substitution Reactions.

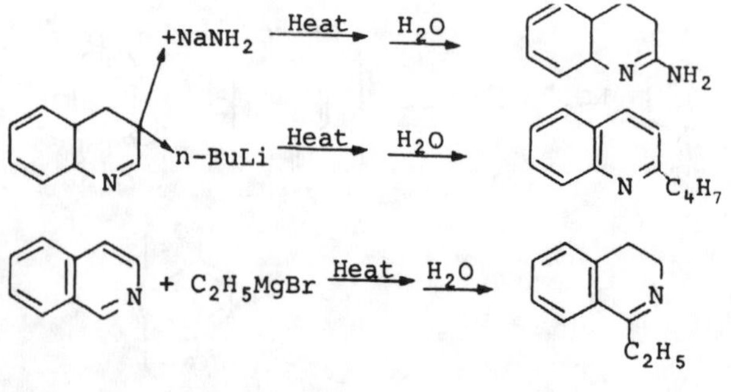
+NaNH2
Heat
H2O
n-BuLi
+ C2H5MgBr

Problem Solving Examples:

 N-Methyl pyridinium chloride is converted by aqueous alkali and a mild oxidizing agent (ferricyanide) to a solid compound of formula C_6H_7NO, which shows no hydroxyl absorption in the infrared. What is a likely structure for this substance and how is it formed? Would you expect a similar reaction with pyridine?

A A nucleophilic substitution reaction will occur:

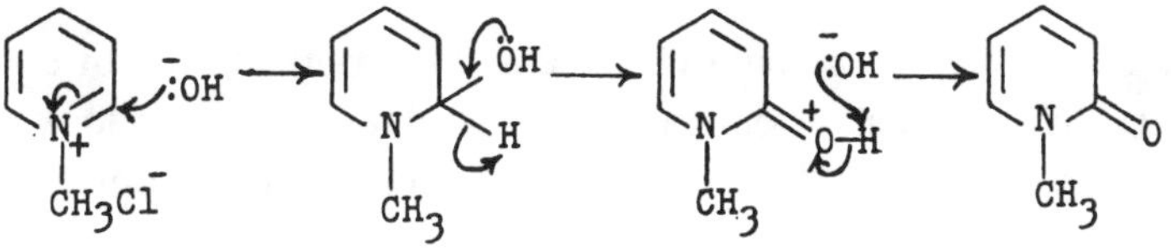

This structure is consistent with the molecular formula and infrared data. With pyridine, a similar reaction will occur at high temperatures with KOH. The product, however, will be a tautomeric system of 2-pyridine with 2-pyridinol:

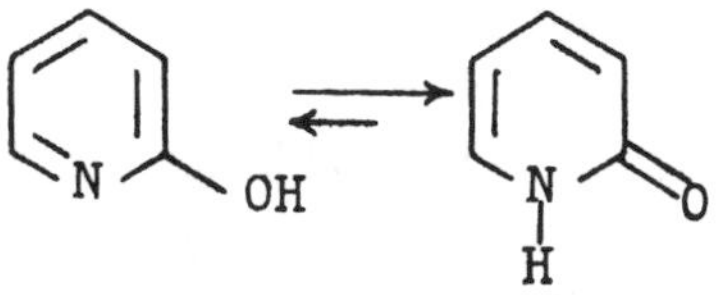

2-pyridinol   2-pyridone

Note that *[structure]* will not exist in the hydroxyl form because

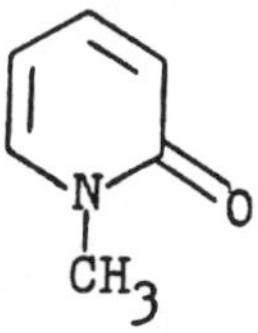

that would put a positive charge on the molecule.

(a) Explain why electrophilic substitution reactions proceed more slowly in pyridine relative to benzene.

(b) Why does substitution occur more readily at the 3 position?

In order to explain the reactivity and orientation of electrophilic substitution reaction in pyridine, it helps to look at the intermediates involved. Let us consider attack at the 4 position. We obtain the following intermediates:

A   B   C

Structure C is especially unstable because the nitrogen atom has only a sextet of electrons. Similar intermediates can be drawn for ortho attack. All these structures are less stable than the corresponding intermediates for substitution on the benzene ring due to the electron withdrawing effect of the nitrogen.

Substitution occurs more readily at the 3 position because at this position a positively charged nitrogen is not produced. The intermediates in this case are:

6.9 Condensed Furans, Pyrroles, and Thiophenes

Structure

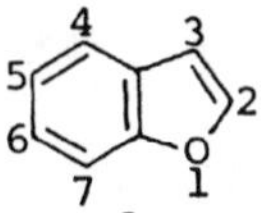

Benzofuran 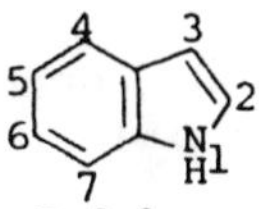 Indole 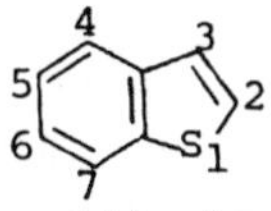 Benzothiophene

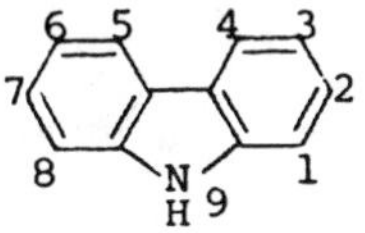

Carbazole

The rings are numbered starting with the heteroatom, except for carbazole.

Synthesis

A) Fisher indole synthesis. The phenyl hydrazone of an aldehyde or ketone is treated with a catalyst such as BF_3, $ZnCl_2$, or polyphosphoric acid.

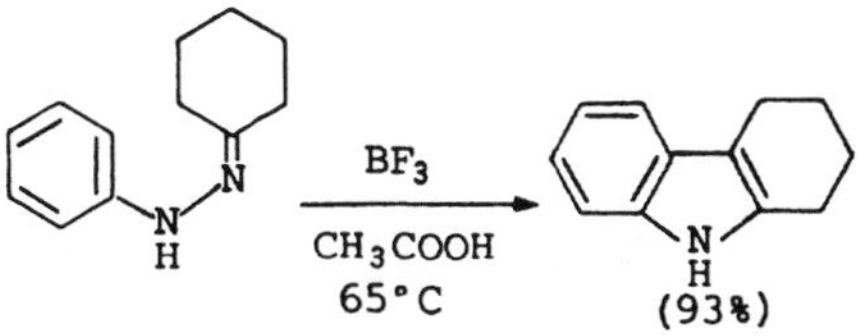

$$\xrightarrow[\substack{CH_3COOH \\ 65°C}]{BF_3}$$

(93%)

1,2,3,4-Tetrahydrocarbozole

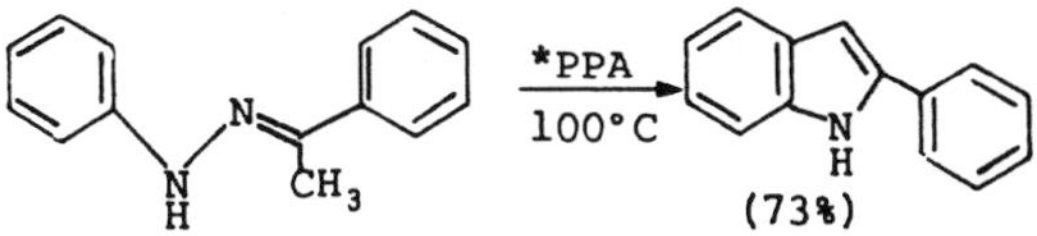

$$\xrightarrow[100°C]{*PPA}$$

(73%)

*PPA=polyphosphoric acid

2-Henylindole

B) Phenylhydrazone of pyruvic acid reacts to yield indole-2-carboxylic acid, which can be decarboxylated to produce indole.

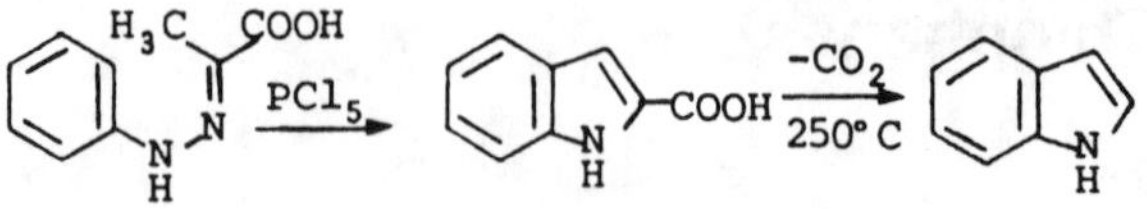

C) Benzofuran is prepared from coumarin, which is itself prepared from salicylaldehyde by the Perkin synthesis.

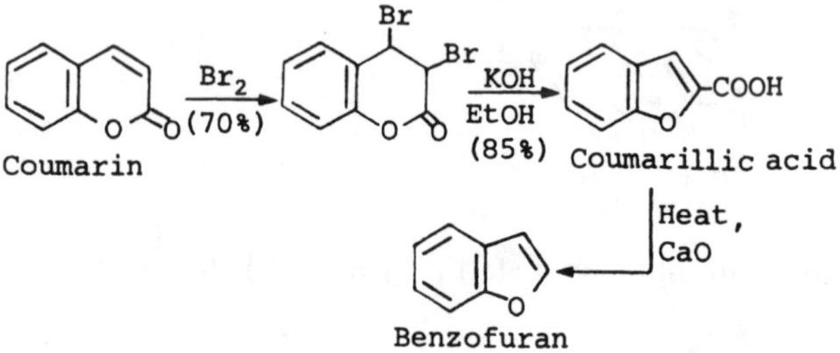

Coumarin   (70%)   (85%) Coumarillic acid

Benzofuran

6.10 Alkaloids

Alkaloids are compounds of vegetable origin with heterocyclic ring system containing one or more basic nitrogen atoms. Most alkaloids are optically active and have a variety of structures. They are toxic and some act as narcotics.

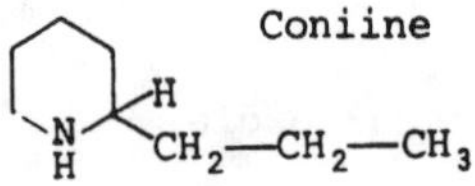

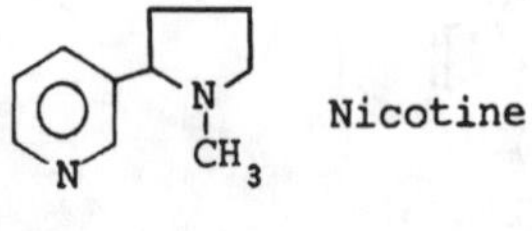

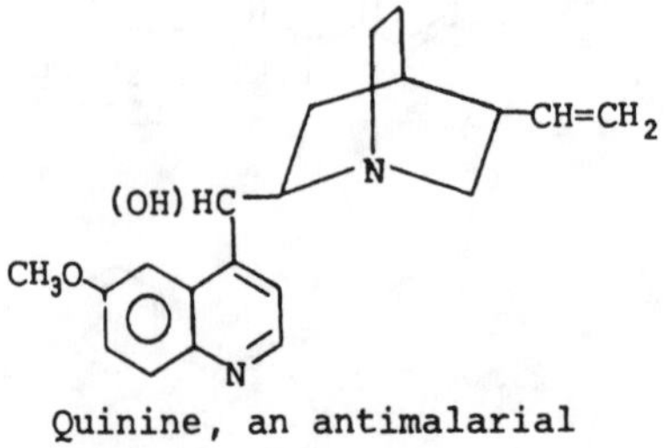

Coniine

Nicotine

Quinine, an antimalarial

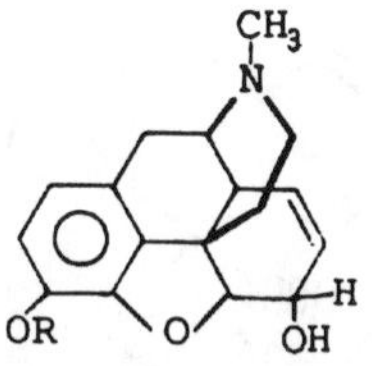

R=H :Morphine
R=CH₃ : Codeine
Narcotics

Problem Solving Example:

Q A synthesis of the alkaloid morphine was completed by Gates and Tschudi in 1952 by way of the following key intermediates starting from naphthalene. Show the reagents, conditions, and important reaction intermediates that you expect would be successful in achieving each of the indicated transformations, noting that more than one synthetic step may be required between each key compound and considering carefully the order in which the operations should be carried out. Indicate those reactions that might be expected to give a mixture of stereo- or position-isomers.

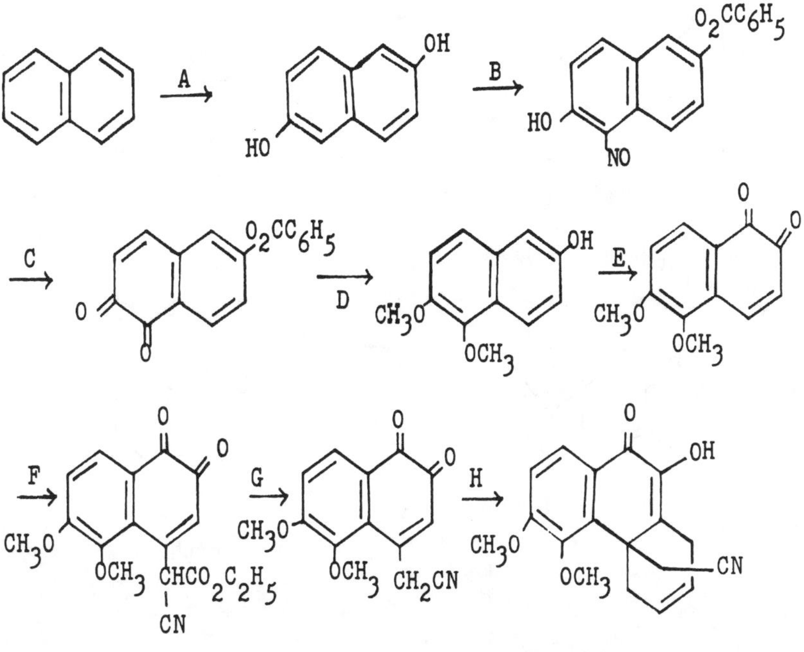

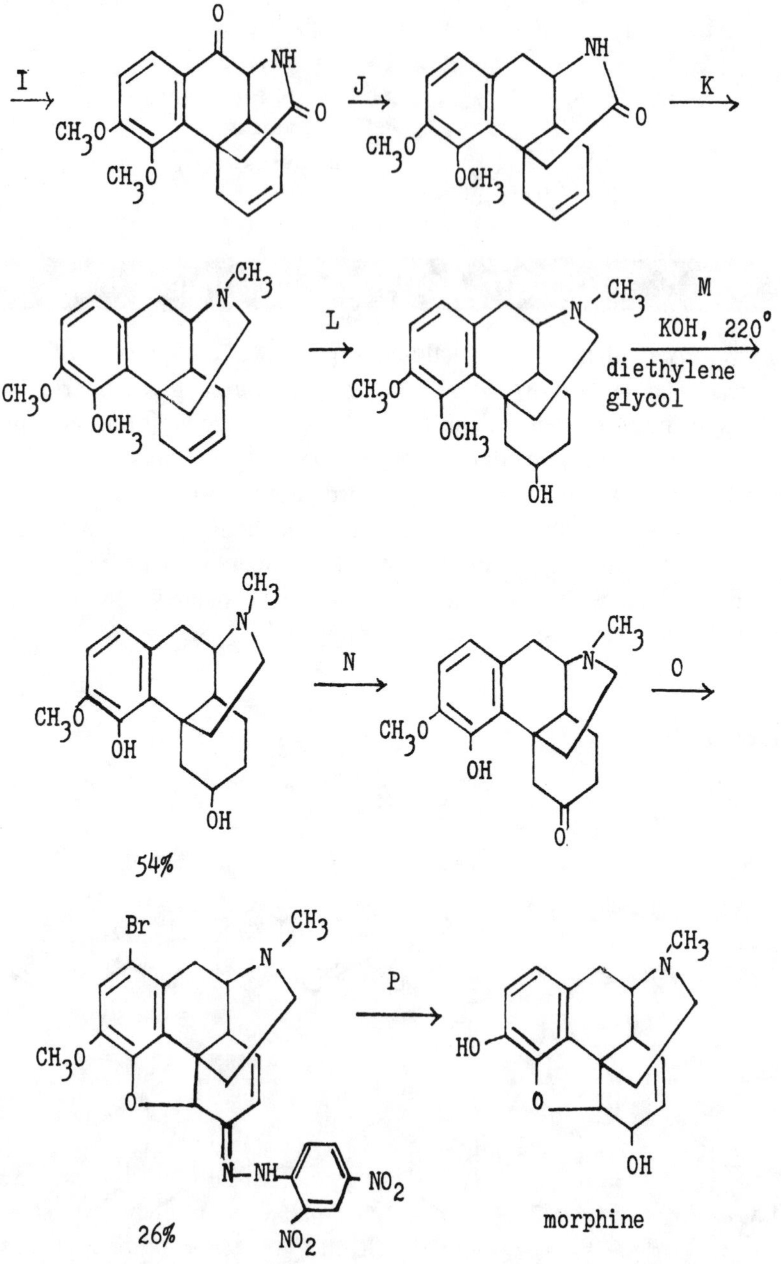
I
J
K
L
M
KOH, 220°
diethylene glycol
N
O
P
54%
26%
morphine

A Morphine is the principle alkaloid of opium and the constituent primarily responsible for opium's physiological effects. Its analgesic and euphoric properties have been a source of study for over hundreds of years.

We should note that morphine's synthesis, as has often been the case with syntheses, served as the final and definitive proof of the structure of morphine. Some transformations used to synthesize morphine will be familiar while others will seem novel. It is often necessary to devise reactions unique to the particular reactants in order to achieve the required transformations. We will attempt to note the significance of each transformation used in the original undertaking of this synthesis. Details of slight modifications and the actual yields of the synthesis can be obtained from the original literature.

Morphine has the following structure and numbering system:

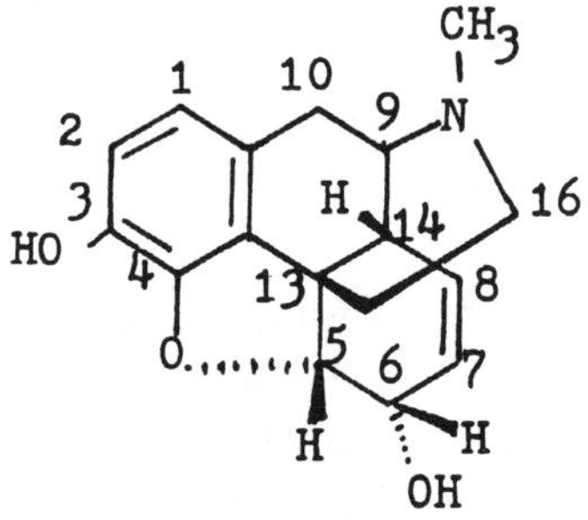

We can already see some of the difficulties in producing this molecule by examining the stereochemistry at its several chiral center bridgeheads. On the other hand, morphine can be seen as a combination of two phenylalanine units and its synthesis seemed quite possible.

(A) Naphthalene, 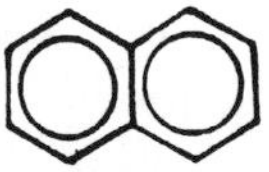 , is a major constituent of coal tar.

2,6-dihydroxynaphthalene can be obtained first by producing β-naphthol via fusion of β-naphthalene-sulfonic acid, which was made in turn from the high temperature sulfonation of naphthalene. β-naphthol (whose hydroxyl group will be protected) can be sulfonated at the 6 position and a second fusion performed. Minor products would lessen the yield of 2,6-dihydroxynaphthalene.

β-napthalenesulfonic acid

β-napthol

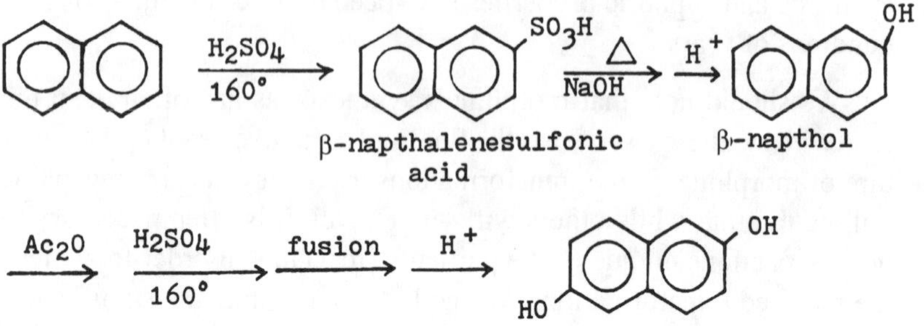

(B) The dihydroxynaphthalene is converted to the monobenzene in high yield in the exclusion of the dibenzoate by making use of the latter's low solubility in boiling alcohol. The mono-benzoate can be converted to 6-benzyloxy-1-nitroso-2-naphthol by nitrosating the reactant via the use of $NaNO_2$ in CH_3COOH. The attacking electrophile can be $[NO]^+$ itself or some more complicated species. We know that normally the electrophile attacks at the α-position of naphthalene due to the formation of a more stable carbocation than that which would have resulted from β-attack. This greater stability can be explained by the fact that the carbocation in the α attack has more resonance contributing structures in which the aromatic ring remains intact (2-α, 1-β).

more activated ring

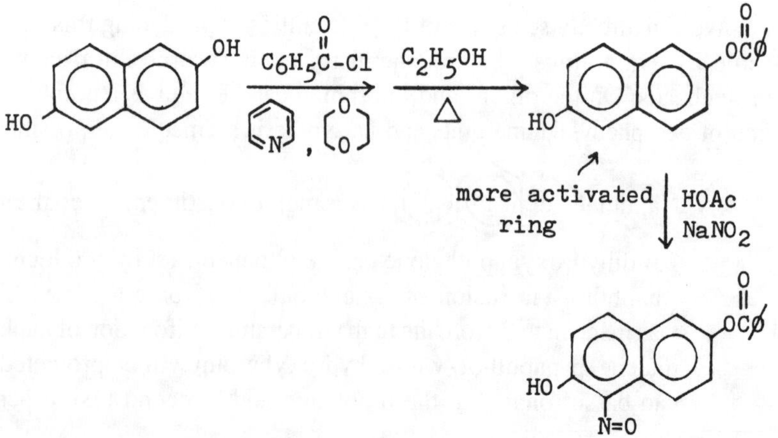

(C) The oxidation potential of the nitroso group is between that of a nitro group and an amino group, and therefore, it can be catalytically reduced to the latter. FeCl$_3$ is an oxidant that promotes a one-electron transfer which converts the α-aminonaphthol into 6-benzyloxy-1,2-naphthoquinone.

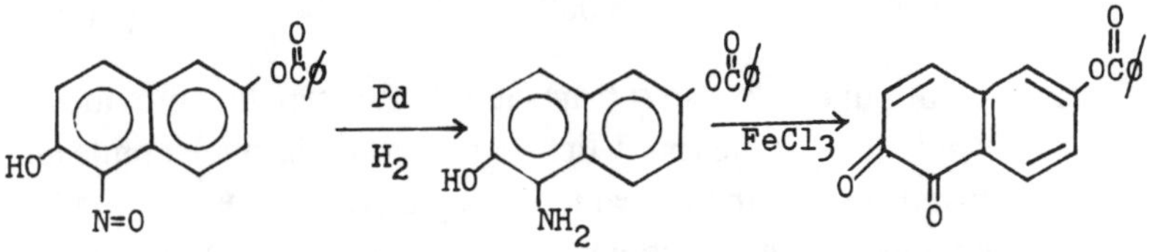

In practice the aminonaphthol was not actually isolated.

(D) The diquinone is reduced by sulfur trioxide, and the resultant hydroquinone is protected in the standard way by treatment with dimethylsulfate in the presence of base (a weak base is used to prevent hydrolysis). The ester linkage (at the 6 position) is subsequently hydrolyzed by alkali.

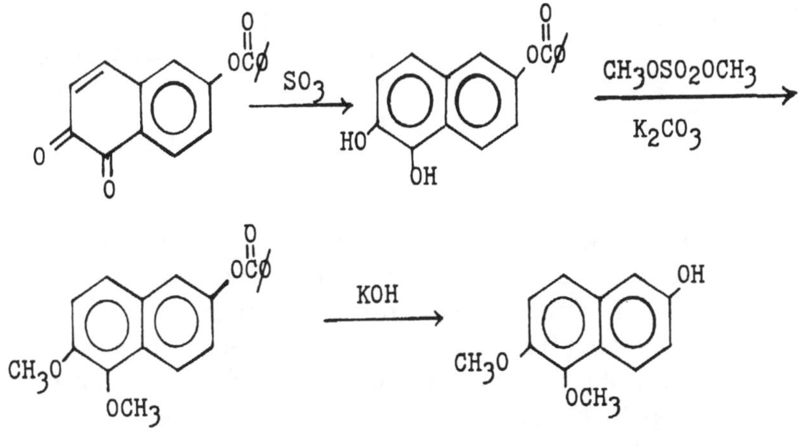

(E) The more activated ring of the dimethoxynaphthol is nitrosated using the same method as before. The nitroso group is catalytically reduced and the subsequent α-aminonaphthol is oxidized by FeCl$_3$ to the diquinone.

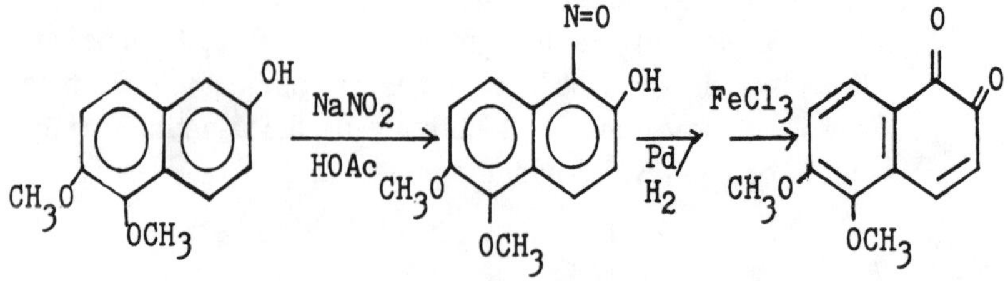

(F) The diquinone has an α,β-unsaturated system that provides an excellent substrate for a Michael reaction. The nucleophile employed here is the carbanion of cyano-acetic ester, which is generated by the combination of the ester and triethylamine, a tertiary amine and therefore a poor nucleophile.

The hydroquinone thus formed is immediately oxidized by ferricyanide incorporated in the reaction mixture.

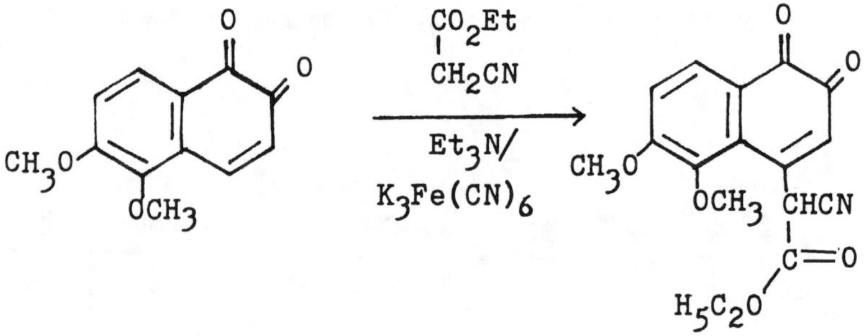

(G) Under the influence of alkali, the ester will be hydrolyzed to a carboxylate salt. This in turn can be easily decarboxylated due to the fact that it is both an α-cyanoester and a vinylogous α,β-ketoester. (The carbonyl group has its influence extended by the double bond.) Water then donates a proton to neutralize the negative charge.

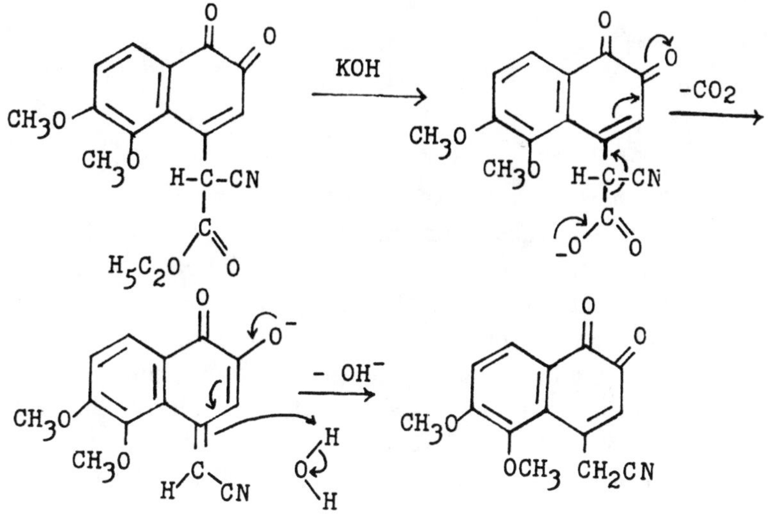

(H) The double bond of the nonaromatic ring may act as a dieno-
phile in a Diels-Alder reaction in which it reacts with butadiene.

Note the usefulness of this reaction information of the six-carbon ring. It is also a stereospecific process, but its stereospeci-
ficity is not evident in the product, since the proton adjacent to
the carbonyl group is lost in the subsequent tautomerization of
the 1,2-diketone to form the α-hydroxy-α, β-unsaturated ketone.

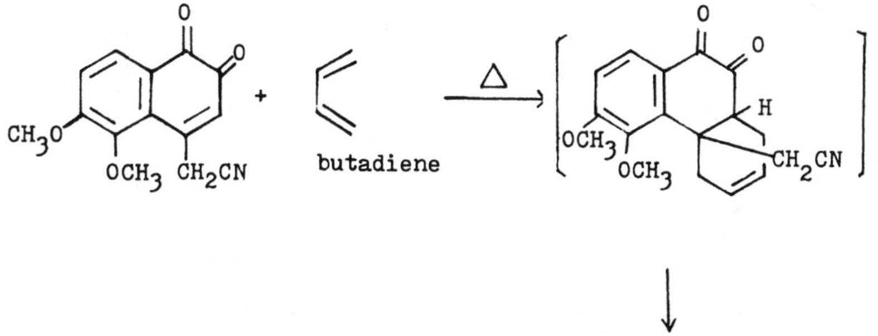

butadiene

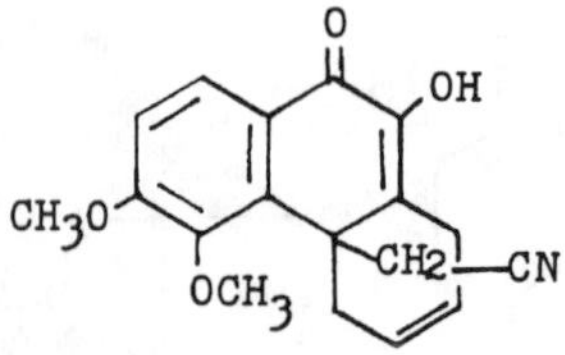

(I) The next step of the synthesis is such a surprising one that its actual mechanism was originally never considered. However, its product was finally verified using infrared spectroscopy. The reaction can be best termed a reductive cyclization. A possible mechanism involves the formation of an amide from the reduction of the nitrile. The amine portion of the amide then condenses with the carbonyl tautomer. Although the lactam thus formed has three chiral centers, only one pair of enantiomers is obtained. This is due to the fact that the hydrogens are delivered to the less hindered side of the bridged compound (opposite the cyanomethyl group), giving C_{14} only one possible configuration in which the new ring could be formed.

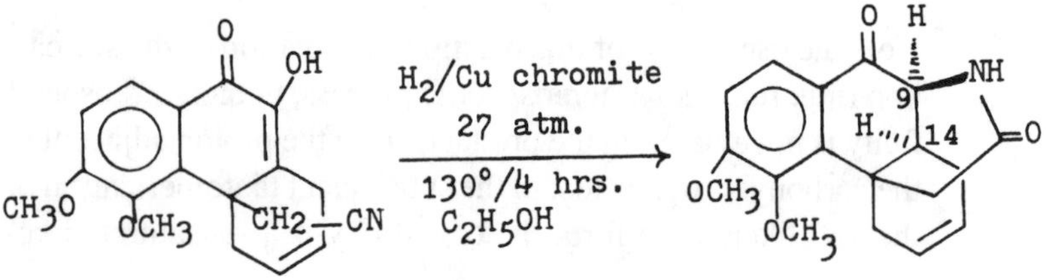

(J) The ketone functionality of the intermediate is removed by Wolff-Kishner reduction. There is no enolization or change in the stereochemistry of C_{14} because of the nature of the ring system.

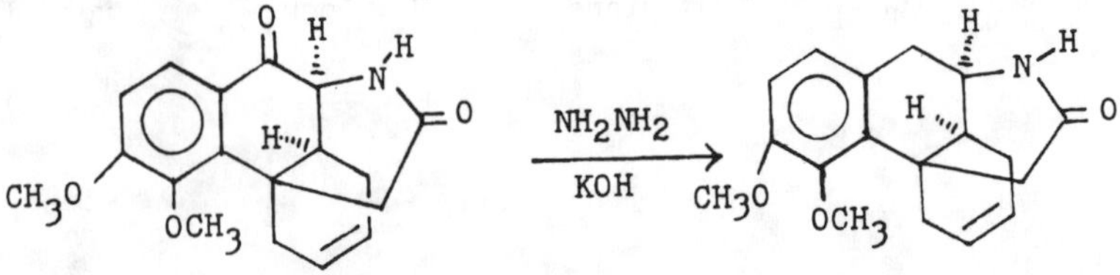

(K) In the presence of a potent base, we may alkylate the lactam and then produce the tertiary amine by reduction with LiAlH$_4$. The amine may now be resolved by employing optically active dibenzoyltartaric acid. After separation and regeneration, one of the enantiomeric amines coincides with a degradation product of morphine.

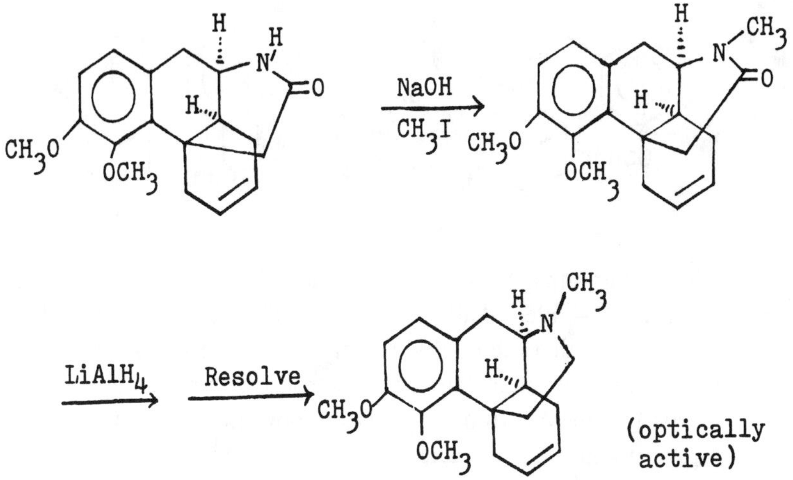

(L) The optically active amine can be hydrated at the isolated double bond, but it is difficult to determine whether the hydroxyl group will be situated at C$_6$ or C$_7$. Its placement at C$_6$ may be rationalized by the fact that the incoming nucleophile would be attracted to the protonated ethylamine bridge. This is a weak argument, however, and there are probably other products formed.

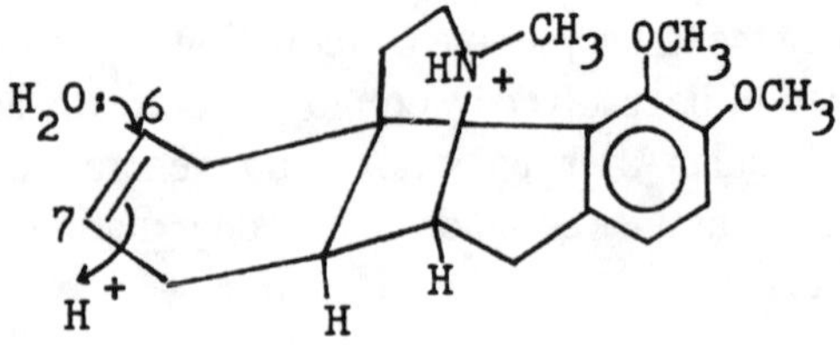

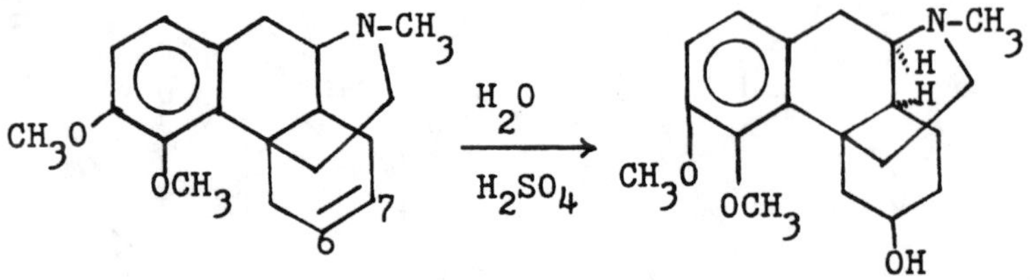

(M) The ether linkage at C_4 may be removed in a somewhat unusual fashion—under alkaline conditions. This can be accounted for by noting that the methyl moiety is an excellent substrate for nucleophilic attack (since its electropositive nature is strengthened by the electron-withdrawing powers of the ring) and the leaving group is a stable phenoxide anion.

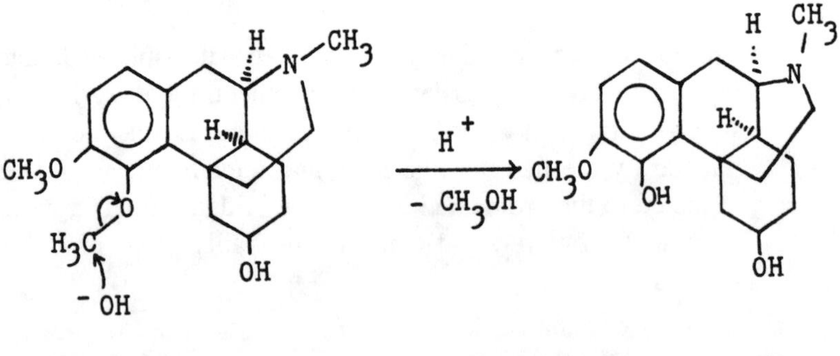

Note, however, that base-catalyzed ether cleavage is still not a facile occurrence and thus vigorous conditions are necessary.

(N) The hydroxyl group is selectively oxidized to the ketone by a variation of the Oppenauer oxidation, namely a disproportionation process similar to the Cannizarro reaction. The alcohol is oxidized to ketone while an alcohol is formed from the reduction of another ketone. The reaction is reversible depending on the concentration of the reactants.

(O) At this junction in the synthesis, we must deduce a means to alter the stereochemistry at C_{14} because it is the inverse of that of morphine. This is done by introducing bromine into the molecule at C_7.

The yield was diminished because the bromide did not go exclusively to C_7, but also to C_1.

α-bromo ketones are known to produce α,β-unsaturated ketones after treatment with 2,4-dinitrophenylhydrazine, followed by hydrolysis. Such formation of a conjugated double bond (between C_7 and C_8) allows the chiral center at C_{14} to invert its configuration through a sequence of enolization and reprotonation. The hydrazone was found to contain the opposite configuration at C_{14} than before and this was known to be the more stable of the two.

After the necessary stereochemical transformation is performed, the newly formed double bond is reduced and the ketone converted to a tribromo derivative. The bromine at C_5 will be displaced by the adjacent hydroxyl group in a nucleophilic reaction, forming a stable five-membered ring. The double bond between C_7 and C_8 was formed in the same way as before. A possible mechanism for this transformation is as follows:

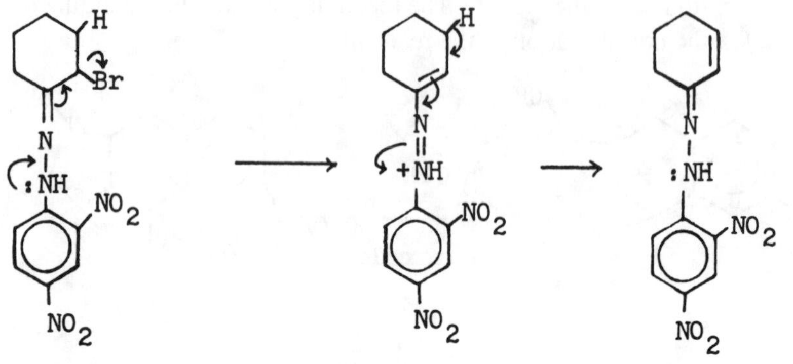

The sequence of the synthesis just discussed is:

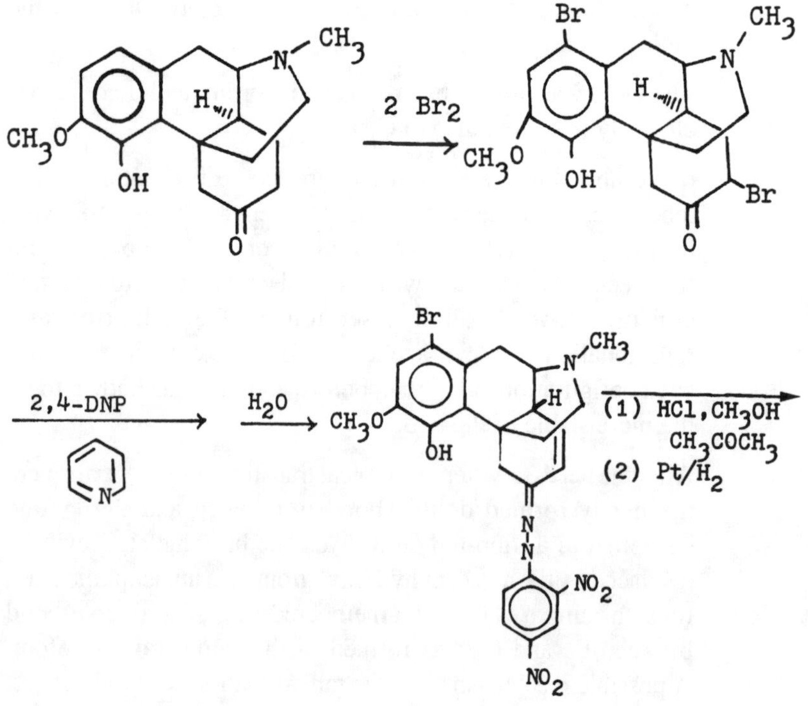

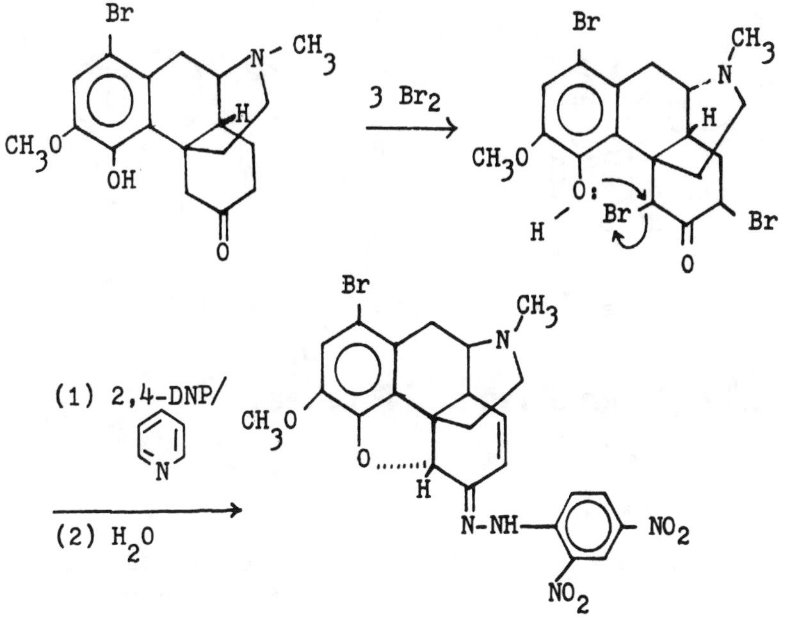

(P) The hydrazone can now be hydrolyzed. The carbonyl group thus formed at C_6 may now be reduced, and the bromine removed from C_1, both by treatment with $LiAlH_4$ This reagent gives the correct stereochemistry in the reduction. The remaining methyl ether linkage is removed by acid treatment to give morphine.

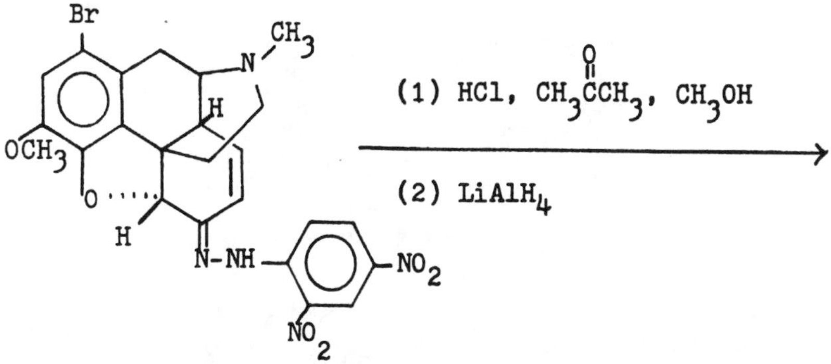

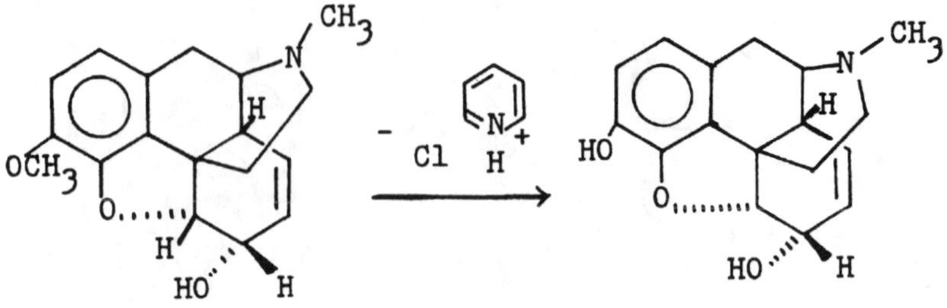

codeine   morphine

6.11 Pyrimidines and Purines

Structure

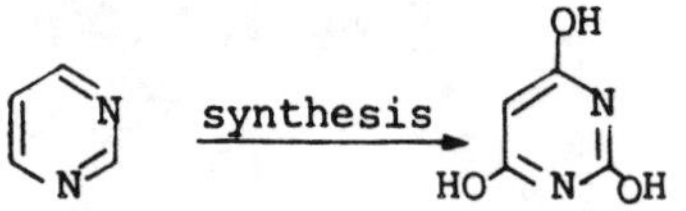

Barbituric acid
(2,4,6-trihydro-
xypyrimidine)

Derivatives of pyrimidine based on 2,4,6-trihydroxypyrimidine are called barbituric acids. The above structure is only one of several tautomeric forms.

Barbituric acids can be synthesized from urea and alkyl- or aryl-substituted diethyl malonate.

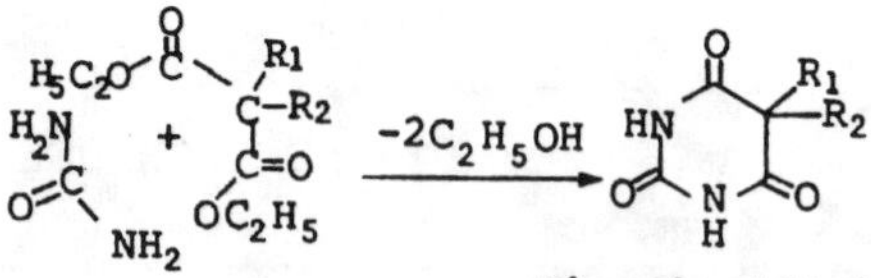

Phenobarbital

$R_1 = C_6H_5$
$R_2 = C_2H_5$

Adenine, a derivative of purine, 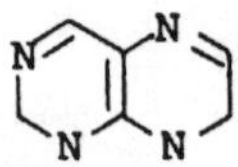 is synthesized by the cyclization of formamidine and phenylazomalodinitrine by way of 4,5,6-triaminopyrimidine.

4,5,6-Triamino-
pyrimidine

Adenine

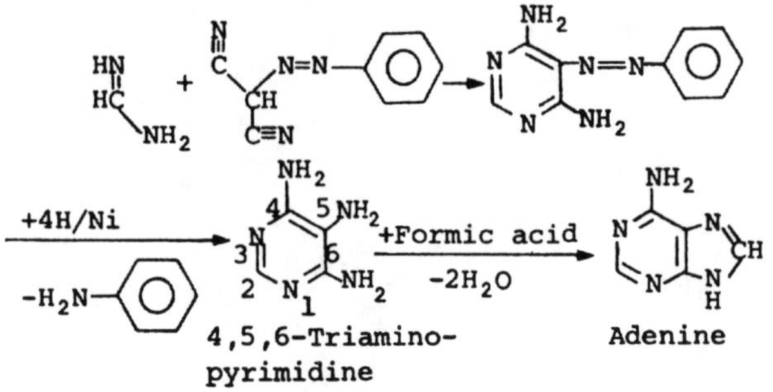

Uric acid is 2,4,6-trihydroxypurine; it exists in several tautomeric lactam forms: caffeine, the alkaloid of the coffee bean and tea plant, is 3,5,7-tri-N-methyl-4,6-dioxopurine.

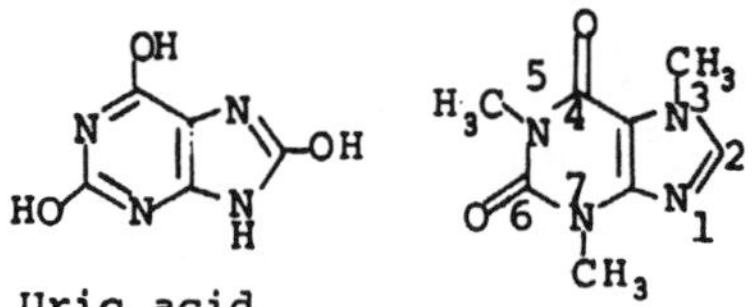

Uric acid

Caffeine

Problem Solving Examples:

Outline a synthesis for adenine, one of the purines, from hydrogen cyanide and ammonia.

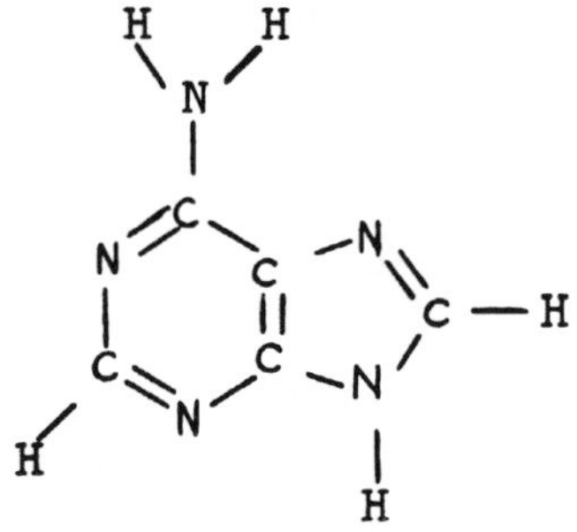

Adenine

A It has been proposed that adenine can be synthesized from hydrogen cyanide and ammonia through a series of additions and "enolizations." If we dissect adenine, we will see that it can be separated into five cyanide groups:

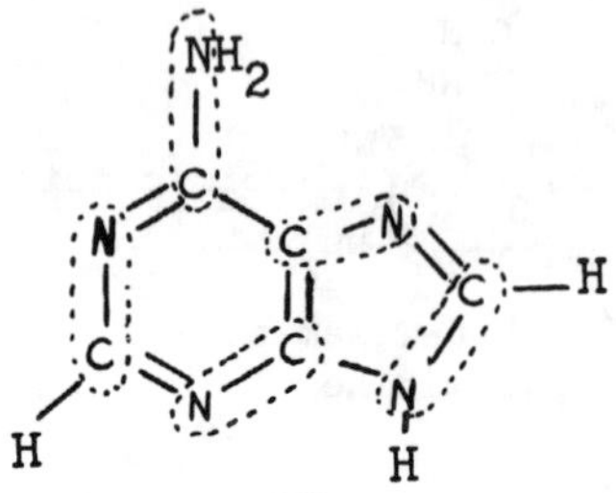

Adenine

Before we start the synthesis, we must review certain facts:

(1) Hydrogen cyanide is a weak acid.

(2) The polarization of HCN is:

$$\overset{\delta+}{H} - \overset{\delta+}{C} \equiv \overset{\delta-}{N}:$$

(3) Ammonia is a weak base.

(4) Imine-enamine tautomerism:

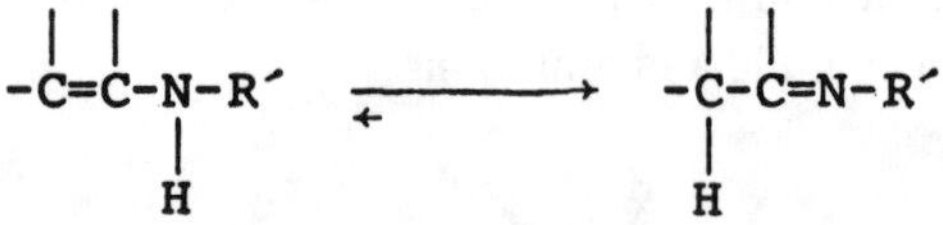

Enamine   Imine (more stable form)

When synthesizing adenine, we will first form the five-member ring, with reactive groups extending out; this way, we will be able to form the attached six-member ring:

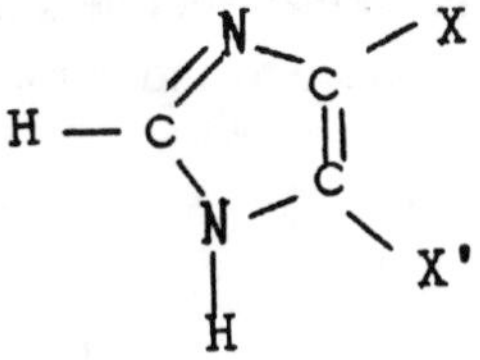

Note: X and X' are the reactive groups.

In synthesizing adenine, we must work systematically, keeping in mind our first goal, formation of the five-membered ring.

(i)

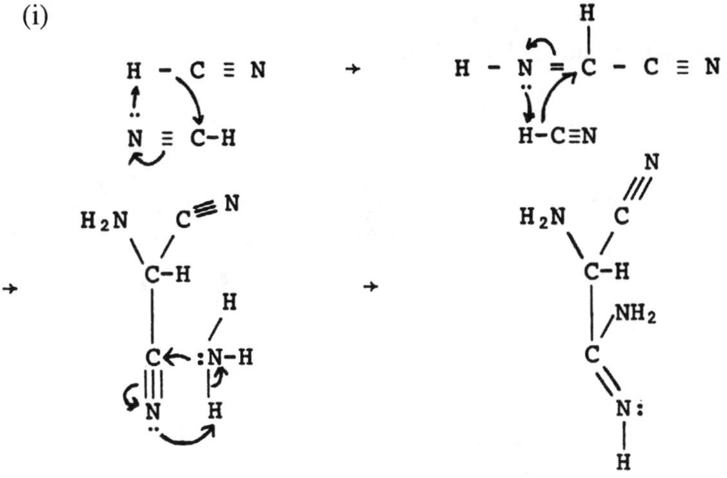

I

Note that this first step, and all subsequent steps, are dependent on the polarity of HCN, the acidity of HCN, and the basic properties of the line-pair electrons of nitrogen.

(ii) In the reaction vessel, we also have NH_3 reacting with HCN to form structure II:

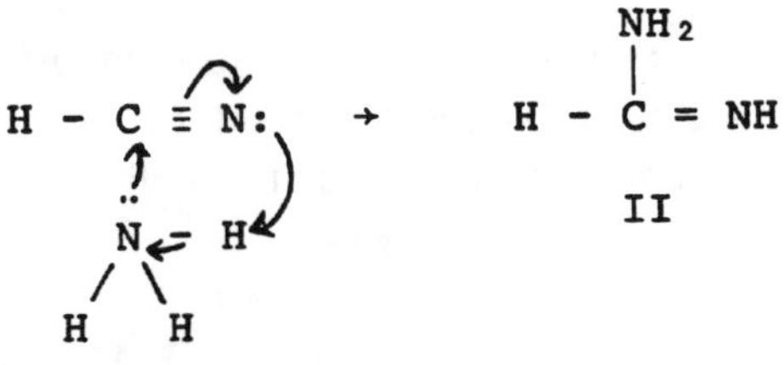

II

Note that the carbon in structure II is a good electrophile since it is singly bonded to one nitrogen and doubly bonded to another nitrogen:

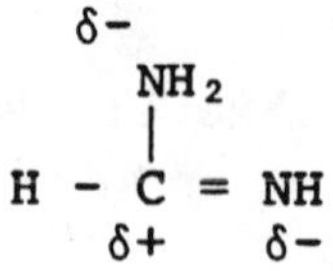

$$\delta-$$
$$NH_2$$
$$|$$
$$H - C = NH$$
$$\delta+ \qquad \delta-$$

II

When structures I and II react:

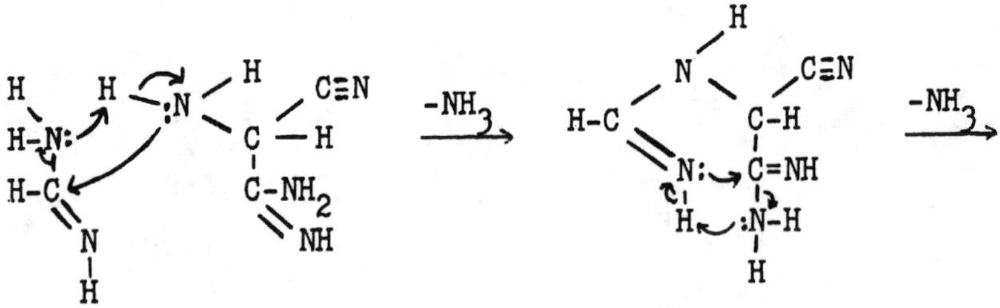

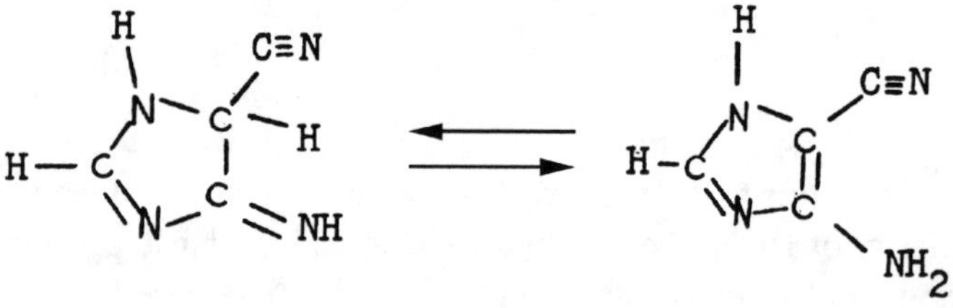

$$\xrightarrow{-NH_3}$$

$$\xrightarrow{-NH_3}$$

imine   enamine
(more stable form)

Note that the enamine is the *more* stable form in this case due to the stability of the aromatic five-member ring; the lone-pair electrons of the top nitrogen is able to participate in the π cloud.

The reactive groups extending out of the ring ("panhandles") are $-C \equiv N$ and $-NH_2$. When structure III reacts with another molecule of structure II:

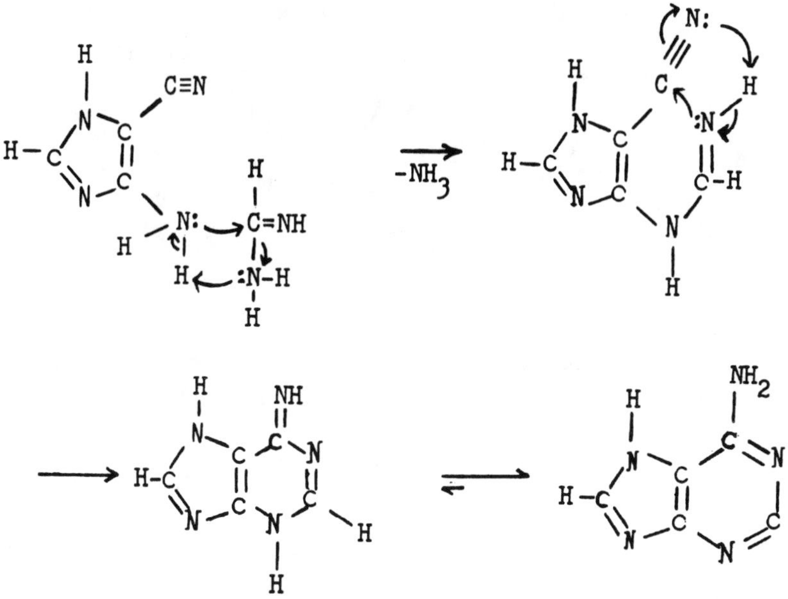

imine

enamine (more stable)

Again, the enamine is the more stable form due to the stabilization of the aromatic ring. Structure IV, we recognize, is adenine.

Q Write equations for the steps involved in hydrolysis of adenine deoxyribonucleoside to deoxyribose and adenine. Would you expect the reaction to occur more readily in acidic or basic solution?

The structure of adenine deoxyribonucleoside is:

Deoxyribose

Adenine

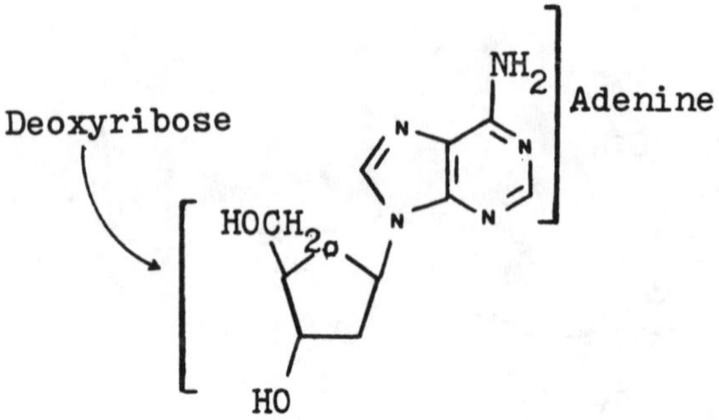

Since ethers and amines are stable toward base, hydrolysis should occur more readily in acidic conditions. The mechanism involved in hydrolysis of adenine deoxyribonucleoside is:

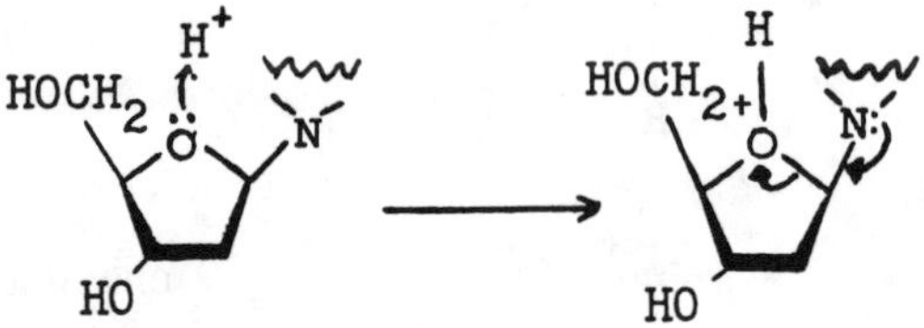

Note: $-N \lessgtr$ represents the adenine portion.

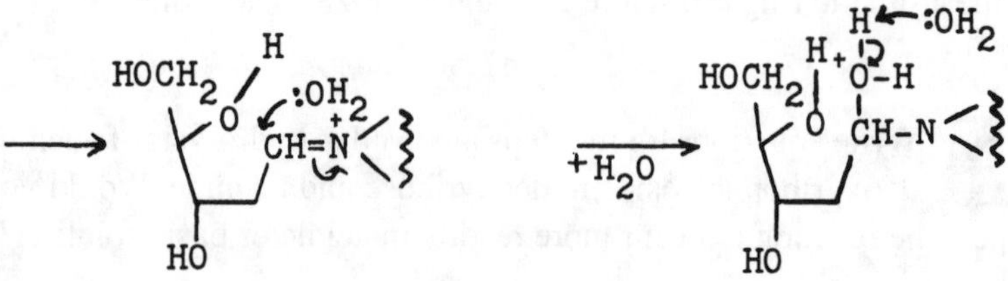

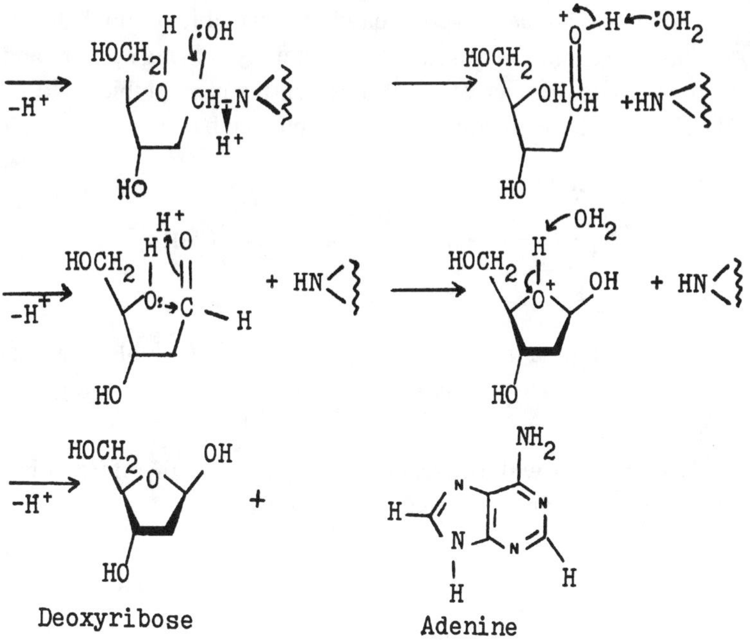

Deoxyribose   Adenine

The alkaloid hygrine is found in the coca plant. Suggest a structure for it on the basis of the following evidence:

Hygrine ($C_8H_{15}ON$) is insoluble in aqueous NaOH but soluble in aqueous HCl. It does not react with benzenesulfonyl chloride. It reacts with phenylhydrazine to yield a phenylhydrazone. It reacts with NaOI to yield a yellow precipitate and a carboxylic acid ($C_7H_{13}O_2N$). Vigorous oxidation by CrO_3 converts hygrine into hygrimic acid ($C_6H_{11}O_2N$).

Hygrinic acid can be synthesized as follows:

$BrCH_2CH_2CH_2Br + CH(COOC_2H_5)_2^- Na^+ \rightarrow A(C_{10}H_{17}O_4Br)$

$A + Br_2 \rightarrow B(C_{10}H_{16}O_4Br_2)$

$B + CH_3NH_2 \rightarrow C(C_{11}H_{19}O_4N)$

$C + aq. Ba(OH)_2 + \xrightarrow{heat} D \xrightarrow{HCl} E \xrightarrow{heat}$ hygrimic acid $+ CO_2$

A Alkaloids are natural compounds of vegetable origin that contain nitrogen. They usually have heterocyclic ring systems and one or more basic nitrogen atoms. Since hygrine is insoluble in aqueous NaOH but soluble in aqueous HCl, it must be basic:

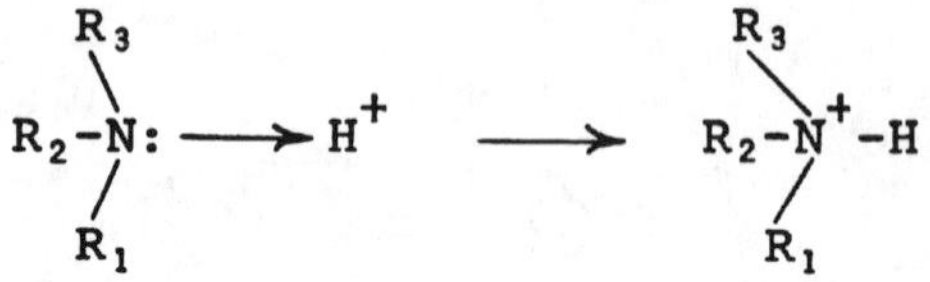

$$R_2-\overset{\overset{\displaystyle R_3}{|}}{N}: \longrightarrow H^+ \longrightarrow R_2-\overset{\overset{\displaystyle R_3}{|}}{\underset{\underset{\displaystyle R_1}{|}}{N}}^+-H$$

(insoluble in water)   ionic (soluble in water)

$\downarrow OH^-$

It does not react with benzene-sulfonyl chloride; therefore, it is a tertiary amine. This analysis is known as the Hinsberg test:

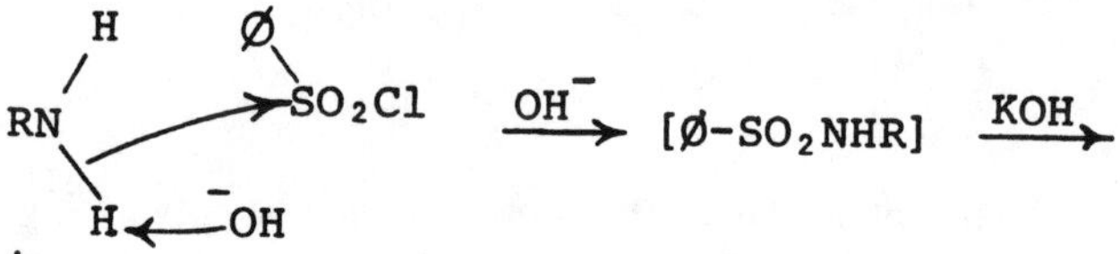

$$RN\overset{H}{\underset{H\longleftarrow \,^-OH}{\diagdown}}\!\!\!\!\longrightarrow \overset{\emptyset}{SO_2Cl} \xrightarrow{OH^-} [\emptyset-SO_2NHR] \xrightarrow{KOH}$$

1° amine

$\emptyset-SO_2NR^-K^+$

clear solution

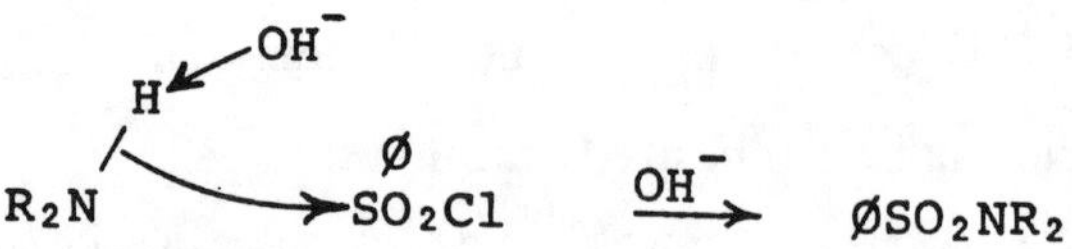

$$R_2N\overset{H\overset{\displaystyle \longleftarrow OH^-}{\diagup}}{\diagup}\!\!\!\longrightarrow \overset{\emptyset}{SO_2Cl} \xrightarrow{OH^-} \emptyset SO_2NR_2$$

2° amine   insoluble

$$R_3N + \emptyset SO_2Cl \xrightarrow{OH^-} \text{no reaction}$$

3° amine

The tertiary amine does not react because there are no hydrogens attached to nitrogen.

Since hygrine reacts with phenylhydrazine to yield a phenyl-hydrazone, it must have a carbonyl function:

$$\underset{O}{\overset{\diagdown \diagup}{\underset{\|}{C}}} \;+\; :NH_2N\overset{\diagup H}{\underset{\diagdown \emptyset}{}} \;\overset{H^+}{\longrightarrow}\; \left[\; \underset{OH}{\overset{|}{\underset{|}{-C}}}-NHNH\emptyset \;\right] \;\rightarrow\; \overset{\diagdown}{\underset{\diagup}{C}}=NNH\emptyset \;+\; H_2O$$

phenylhydrazine   phenylhydrazone

This fact together with the subsequent iodoform test tells us that the carbonyl function is a methyl ketone:

$$\underset{\underset{O}{\|}}{R-C-CH_3} \;\overset{OI^-}{\longrightarrow}\; \overset{O}{\overset{\|}{R-C}}-O^- \;+\; CHI_3$$

A   Yellow precipitate

Structure A has the molecular formula of $C_7H_{13}O_2N$. Vigorous oxidation by CrO_3 removes one more carbon and two more hydrogens than the haloform reaction; therefore, a tentative conclusion is a

$-CH_2\overset{O}{\overset{\|}{C}}-CH_3$ group attached to a resistant group:

$$\underset{R-CH_2-\overset{O}{\overset{\|}{C}}-CH_3}{} \;\overset{C_2O_3}{\longrightarrow}\; \overset{O}{\overset{\|}{RC}}-OH$$

By following the synthesis for hygrinic acid, we can deduce the structure of hygrine. The first step is the malonic ester synthesis:

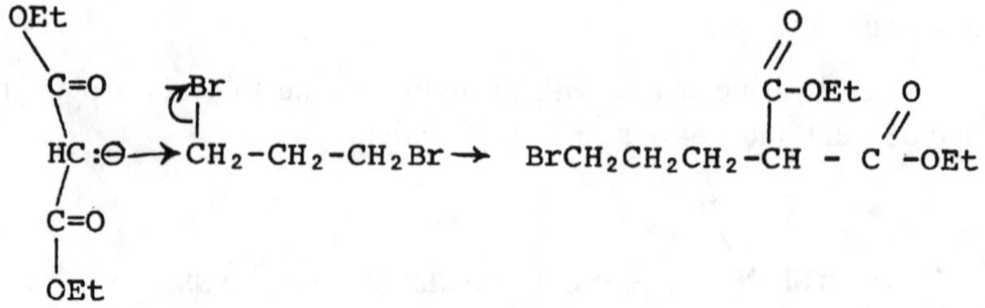

$$\begin{array}{l} \text{OEt} \\ | \\ \text{C=O} \\ | \\ \text{HC}:\ominus \\ | \\ \text{C=O} \\ | \\ \text{OEt} \end{array} \quad \overset{\curvearrowright \text{Br}}{\underset{|}{\text{CH}_2-\text{CH}_2-\text{CH}_2\,\text{Br}}} \rightarrow \quad \begin{array}{c} \overset{O}{\overset{\|}{\text{C}-\text{OEt}}} \quad \overset{O}{\overset{\|}{}} \\ | \\ \text{BrCH}_2\text{CH}_2\text{CH}_2-\text{CH} \;-\; \text{C} \;-\text{OEt} \end{array}$$

The next step is alpha-halogenation of the diester, known as the Hell-Volhard-Zelinsky reaction:

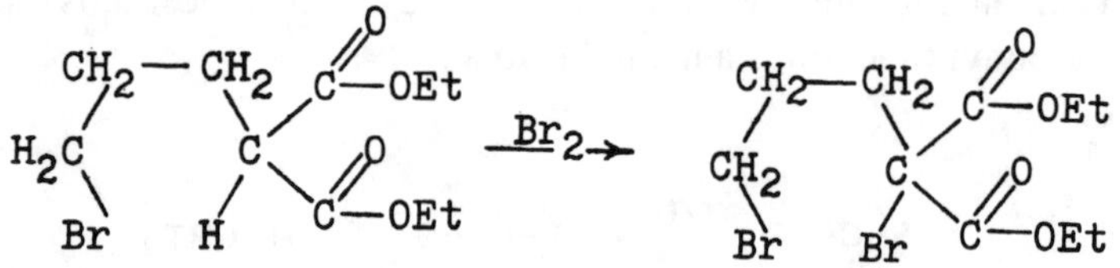

A   B

The next step is an S_N2 displacement by the nucleophile $CH_3 - NH_2$, and then an intra molecular S_N2 displacement of the remaining bromide.

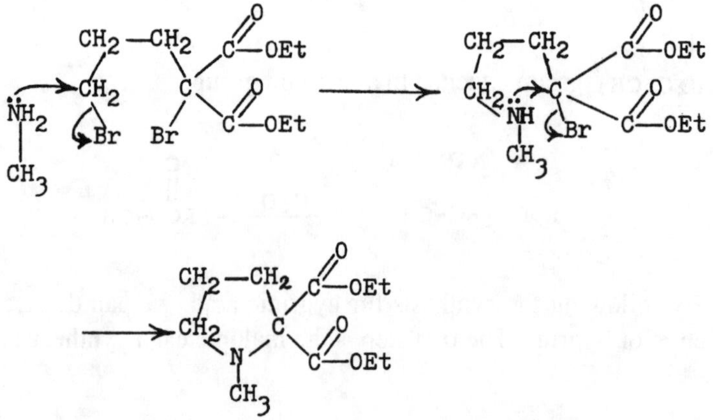

C

The next step is hydrolysis of the diester and subsequent acidification:

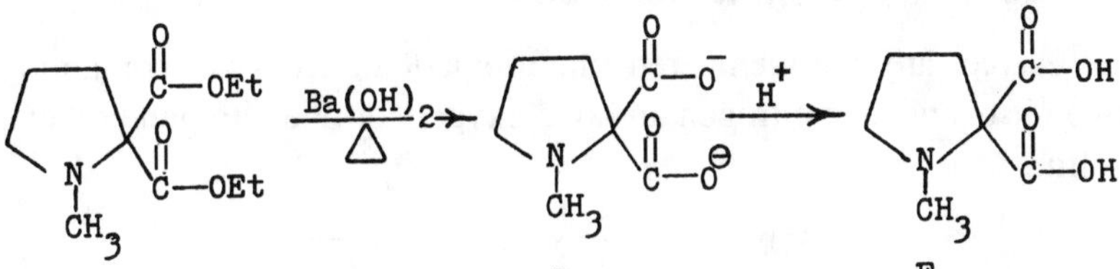

C   D   E

The next step is decarboxylation by heat to form hygrinic acid:

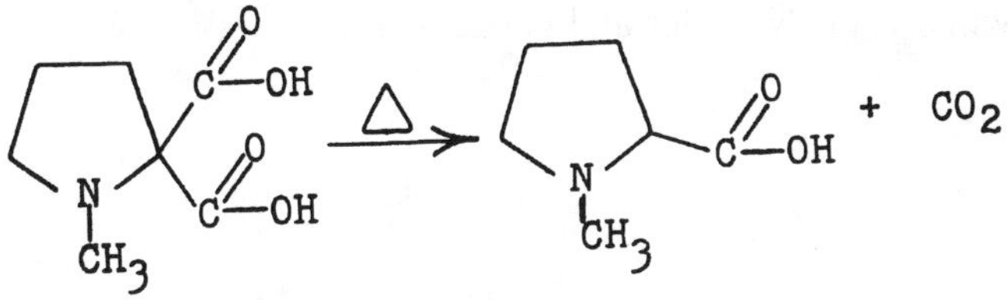

E   hygrinic acid

From our previous conclusion, we know:

$$RCH_2-\overset{\overset{\textstyle O}{\|}}{C}-CH_3 \quad \xrightarrow{[O]} \quad R\overset{\overset{\textstyle O}{\|}}{C}-OH$$

hygrine   hygrinic acid

Therefore, R = 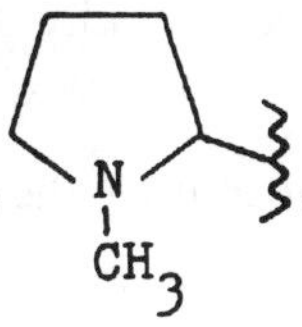 and the structure for hygrine is:

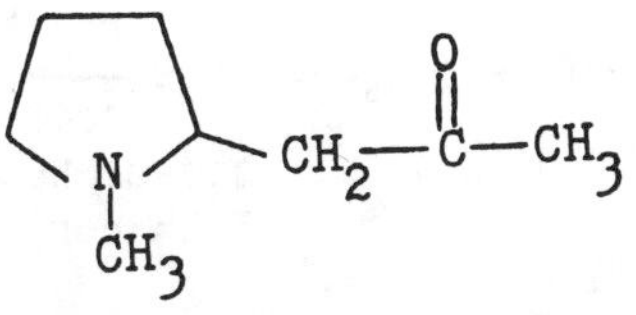

6.12 Azoles

Structures and Nomenclatures

Azoles are five-membered-ring aromatic heterocycles containing two nitrogens, one nitrogen and one oxygen, or one nitrogen and one sulfur.

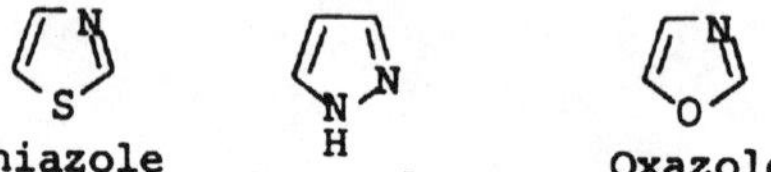

Thiazole   **Pyrazole**   **Oxazole**

Synthesis of Pyrazoles and Isoxazoles:

A)

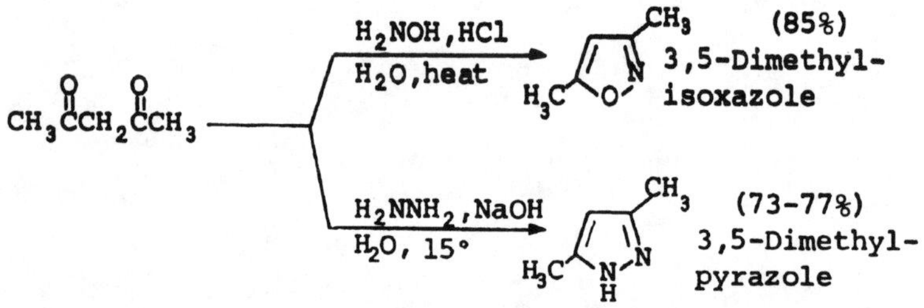

$$CH_3\overset{O}{\overset{\|}{C}}CH_2\overset{O}{\overset{\|}{C}}CH_3$$

$\xrightarrow[\text{H}_2\text{O, heat}]{\text{H}_2\text{NOH, HCl}}$ (85%) 3,5-Dimethyl-isoxazole

$\xrightarrow[\text{H}_2\text{O, 15°}]{\text{H}_2\text{NNH}_2, \text{NaOH}}$ (73–77%) 3,5-Dimethyl-pyrazole

The above reaction occurs by the action of hydrazine or hydroxylamine with 1,3-dicarbonyl compound or its equivalent.

B) Synthesis of isoxazoles can also occur by the cycloaddition of a nitrile oxide to an acetylene.

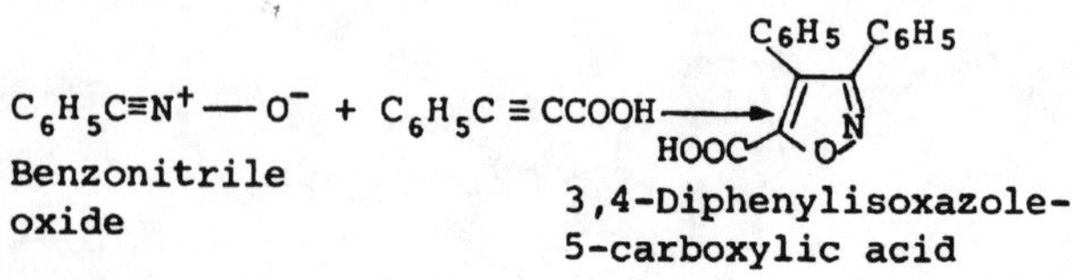

$$C_6H_5C\equiv N^+ \!-\! O^- + C_6H_5C \equiv CCOOH \longrightarrow$$

C_6H_5 C_6H_5

HOOC

3,4-Diphenylisoxazole-5-carboxylic acid

Benzonitrile oxide

C)   General synthesis of 1,3-azoles involves the dehydration of 1,4-dicarbonyl compounds (a form of Paal-Knorr cyclization).

$$C_6H_5\overset{O}{\underset{}{C}}-\overset{H}{\underset{}{N}}CH_2\overset{O}{\underset{}{C}}C_6H_5 \xrightarrow[\text{heat}]{H_2SO_4}$$

2,5-Diphenyloxazole

$$C_6H_5\overset{O}{\underset{}{C}}-\overset{H}{\underset{\underset{C_6H_5}{|}}{N}}CH\overset{O}{\underset{}{C}}C_6H_5 \xrightarrow[\text{HOAc 120°C}]{NH_4^+\,OAc^-}$$

(93%) 2,4,5-Triphenyl-imidazole

$$CH_3\overset{O}{\underset{}{C}}CH_2-\overset{H}{\underset{}{N}}-\overset{O}{\underset{}{C}}CH_3 + P_2S_5 \xrightarrow{120°C}$$

2,5-Dimethylthiazole

Reactions of Azoles

The azoles are significantly less reactive than furan, pyrrole, and thiophene. Their order of reactivity of 1,2-azoles is

$$\underset{\overset{|}{H}}{\boxed{N}}N \quad > \quad \boxed{S}N \quad > \quad \boxed{O}N$$

A)   Electrophilic substitution reaction.

$$\boxed{S}N \xrightarrow[\text{115°C,19hrs.}]{HNO_3,H_2SO_4}$$

(97%)4-Nitroisothiazole

B)   Substitution reaction for imidazoles.

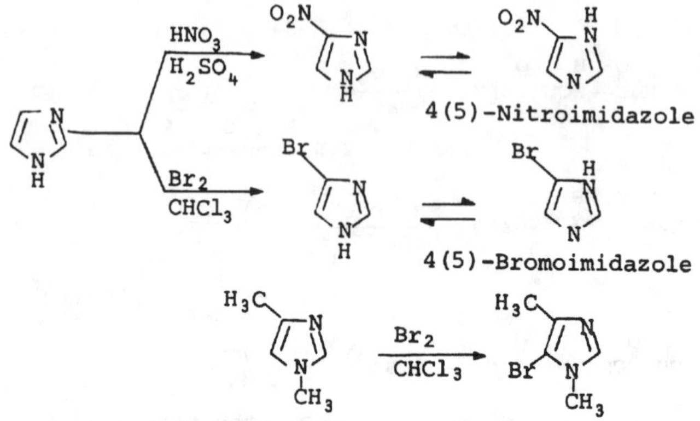

4(5)-Nitroimidazole

4(5)-Bromoimidazole

1,4-Dimethylimi-   5-Bromo-1,4-dimethy-

Problem Solving Example:

(a) Account for the aromatic properties of the imidazole ring.

(b) Arrange the nitrogen atoms of histamine (the substance responsible for many allergenic reactions) in order of their expected basicity, and account for your answer.

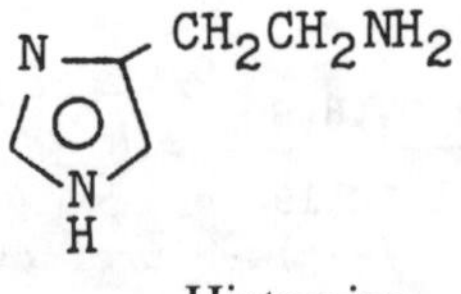

$$N \overbrace{}^{} CH_2CH_2NH_2$$

Histamine

(a) The aromatic properties of the imidazole ring can best be accounted for by looking at its orbital structure.

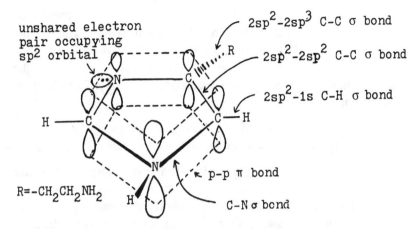

Every atom of the ring is sp² hybridized and has three bonding orbitals. These orbitals lie in a common plane and are 120° apart. Each carbon atom contributes one electron to each of its sp² orbitals, thereby forming σ bonds with three other atoms, and has one electron left over, which occupies a 2p orbital.

The "pyridine" nitrogen contributes one electron to each of its two s bonds with carbon, and its unshared (lone) pair of electrons occupies the third sp² orbital. It has one electron left over, then, which occupies a 2p orbital.

The "pyrrole" nitrogen differs from the "pyridine" nitrogen in that it is bonded not only to two carbons but also to a hydrogen atom. While contributing an electron to each of its three s bonds, it still has an unshared pair of electrons left. These electrons will occupy a 2p orbital. This p orbital and the other p orbitals of the ring are situated close enough to overlap, giving rise to two π clouds, one above and one below the plane of the ring. The six electrons are delocalized in these clouds around the ring and stabilize the system.

Note that the imidazole ring satisfies Hückel's rule for aromaticity, which states that in order for a molecule to satisfy the closed-shell stability of the π system, the number of π electrons must be $4n+2$ (where n equals 0, 1, 2, 3, ...). For the imidazole ring, there are 6 π electrons, or $(4n + 2)$ π electrons where $n = 1$.

(b) The basicity of a nitrogen atom is a direct result of its ability to accept a proton or share its free electron pair with an acid.

The aliphatic nitrogen is tetrahedral and its lone electron pair occupies an sp^3 orbital. This is in contrast to the "pyridine" nitrogen, which occupies an sp^2 orbital. Since an sp^2 orbital has more s character than an sp^3 orbital, the lone-pair electrons in the sp^2 orbital will be held more tightly; therefore, they are less available for sharing, resulting in the "pyridine" nitrogen being less basic than the aliphatic. A "pyrrole" nitrogen atom is still less basic than a "pyridine" one, since its lone electron pair is involved in the formation of the π cloud, and is thus not able to be shared with an acid.

Whereas in general the following order of basicity holds:

aliphatic $-NH_2$ > "pyridine" $-N$ > "pyrrole" $-N$,

the "pyridine" nitrogen in the imidazole ring in histamine is actually indistinguishable from the "pyrrole" nitrogen. This is because of the presence of tautomeric shifts:

Therefore, the two nitrogen atoms in the ring will display the same degree of basicity.

Quiz: Heterocyclic Compounds

1. The following compound is named

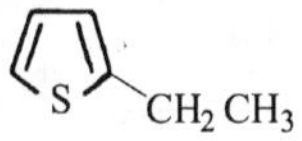

   (A) 1-ethylpyrrole.   (D) 2-ethylthiophene.

   (B) 2-ethylfuran.   (E) 1-sulfurfuran.

   (C) 5-ethylpyrrole.

2. During Knorr pyrrole synthesis a substituted pyrrole is produced from the reaction of

   (A) a ketone and an aldehyde.

   (B) an α-amino ketone and a ketone.

   (C) a β-ketoester and an alcohol.

   (D) a 1,4 diketone.

   (E) None of the above.

3. Electrophilic substitution of heterocyclic compounds occurs at the

   (A) α-position.   (D) All of the above.

   (B) β-position.   (E) None of the above.

   (C) γ-position.

4. The Skraup method of quinoline synthesis involves heating a mixture of

   (A) aniline, cinnamic aldehyde, and peroxide.

   (B) pyridine, glycerol, and α,β-unsaturated ketone.

(C) o-aminobenzaldehyde, glycerol, ketone, and hydro-chloric acid.

(D) alkyl halide, phosphorus pentoxide, and lepidine.

(E) aniline, glycerol, nitrobenzene, and sulfuric acid.

5. Pyridines undergo all of the following reactions EXCEPT

    (A) bromination.   (D) electrophilic substitution.

    (B) amination.   (E) nucleophilic substitution.

    (C) acylation.

6. The product of the following reaction is

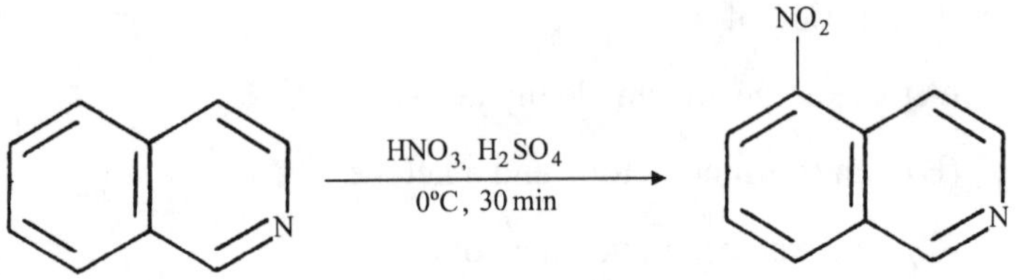

(A) 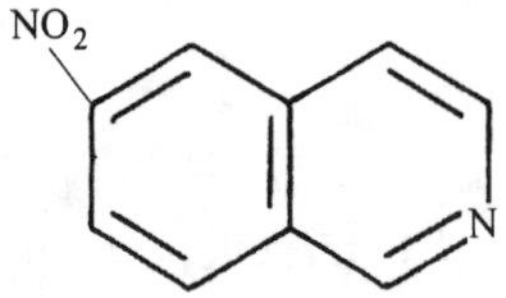   (D)   All of the above.

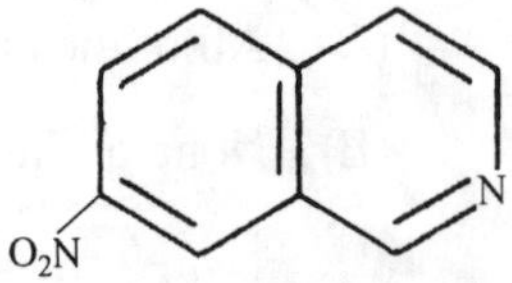

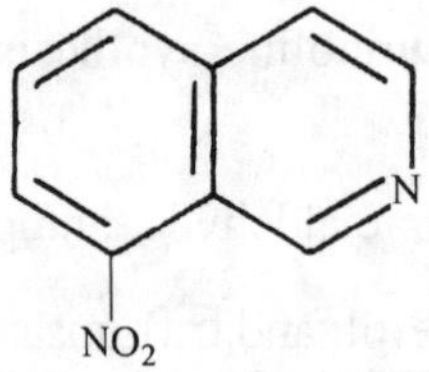

(B)   (E)   None of the above.

(C)

7. The following structure is

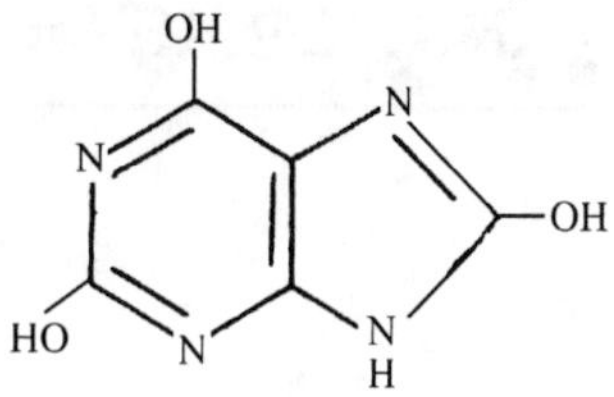

    (A) caffeine.   (D) morphine.

    (B) uric acid.   (E) phenobarbital.

    (C) adenine.

8. All of the following are alkaloids EXCEPT

    (A) nicotine.   (D) coniline.

    (B) quinine.   (E) pyrimidine.

    (C) morphine.

9. Azoles are five-membered-ring aromatic heterocycles containing

    (A) two nitrogens.

    (B) one nitrogen and one oxygen.

    (C) one nitrogen and one sulfur.

    (D) All of the above.

    (E) None of the above.

10. Azoles may be synthesized using all of the following reactants
    EXCEPT

    (A) hydrazine.   (D) barbituric acid.

    (B) nitrile oxide.   (E) 1,4-dicarbonyl.

    (C) acetylene.

ANSWER KEY

1.	(D)	6.	(C)
2.	(B)	7.	(B)
3.	(D)	8.	(E)
4.	(E)	9.	(D)
5.	(C)	10.	(D)

CHAPTER 7

Carbohydrates

Simple carbohydrates are polyhydroxy aldehydes and polyhydroxy ketones existing in cyclic hemiacetal and hemiketal forms. They are, actually or potentially, hydroxy or polyhydroxy oxoderivatives of the hydrocarbons.

Structural formula: $n\left[C_x(H_2O)_x\right] - (n-1)H_2O$

x = number of carbon atoms in a building unit.

n = number of building units per molecule.

7.1   Nomenclature and Classification

Carbohydrates with an aldehyde group are called aldoses and those with a keto group are called ketoses.

```
        CHO                    CH₂OH
        |                      |
     (CHOH)                    C = O
           n                   |
        CH₂OH               (CHOH)
                                  n
                              CH₂OH
```

An aldose   A ketose
(Polyhydroxy (Polyhydroxy ketone)
aldehyde)

Aldoses and ketoses are further characterized as aldo- or keto-, trioses, -tetroses, -pentoses, -hexoses, etc., according to the number of carbon atoms they have.

Example

$$
\begin{array}{cc}
\text{CHO} & \text{CH}_2\text{OH} \\
| & | \\
\text{CHOH} & \text{C} = \text{O} \\
| & | \\
\text{CH}_2\text{OH} & (\text{CHOH})_4 \\
& | \\
& \text{CH}_2\text{OH}
\end{array}
$$

An aldotriose   A 2-ketoheptose

Carbohydrates are classified as:

A) Monosaccharides: single aldoses and ketoses that do not hydro-lyze to simpler sugars.

B) Oligosaccharides: di-, tri-, tetra-, etc., saccharides that can be hy-drolyzed to monosaccharides (up to 10).

C) Polysaccharides: macromolecule consisting of higher numbers (up to thousands) of monosaccharides.

Saccharides containing a nonsugar component are called glyco-sides.

Problem Solving Example:

Answer the following as Fischer did in his famous monosac-charide structure determination.

(a) Write projection formulas for all the theoretically possible D-aldopentoses, $HOH_2C(CHOH)_3CHO$.

(b) One of the D-aldopentoses is the naturally occurring D-arabi-nose, enantiomeric with the more abundant L-arabinose. Oxi-dation of D-arabinose with nitric acid gives an optically active five-carbon trihydroxydicarboxylic acid. Which of the D-aldopentoses could be the D-arabinose?

(c) D-arabinose is converted by the following transformations into
G-glucose and D-mannose. (This is the classic Kiliani-Fischer
cyanohydrin synthesis of sugars.) What do these transformations tell about the relationship between the configurations of
mannose and glucose?

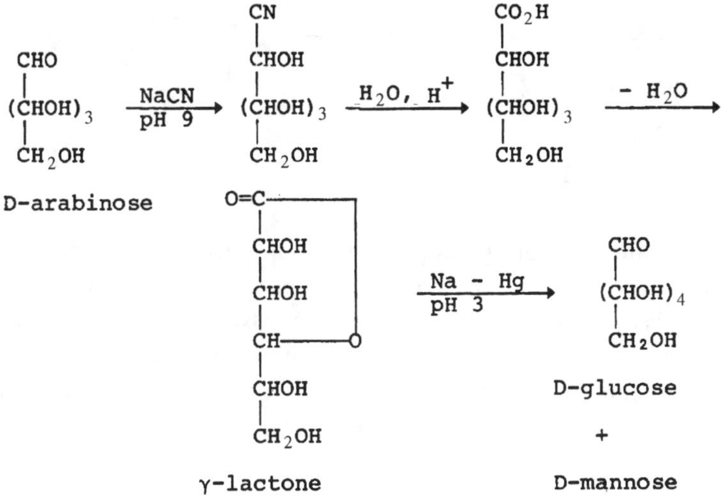

(d) Oxidation of D-glucose and D-mannose gives the six-carbon,
tetrahydroxydicarboxylic acids, glucaric and mannaric acids,
respectively. Both are optically active. What then are the configurations of the D- and L-arabinoses?

(e) D-glucaric acid can form two different γ-monolactones, whereas
D-mannaric acid can form only one monolactone. What then
is the configuration of D-glucose and D-mannose?

A Carbohydrates are a group of extremely important naturally
occurring compounds. They have the general structural formula $C_n(H_2O)_n$ with n being an integer greater than or equal to three.
The most basic type of carbohydrates are the sugars (saccharides), which
are polyhydroxyaldehydes or polyhydroxyketones that exist almost

exclusively in cyclic form. D-glucose, the ubiquitous natural product, is a diastereomer of 2,3,4,5,6-pentahydroxylhexanol.

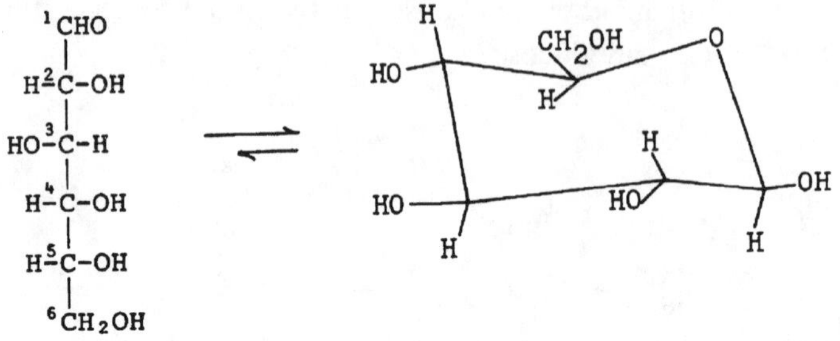

D-glucose

Note the numbering system of the monosaccharides in which the carbonyl carbon (or the terminal carbon nearest the carbonyl group in a ketose) is considered carbon number one. Also note that the "D" configuration refers to the stereochemistry about the chiral center farthest from the carbonyl group. The D-family of sugars is based upon the three-carbon sugar D-(+)-glyceraldehyde:

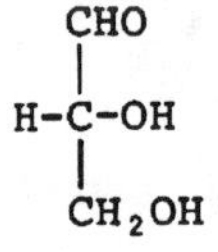

Actually, Fischer did not know the absolute configuration of this carbohydrate and was lucky in designating this compound correctly as a D-sugar.

(a) Carbohydrates are classified in two general ways by designating the number of carbons in the sugar and the type of carbonyl group it possesses (ketone or aldehyde). Thus, a five-carbon sugar with an aldehyde carbonyl is termed an aldopentose. The general structural formula for an aldopentose is:

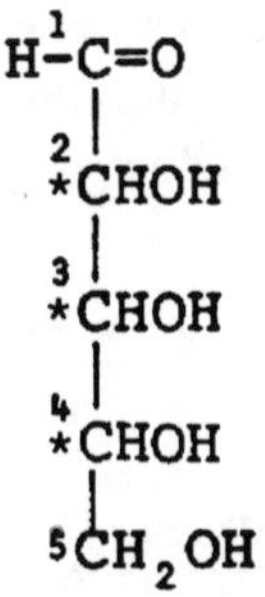

$$H-\overset{1}{C}=O$$
$$\overset{2}{\underset{*}{C}HOH}$$
$$\overset{3}{\underset{*}{C}HOH}$$
$$\overset{4}{\underset{*}{C}HOH}$$
$$\overset{5}{C}H_2OH$$

Three of the carbons are asymmetric (bonded to four different groups), and therefore, they are chiral centers, denoted with * allowing for the existence of eight (2^n where n = number of chiral carbons = 3) different stereoisomers. We can denote the stereochemistry of these different isomers through the use of Fischer projection formulas, remembering its convention of horizontal lines representing bonds coming out of the plane of the paper and vertical lines representing bonds behind this plane. For example, in D-(+)-glyceraldehyde

We are asked to list all the possible D-aldopentoses that require the stereochemistry at C_4 be fixed. This means that there are only four different D-aldopentoses. The diastereomers are:

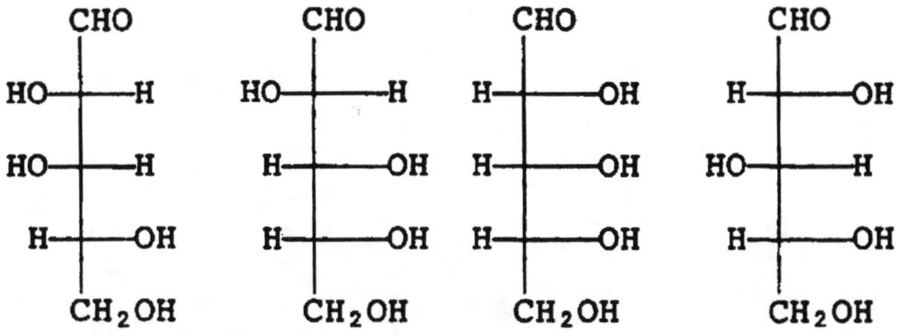

(b) Nitric acid is a potent oxidizing agent that will oxidize both the terminal carbonyl and hydroxyl group to carboxylic acid functionalities. The product is a dicarboxylic acid, which is termed an aldaric acid.

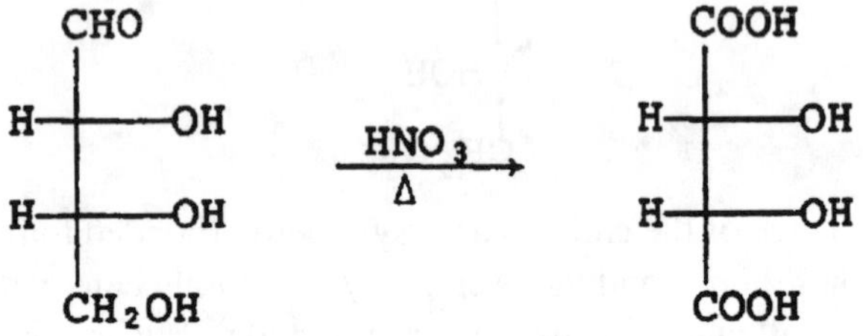

$$\text{D-erythrose} \xrightarrow[\Delta]{\text{HNO}_3} \text{D-erythric acid}$$

CHO
H——OH
H——OH
CH$_2$OH

D-erythrose

COOH
H——OH
H——OH
COOH

D-erythric acid

The two terminal groups of an aldaric acid are identical, and if the acid possesses a plane of symmetry, it will be an optically inactive meso compound. Two of the D-aldopentoses will give meso aldaric acids while two will yield optically active aldaric acids.

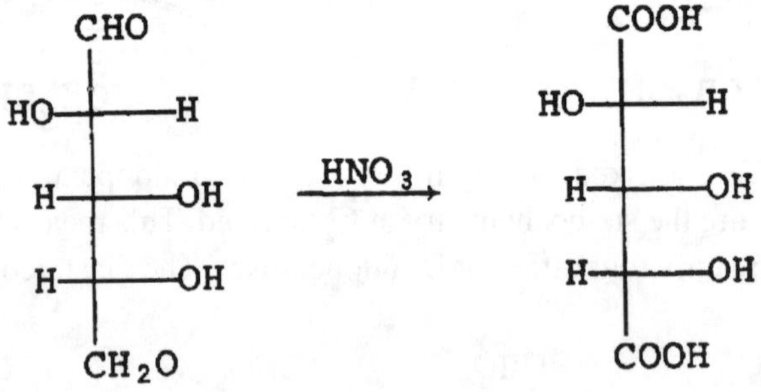

CHO
HO——H
H——OH
H——OH
CH$_2$O

$$\xrightarrow{\text{HNO}_3}$$

COOH
HO——H
H——OH
H——OH
COOH

optically active

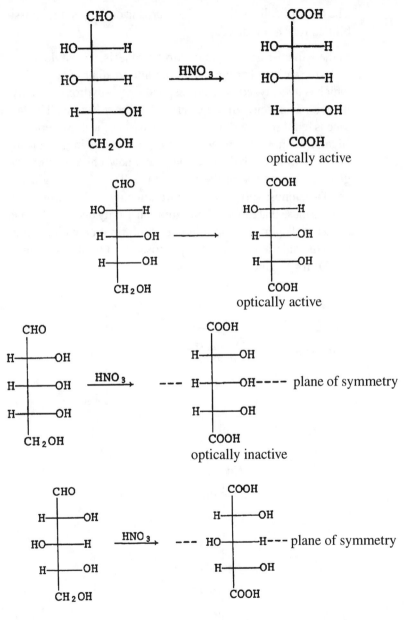

optically active

optically active

optically inactive

optically inactive

The two optically active products are the only structural possibilities for D-arabinose.

(c) In the Kiliani-Fischer cyanohydrin synthesis, a cyanide ion attacks the carbonyl group to form an intermediate cyanohydrin, which is hydrolyzed to its corresponding γ,δ-hydroxy carboxylic acid. This acid will be esterified to the γ-lactone. The lactone is reduced by sodium amalgam to form the diastereomeric sugars. We note that none of the chiral centers in the starting material were altered. Also, only one new chiral center was generated by attack of the nucleophile. Therefore, D-glucose and D-mannose differ in configuration only at C_2. Compounds such as these that differ in configuration at only one chiral center are termed epimers. Combining this information with our data from the aldaric acid, we can propose four possible structures (A–D) for D-glucose and D-mannose.

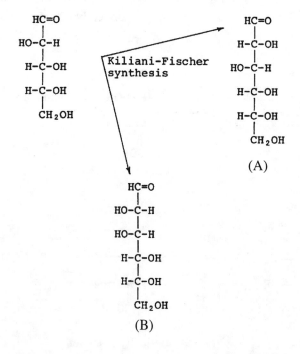

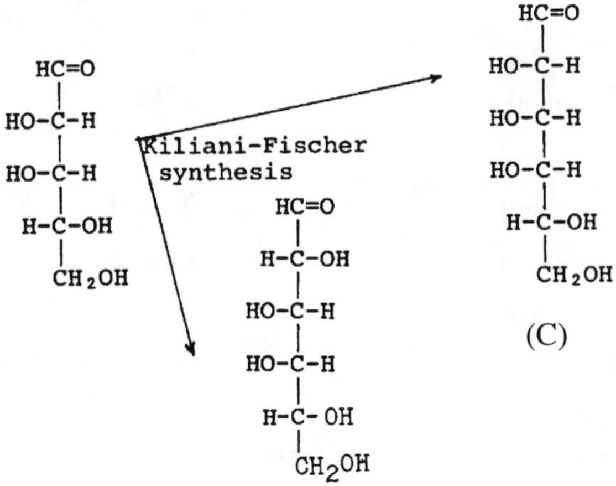

HC=O
|
HO-C-H
|
HO-C-H
|
H-C-OH
|
CH₂OH

Kiliani-Fischer
synthesis

HC=O
|
HO-C-H
|
HO-C-H
|
HO-C-H
|
H-C-OH
|
CH₂OH

(C)

HC=O
|
H-C-OH
|
HO-C-H
|
HO-C-H
|
H-C-OH
|
CH₂OH

(D)

(d) Nitric acid will oxidize a carbohydrate to an aldaric acid. Realizing that a dicarboxylic acid with a plane of symmetry lacks optical activity, we can limit the structural possibilities for the aldohexoses to two Fischer projections.

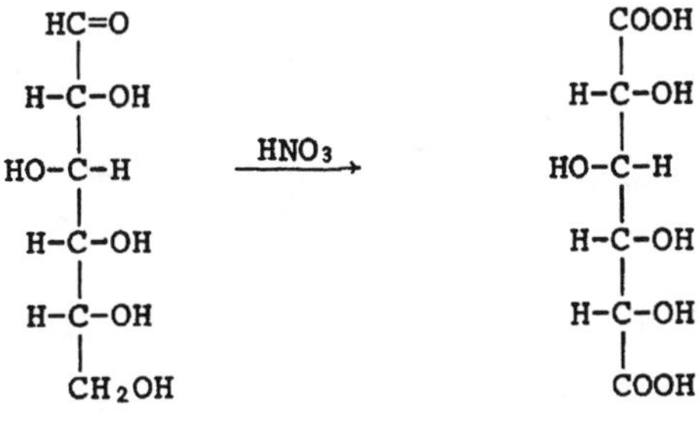

HC=O
|
H-C-OH
|
HO-C-H
|
H-C-OH
|
H-C-OH
|
CH₂OH

(A)

$\xrightarrow{\text{HNO}_3}$

COOH
|
H-C-OH
|
HO-C-H
|
H-C-OH
|
H-C-OH
|
COOH

optically active

(A)   optically active

```
      HC=O                                    COOH
       |                                       |
   HO-C-H                                   HO-C-H
       |                                       |
   HO-C-H          HNO₃                     HO-C-H
       |         ──────────→                   |
    H-C-OH                                   H-C-OH
       |                                       |
    H-C-OH                                   H-C-OH
       |                                       |
     CH₂OH                                    COOH
```

(B)   optically active

```
      HC=O                                    COOH
       |                                       |
   HO-C-H                                   HO-C-H
       |                                       |
   HO-C-H          HNO₃                     HO-C-H
       |         ──────────→                   |
   HO-C-H                                   HO-C-H
       |                                       |
    H-C-OH                                   H-C-OH
       |                                       |
     CH₂OH                                    COOH
```

(C)   optically active

```
      HC=O                                    COOH
       |                                       |
    H-C-OH                                   H-C-OH
       |                                       |
   HO-C-H          HNO₃                     HO-C-H
       |         ──────────→          ---------+--------  plane of
   HO-C-H                                   HO-C-H        symmetry
       |                                       |
    H-C-OH                                   H-C-OH
       |                                       |
     CH₂OH                                    COOH
```

(D)   optically inactive

Compound C yields an optically active aldaric acid upon treatment with HNO_3 but it can be eliminated as a possible structure for D-glucose or D-mannose because it is an epimer of compound D which yielded a meso aldaric acid. We recall that D-glucose and D-mannose are epimers, differing only at C-2. However, examining the configurations of C_3, C_4, C_5 of the

aldohexose A and B, we can depict D-arabinose and its enanti-
omer L-arabinose.

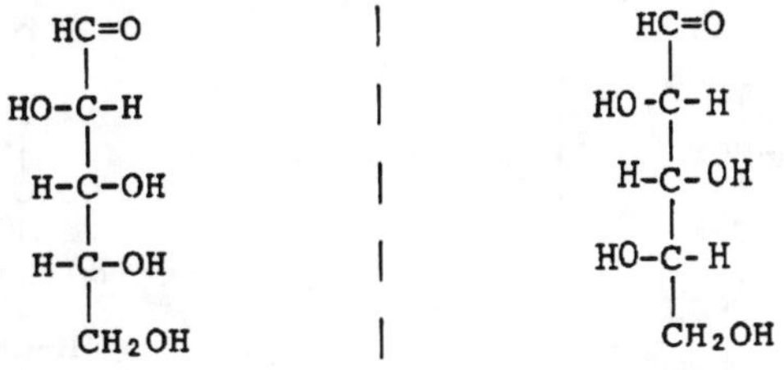

HC=O

HO–C–H

H–C–OH

H–C–OH

CH₂OH

D-arabinose

HC=O

HO–C–H

H–C–OH

HO–C–H

CH₂OH

L-arabinose

Thus, compounds A and B must be differentiated as D-glucose
and D-mannose.

(e) Examining the aldaric acid of compound A, we see that
two distinct γ-monolactones can be derived from it.

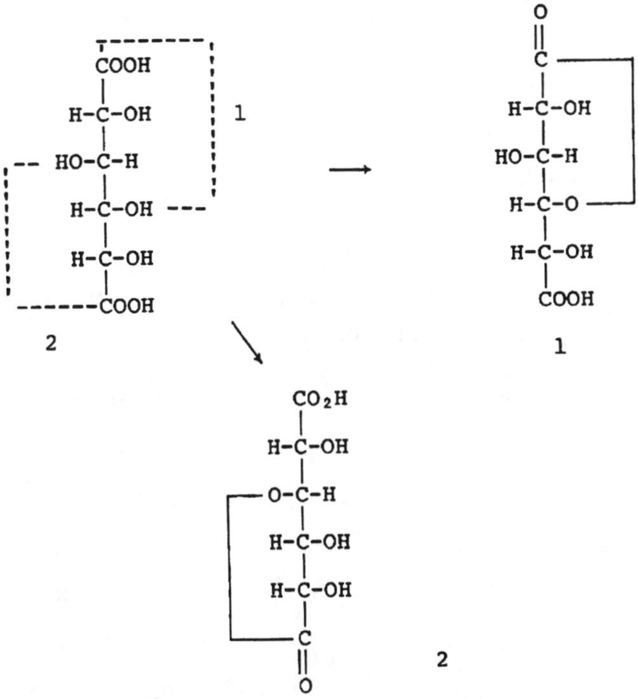

COOH

H–C–OH

HO–C–H

H–C–OH

H–C–OH

COOH

2

O
‖
C

H–C–OH

HO–C–H

H–C–O

H–C–OH

COOH

1

⟶

1

CO₂H

H–C–OH

O–C–H

H–C–OH

H–C–OH

C
‖
O

2

On the other hand, compound B's aldaric acid can form only one lactone. This can be verified by rotation of the Fischer projections by 180°.

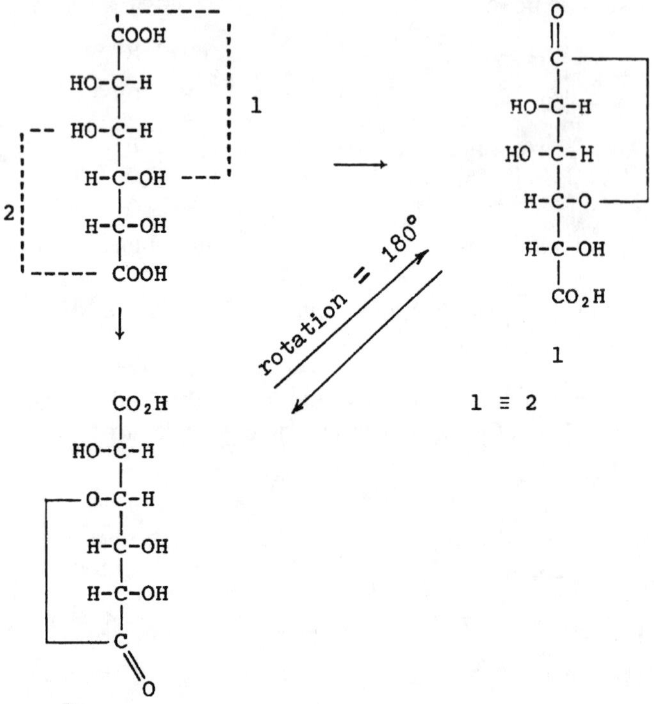

Therefore, we can now designate the correct structures for D-glucose and D-mannose. They are:

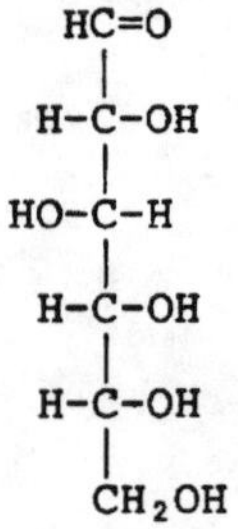

D-glucose

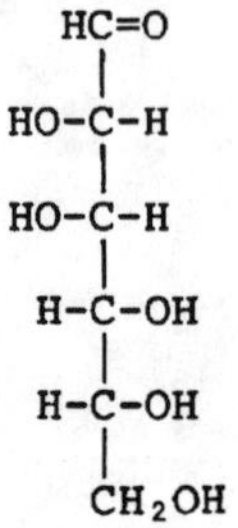

D-mannose

7.2 Physical Properties

A) Monosaccharides are readily soluble in water, have a sweet taste, which increases with the number of –OH groups, and are usually crystalline.

B) Glycolic aldehyde is sweet and soluble in cold water and alcohol, but insoluble in ether. It can be obtained in crystalline form (m.p. 95–97°C).

C) Glyceraldehyde (m.p. 138°C) is a white powder, not very sweet, soluble in water, slightly soluble in alcohol, and insoluble in ether.

D) Dihydroxyacetone is sweet, very soluble in cold water, slightly soluble in ether, almost insoluble in hot acetone, and insoluble in ligroin.

E) Arabinose is precipitated by alcohol and yields an orange-yellow osazone.

F) D-fructose forms anhydrous crystals that are somewhat hygroscopic. It forms an insoluble methylphenylhydrazone, whereas the corresponding derivative of D-glucose is soluble.

G) D-mannose may be precipitated by alcohol.

H) Most of the di- and trisaccharides are soluble in water and crystallize quite readily.

I) The sweetness of sucrose is taken as 100 (standard basis). The relative sweetness of common sugars is indicated as:

D-fructose	173.3	Rhamnose	32.5
D-glucose	74.3	D-galactose	32.1
D-xylose	40.0	Raffinose	22.6
Maltose	32.5	Lactose	16.0

J) Polysaccharides, as a rule, are not soluble in water and are not readily obtained in crystalline form.

K) Dextrin has an insipid taste and gives a red to brown coloration.

L) Glycogen tends to give a milky solution that may be clarified by treating with acetic acid. With iodine in potassium iodide, glycogen gives a reddish-brown coloration.

M) Ordinary air-dried starch is tasteless. It contains water but becomes anhydrous on heating to 110°C.

N) On boiling starch with water, the granules break to give a colloidal solution which gives a blue coloration with iodide in potassium iodide. Coloration disappears upon heating and reappears upon cooling.

O) Cellulose may be fibrous, cellular, or woody.

7.3 Preparation of Monosaccharides

Aldol condensation

$$\overset{H}{\underset{H}{RC}} = O + R'CH - \overset{H}{\underset{H}{C}} = O \xrightarrow{-OH} R - \overset{H}{\underset{OH}{C}} - CH_2 \overset{H}{C} = O$$

Kiliani-Fischer Synthesis

Ra is $CH_2OH-HCOH-HCOH-HCOH-$

$$2H-\underset{Ra}{C} = O \xrightarrow[\text{trace } NH_3]{\text{3\% HCN}} \quad \underset{Ra}{\overset{CN}{\underset{|}{H-C-OH}}} + \underset{Ra}{\overset{CN}{\underset{|}{HO-C-H}}}$$

$$\xrightarrow[\text{dil. } H_2SO_4, \text{boil}]{\text{hydrolysis}} \quad H-\underset{Ra}{\overset{CO-OH}{\underset{|}{C-OH}}} + HO-\underset{Ra}{\overset{CO-OH}{\underset{|}{C-H}}}$$

After the polyhydroxy monacids are isolated in the form of their lactones, then

$$
\begin{array}{ccc}
\begin{array}{l}
O=C \\
\;\;| \\
H-C-OH \\
\;\;| \\
HO-C-H \\
\;\;| \\
H-C-O \\
\;\;| \\
H-C-OH \\
\;\;| \\
CH_2OH
\end{array}
&
+ 2Na/(Hg)/2H_2SO_4, aq
&
\begin{array}{l}
O=C-H \\
\;\;| \\
H-C-OH \\
\;\;| \\
HO-C-H \\
\;\;| \\
H-C-OH \\
\;\;| \\
H-C-OH \\
\;\;| \\
CH_2OH
\end{array}
\;\; + NaHSO_4
\end{array}
$$

b-glyconolactone   D-glucose

Mild oxidation or more vigorous oxidation followed by reduction of the polyhydroxy alcohols:

A) Direct oxidation of corresponding alcohols.

$$
\begin{array}{l}
CH_2OH + Br_2, aq/2C_6H_5CO-ONa \rightarrow H-C=O + 2NaBr \\
\;\;| \qquad\qquad\qquad\qquad\qquad\qquad\qquad\;\;\; | \\
Ra \qquad\qquad\qquad\qquad\qquad\qquad\qquad\qquad Ra
\end{array}
$$

$$+ 2C_6H_5-CO-OH$$

B) Oxidation of a polyhydroxy alcohol to the corresponding monoacid and subsequent reduction of the monoacid lactone to the corresponding aldose.

$$
\begin{array}{l}
CH_2OH + 2Br_2/H_2O/4C_6H_5CO-ONa, aq \\
\;\;| \\
Rg \qquad\qquad \rightarrow \;\; O=C-OH + 4NaBr + 4C_6H_5-CO-OH \\
\qquad\qquad\qquad\qquad\;\;| \\
\qquad\qquad\qquad\qquad\; Rg
\end{array}
$$

Rg is $CH_2OH - HCOH - HCOH - HCOH - HCOH-$

Degradation reaction. D-glucose may be degraded to D-arabinose.

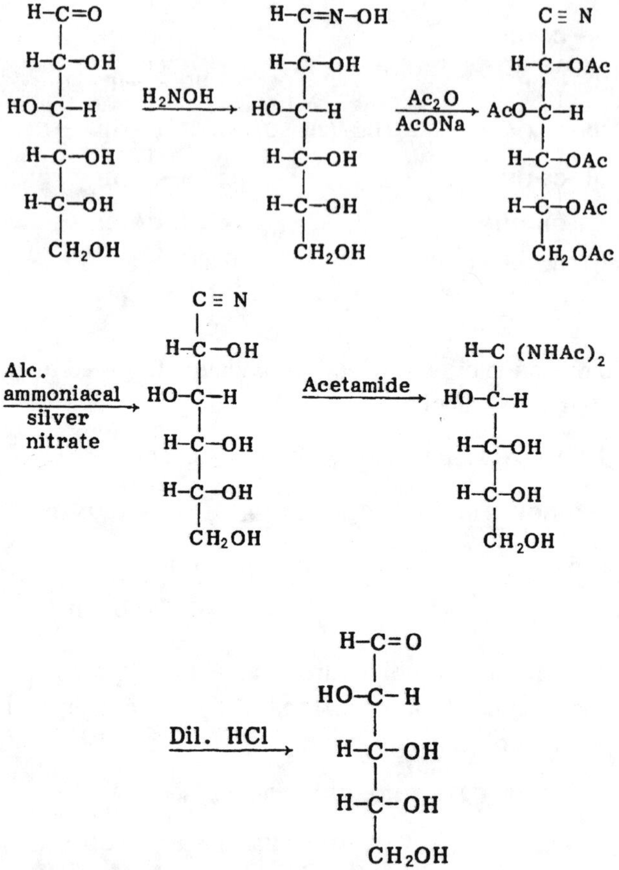

H-C=O   H-C=N-OH   C ≡ N
H-C-OH   H-C-OH   H-C-OAc
HO-C-H $\xrightarrow{H_2NOH}$ HO-C-H $\xrightarrow[\text{AcONa}]{Ac_2O}$ AcO-C-H
H-C-OH   H-C-OH   H-C-OAc
H-C-OH   H-C-OH   H-C-OAc
CH_2OH CH_2OH CH_2OAc

C ≡ N
H-C-OH   H-C $(NHAc)_2$

Alc.   HO-C-H   Acetamide   HO-C-H
ammoniacal $\xrightarrow{\hspace{1cm}}$
silver   H-C-OH   H-C-OH
nitrate
H-C-OH   H-C-OH
CH_2OH CH_2OH

H-C=O
HO-C-H
$\xrightarrow{Dil.\ HCl}$ H-C-OH
H-C-OH
CH_2OH

Hydrolysis of Polysaccharides

   Example

Arabinosans	yield	Arabinose
Xylosans	yield	Xylose
Dextrosans	yield	Glucose
Mannosans	yield	Mannose

Starch is hydrolyzed commercially to give a variety of products, such as corn syrup, grape sugar, and dextrin, by controlling the extent of the hydrolysis.

By the condensation of aldoses with nitromethane, conversion to the salt with alkali, and subsequent hydrolysis with acid.

Problem Solving Example:

Q D-arabinose and D-ribose give the same phenylosazone. D-ribose is reduced to the optically inactive pentahydric alcohol, ribitol. D-arabinose can be degraded by the Ruff method, which involves the following reactions:

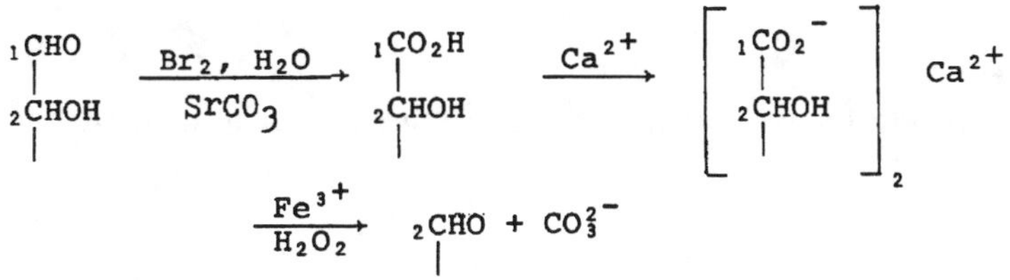

$$\underset{2}{\overset{1}{\text{CHO}}}\ \underset{\text{CHOH}}{|}\ \xrightarrow[\text{SrCO}_3]{\text{Br}_2,\ \text{H}_2\text{O}}\ \underset{2}{\overset{1}{\text{CO}_2\text{H}}}\ \underset{\text{CHOH}}{|}\ \xrightarrow{\text{Ca}^{2+}}\ \left[\ \underset{2}{\overset{1}{\text{CO}_2^-}}\ \underset{\text{CHOH}}{|}\ \right]_2 \text{Ca}^{2+}$$

$$\xrightarrow[\text{H}_2\text{O}_2]{\text{Fe}^{3+}}\ \underset{2}{\text{CHO}} + \text{CO}_3^{2-}$$

The tetrose, D-erythrose, so obtained can be oxidized with nitric acid to mesotartaric acid. What are the configurations of D-arabinose, D-ribose, ribitol, and D-erythrose?

A In an identification problem such as this, it is most advisable to work backwards from the last piece of data given. Mesotartaric acid is a four-carbon (it is derived from a tetrose sugar) dicarboxylic acid, which has the structure:

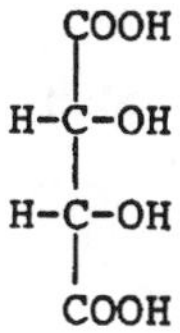

$$\begin{array}{c} \text{COOH} \\ | \\ \text{H}-\text{C}-\text{OH} \\ | \\ \text{H}-\text{C}-\text{OH} \\ | \\ \text{COOH} \end{array}$$

We deduce this from the fact that 1) nitric acid will oxidize the terminal hydroxyl group and the carbonyl group of the sugar to carboxylic acid functionalities, and 2) it is a D-sugar that lacks optical activity in that it has a plane of symmetry (meso compound). This reasoning can be outlined as follows:

$$
\begin{array}{ccc}
\text{C=O} & \text{CO}_2\text{H} & \text{COOH} \\
| & | & | \\
\text{CHOH} & \text{CHOH} & \text{H–C–OH} \\
| \qquad \xrightarrow{\text{HNO}_3} & | & \text{--}\dashv\text{------ plane of} \\
\text{H–C–OH} & \text{H–C–OH} & \text{H–C–OH} \quad \text{symmetry} \\
| & | & | \\
\text{CH}_2\text{OH} & \text{CO}_2\text{H} & \text{COOH} \\
\text{D-sugar} & \text{D-tartaric} & \text{meso-D-tartaric acid} \\
& \text{acid} &
\end{array}
$$

We now know the structure of D-erythrose from its oxidation product, tartaric acid:

$$
\begin{array}{ccc}
\text{CO}_2\text{H} & & \text{CHO} \\
| & & | \\
\text{H–C–OH} & \xleftarrow{\text{HNO}_3} & \text{H–C–OH} \\
| & & | \\
\text{H–C–OH} & & \text{H–C–OH} \\
| & & | \\
\text{CO}_2\text{H} & & \text{CH}_2\text{OH} \\
& & \text{D-erythrose}
\end{array}
$$

D-erythrose is the product of the Ruff degradation of D-arabinose, which significantly did not alter the stereochemistry about the carbons of the pentose. Thus, still ignorant of the stereochemistry, we depict the structure of D-arabinose as:

$$
\begin{array}{c}
\text{CHO} \\
| \\
\text{CHOH} \\
| \\
\text{H–C–OH} \\
| \\
\text{H–C–OH} \\
| \\
\text{CH}_2\text{OH}
\end{array}
$$

d-arabinose

Since phenylosazone formation also does not affect the stereochemistry about C_3 and C_4 of the pentose, the above structure also represents D-ribose whose structure differs only at the C-2 chiral center. Such compounds, differing in stereochemistry only about one carbon, are termed epimers. D-ribose is reduced to an optically inactive alditol, ribitol, and therefore, we can determine the structure of D-ribose and its epimer, D-arabinose.

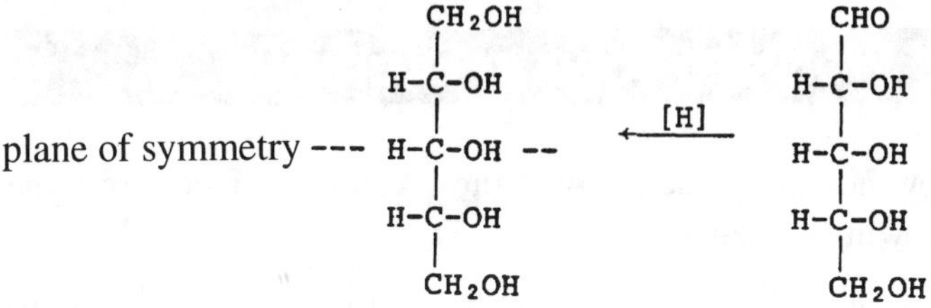

```
                        CH₂OH                        CHO
                          |                           |
                      H-C-OH                      H-C-OH
                          |              [H]          |
plane of symmetry --- H-C-OH  --  ←----------   H-C-OH
                          |                           |
                      H-C-OH                      H-C-OH
                          |                           |
                       CH₂OH                       CH₂OH
```

   (optically inactive
   alditol of D-ribose)   ribitol   D-ribose

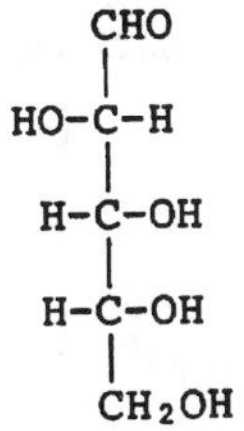

```
              CHO
               |
          HO-C-H
               |
          H-C-OH
               |
          H-C-OH
               |
           CH₂OH
```

           D-arabinose

7.4   Preparation of Disaccharides

7.4.1   Isolation from natural products

A) Sucrose from sugar cane or sugar beet.

B) Maltose from sprouted barley or by partial hydrolysis of starch.

C) Lactose, as by-product in the cheese industry.

Problem Solving Example:

Show how the structure of maltose can be deduced from the following evidence:

(a) The sugar is hydrolyzed by yeast α-D-glucosidase to D-glucose.

(b) Maltose mutarotates and forms a phenylosazone. Methylation with dimethyl sulfate in basic solution followed by acid hydrolysis gives 2,3,4,6-tetra-O-methyl-D-glucopyranose and 2,3,6-tri-O-methyl-D-glucose.

(c) Bromine oxidation of maltose followed by methylation and hydrolysis gives 2,3,4,6-tetra-O-methyl-D-glucopyranose and a tetramethyl-D-gluconic acid, which readily forms a y-lactone.

Maltose is obtained by the partial hydrolysis of starch in aqueous acid. By examining each fragment of information and trying to use this data to determine maltose's constitutional characteristics, we will be able to deduce its structure.

(a) Enzymes are specific in that some will act only on certain glycosidic linkages, while others will only take action with their anomeric counterparts. Here, the enzyme (α-D-glucosidase), is specific in hydrolyzing only α-D-glycosidic linkages. Thus, maltose consists only of D-glucose monosaccharide units and has an α configuration at the anomeric carbon. The anomeric carbon is the C-1 carbon.

(b) Both mutarotation (change in optical rotation upon dissolution of the crystalline compound) and phenylosazone formation (incorporation of two molecules of phenylosazone) are characteristic of carbohydrates with a free aldehyde or ketone carbonyl group. Since maltose possesses these two properties, it has a free carbonyl group, and therefore, both anomeric carbons are not involved in the acetal formation as is the case with a disaccharide such as D-sucrose. We can examine the products of sequential methylation and acid hydrolysis:

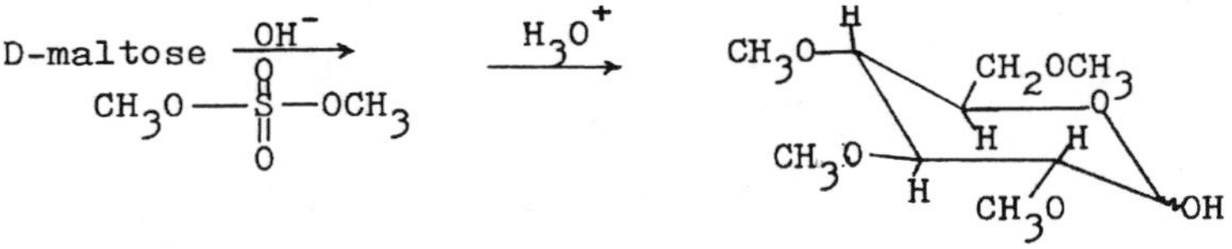

D-maltose $\xrightarrow{OH^-}$ $CH_3O—\overset{O}{\underset{O}{\overset{\|}{\underset{\|}{S}}}}—OCH_3$ $\xrightarrow{H_3O^+}$

2,3,4,6-tetra-O-methyl-D-glucopyranose

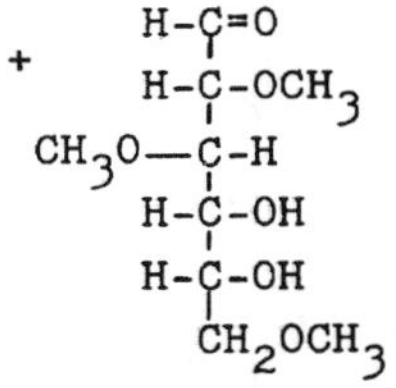

$$
+ \quad
\begin{array}{l}
H-\underset{|}{C}=O \\
H-\underset{|}{C}-OCH_3 \\
CH_3O—\underset{|}{C}-H \\
H-\underset{|}{C}-OH \\
H-\underset{|}{C}-OH \\
CH_2OCH_3
\end{array}
$$

2,3,6-tri-O-methyl-D-
glucose

From this information, we know immediately that at least one of the glucose units was in the pyranose form as seen from the glucopyranose product. The trimethyl glucose infers that the second glucose unit has either its C_4 or C_5 hydroxyl involved in the glucoside formation, while the other forms the intramolecular hemiacetal.

(c) y-lactone formation after bromine oxidation, methylation, and hydrolysis informs us that C_4 of the second glucose unit was involved in the glucoside linkage. A y-lactone has three-ring carbons besides the carbonyl group, which tells us which of the hydroxyl groups was not methylated and therefore was involved in the glucoside formation.

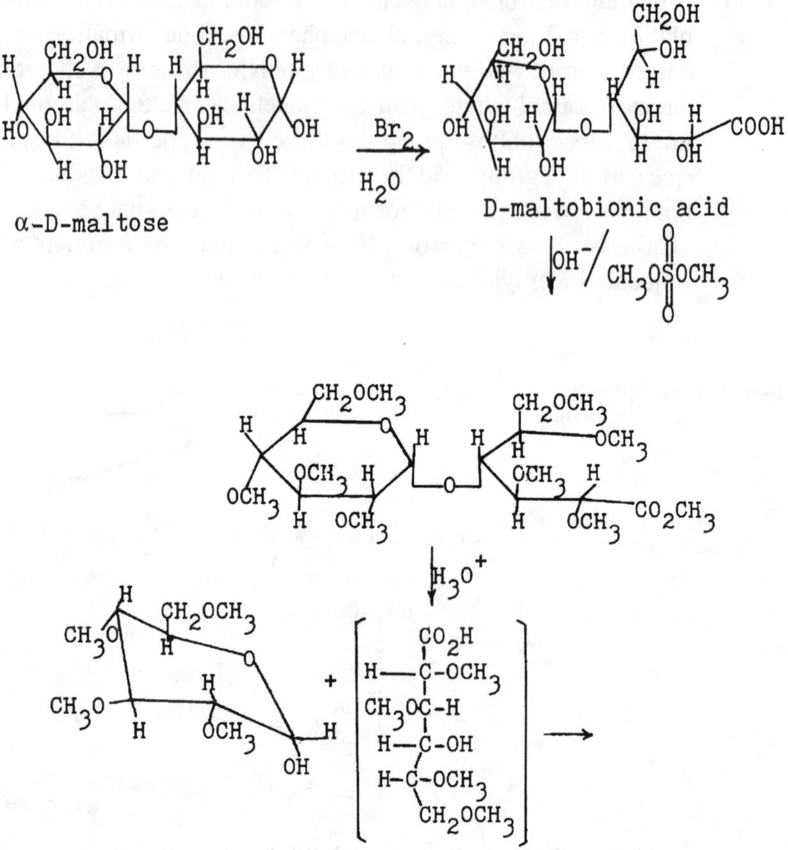

α-D-maltose

$\xrightarrow{\substack{Br_2 \\ H_2O}}$

D-maltobionic acid

$\downarrow OH^- / CH_3O\overset{\overset{\displaystyle O}{\|}}{S}OCH_3$

$\downarrow H_3O^+$

2,3,4,6-tetra-O-methyl-D-glucopyranose   2,3,5,6-tetramethyl-D-gluconic acid

(easily formed)

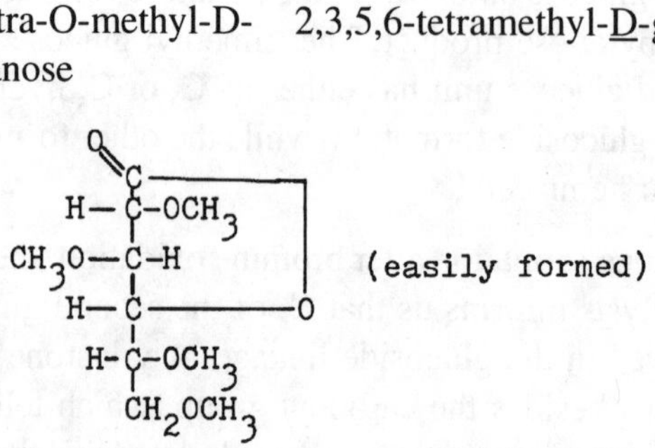

γ-lactone of 2,3,5,6-tetramethyl-D-gluconic acid

7.5   Preparation of Polysaccharides

A) Dextrin is obtained by the controlled hydrolysis of starch.

B) Glycogen is obtained by the extraction of liver tissue.

C) Starch is produced from corn, wheat, potatoes, rice, arrowroot, and the sago palm.

7.6   Reactions of Carbohydrates

Osazone formation by reaction between α or β-D-glucose and 3 moles of phenylhydrazine:

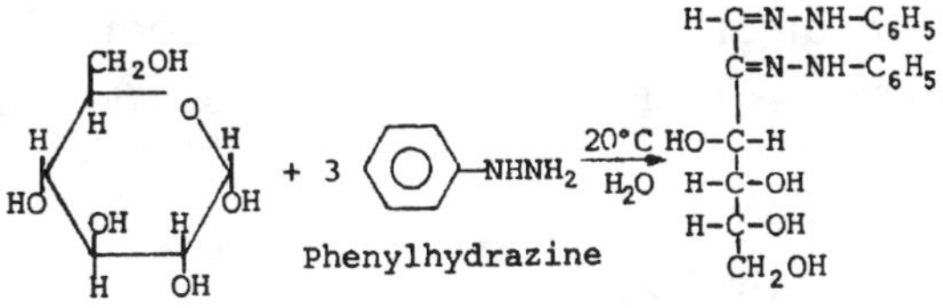

$$H-C=N-NH-C_6H_5$$
$$C=N-NH-C_6H_5$$
$$HO-C-H$$
$$H-C-OH$$
$$H-C-OH$$
$$CH_2OH$$

α–D-Glucose   + 3 Phenylhydrazine $\xrightarrow[H_2O]{20°C}$ Glucosazone

$$+C_6H_5-NH_2+NH_3+2H_2O$$

Reduction. Treatment of aldoses with Na/Hg in alkaline solution is by catalytic hydrogenation:

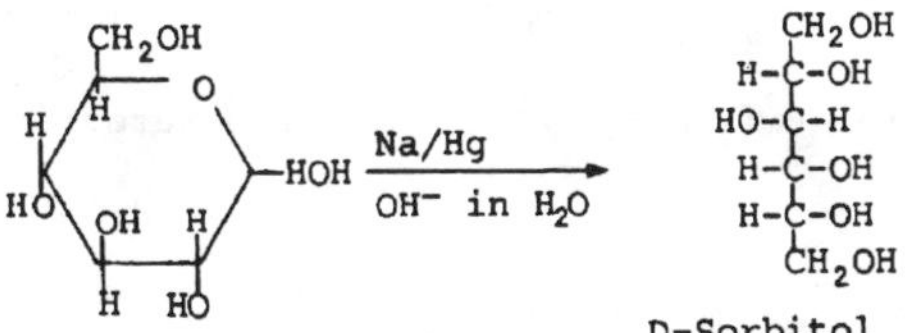

$$CH_2OH$$
$$H-C-OH$$
$$HO-C-H$$
$$H-C-OH$$
$$H-C-OH$$
$$CH_2OH$$

D-Glucose $\xrightarrow[OH^- \ in \ H_2O]{Na/Hg}$ D-Sorbitol (optically active)

Oxidation. Treatment of aldoses with nitric acid to cause the conversion to polyhydroxy-α, ω-dicarboxylic acids:

Example

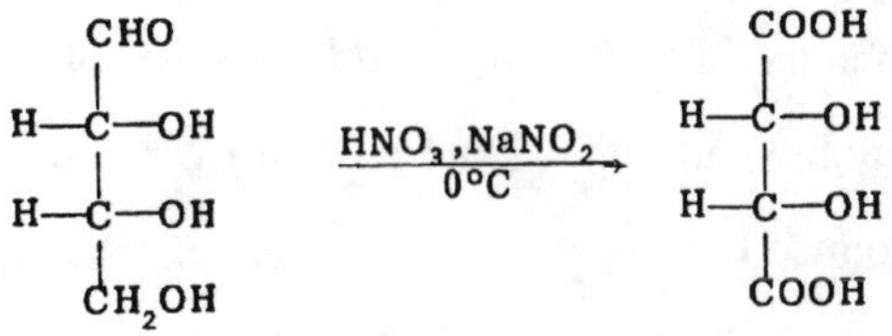

D-Erythrose (active) → meso-Tartaric acid (inactive)

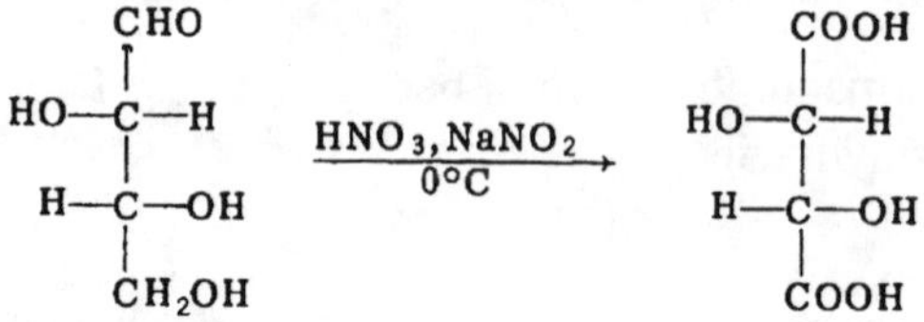

D-Threose (active) → D-Tartaric acid (active)

Kiliani-Fischer Synthesis

The conversion of an aldose to two epimers of the next higher aldose by the application of the cyanohydrin formation of aldehydes:

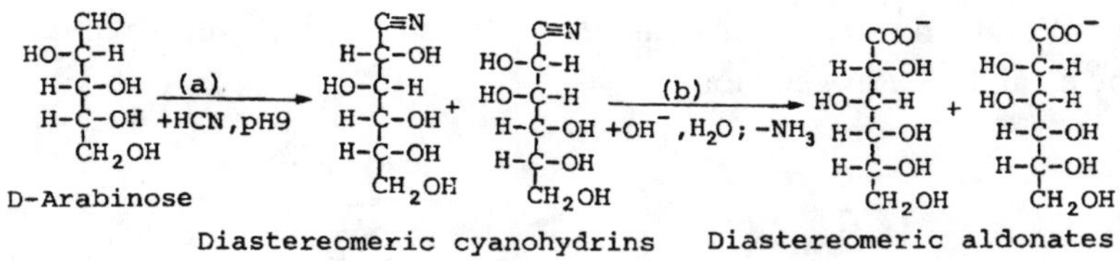

D-Arabinose

Diastereomeric cyanohydrins   Diastereomeric aldonates

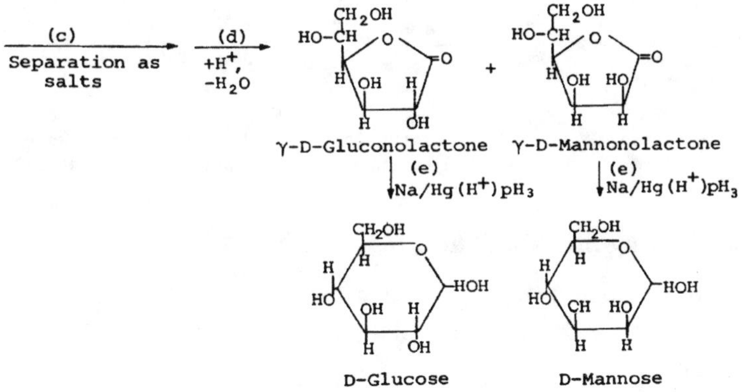

γ-D-Gluconolactone   γ-D-Mannonolactone

D-Glucose   D-Mannose

Formation of glycosides by boiling aldoses in alcohol in the presence of anhydrous hydrogen chloride.

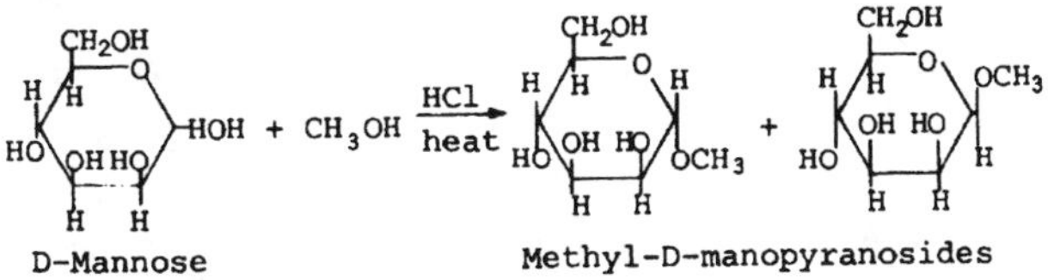

D-Mannose   Methyl-D-manopyranosides

Upon esterification by treatment with acetic anhydride/ZnCl$_2$ or with acetyl chloride, aldohexoses yield diastereomeric α- and β–glycopyranose pentaacetates:

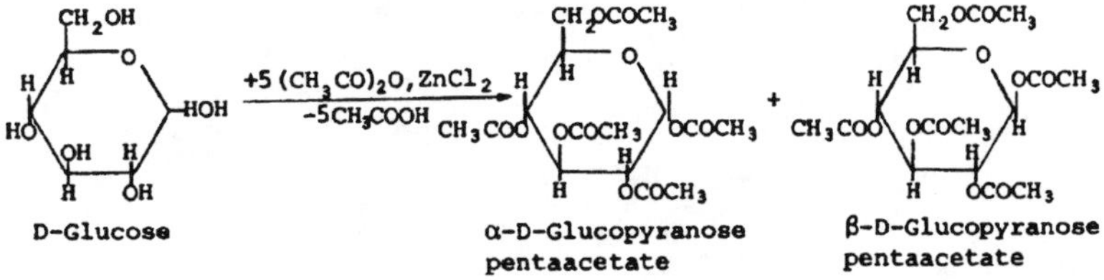

D-Glucose   α-D-Glucopyranose   β-D-Glucopyranose
                 pentaacetate   pentaacetate

Methylation. Polyhydroxy compounds can be converted to ether groups:

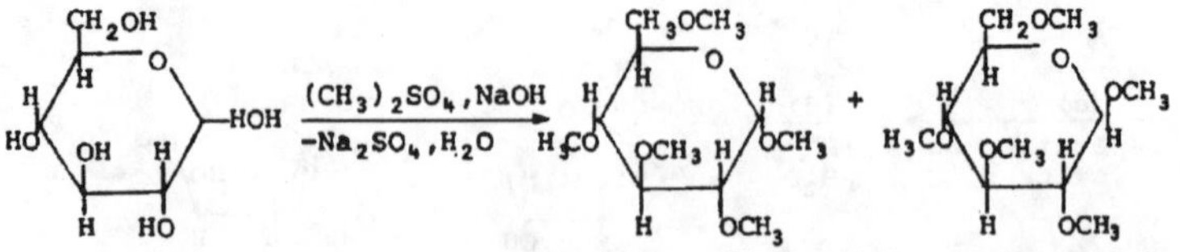

D-Glucose → Methyl-α(and β)-2,3,4,6-tetra-O-methyl-D-glycopyranoside (a permethylated sugar)

Reaction with Alkali and Mineral Acids

By the treatment of glucose with calcium hydroxide in alkaline medium, many products are obtained. This is due to isomerization of α-ketols:

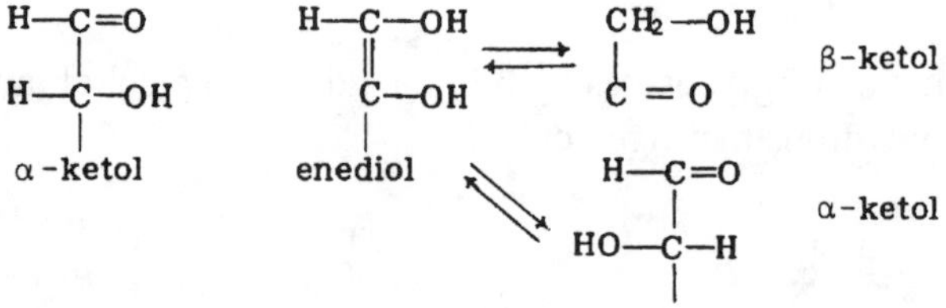

Treatment of aldoses with strong mineral acids causes dehydration, yielding furfural in the case of pentoses. Hexoses yield 5-hydroxymethyl furfural:

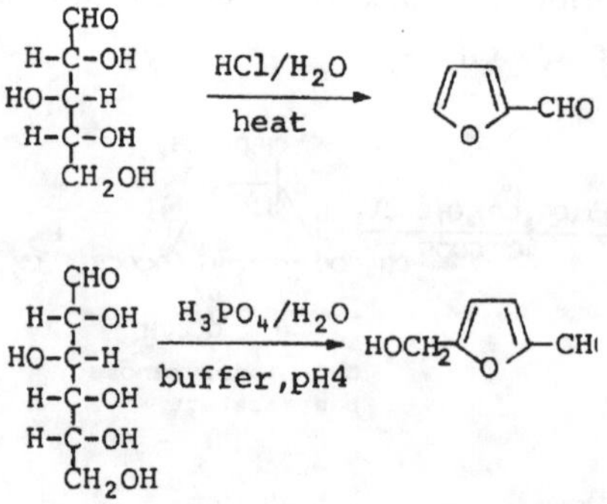

Formation of Acetonide

Monosaccharides containing cis-glycol groups react with acetone in the presence of dehydrating agents to give isopropylidene derivatives called acetonides.

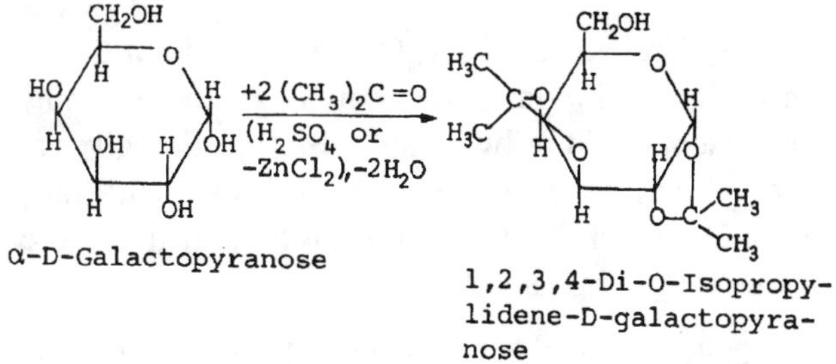

α-D-Galactopyranose

1,2,3,4-Di-O-Isopropylidene-D-galactopyranose

Oligosaccharides

Two monosaccharides condense to yield various types of disaccharides. If the hemiacetal –OH group of one molecule condenses with an alcoholic –OH of the second molecule, a maltose-type disaccharide is formed.

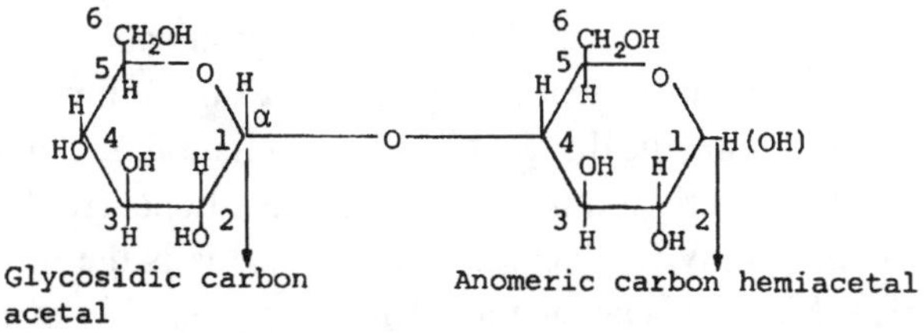

Glycosidic carbon acetal

Anomeric carbon hemiacetal

Polysaccharides

Large numbers of monosaccharides joined by glycosidic linkages are called polysaccharides. They have large molecular weights.

Cellulose is a large linear unbranched natural polymer consisting of 3,000–5,000 D-glucose units per chain. Cellulose and lignin form the cell walls of wood and plants.

Starch is found in granular form in seeds and roots of plants. It has two components, amylose (20%) and amylopectin (80%), both consisting of D-glucose units.

Problem Solving Examples:

(a) (+)-Trehalose, $C_{12}H_{22}O_{11}$, a nonreducing sugar found in young mushrooms, gives only D-glucose when hydrolyzed by aqueous acid or by maltase. Methylation gives an octa-D-methyl derivative that, upon hydrolysis, yields only 2,3,4,6-tetra-O-methyl-D-glucose. What is the structure and systematic name for (+)-trehalose?

(b) (–)-Isotrehalose and (+)-neotrehalose resemble trehalose in most respects. However, isotrehalose is hydrolyzed by either emulsin or maltase, and neotrehalose is hydrolyzed only by emulsin. What are the structures and systematic names for these two carbohydrates?

(a) (+)-Trehalose is a disaccharide since it can be formed by the condensation of two monosaccharide units ($2C_6H_{12}O_6$–H_2O $= C_{12}H_{22}O_{11}$). (+)-Trehalose is a nonreducing sugar which means that it does not have a free carbonyl group and therefore does not mutarotate or react with Benedict's or Fehling's solutions. Furthermore, the disaccharide has each monosaccharide unit's anomeric carbon linked in a glycosidic bond. Acid hydrolysis yielding only D-glucoses imply that this is the monosaccharide. Maltase is an enzyme that hydrolyzes only α linkages as in maltose. C_1 is part of a formyl group and C_5 has a free hydroxyl group after methylation and hydrolysis. Therefore, (+)-trehalose consists of a-D-glucopyranose rings. Its systematic name and structure is:

α-D-glucopyranosyl-α-D-glucopyranoside

(b) (–)-Isotrehalose differs from (+)-trehalose only in that it is hydrolyzed by either emulsin or maltase. Therefore, one ring will have an α glycosidic linkage, while this linkage will be β to the other ring. The systematic name and formula of (–)-Isotrehalose are:

α-D-glucopyranosyl-β-D-glucopyranoside

Similarly, (+)-neotrehalose is hydrolyzed only by emulsin, and therefore, the glycosidic linkage is β to both rings.

β-D-glucopyranosyl-β-D-glucopyranoside

Q The disaccharide melibiose is hydrolyzed by dilute acid to a mixture of D-glucose and D-galactose. Melibiose is a reducing sugar and is oxidized by bromine water to melibionic acid, which is methylated by sodium hydroxide and dimethyl sulfate to octa-O-methylmelibionic acid. Hydrolysis of the latter gives a tetra-O-methylgluconic acid (A) and a tetra-O-methylgalactose (B). Compound A is oxidized by nitric acid to tetra-O-methylglucaric acid. Compound

B is also obtained by the acidic hydrolysis of methyl 2,3,4,6-tetra-O-methylgalactopyranoside. Melibiose is hydrolyzed by an α-galactosidase from almonds. What is the structure of melibiose?

A In a carbohydrate identification problem such as this, one first finds general characteristics of the unknown compound. Utilizing this broad data, we can discover more specific properties of the carbohydrate, eventually deducing the complete structure of the compound. A disaccharide is a carbohydrate that has two monosaccharide components. Here, we know that the monosaccharides making up the disaccharide are glucose and galactose. We are also informed that the disaccharide melibiose is a reducing sugar. This implies that one of the monosaccharides retains a hemiacetal linkage. This means that melibiose, unlike a disaccharide such as sucrose, will be in equilibrium with a carbonyl compound. However, the important questions remain to be answered: (1) Which sugar (glucose or galactose) will contain the hemiacetal linkage? (2) What is the respective size of each of the monosaccharide rings (furanose or pyranose)? (3) What is the nature of the acetal linkage between the monosaccharides (α or β) Bromine water is a reagent that specifically oxidizes the carbonyl group of a reducing sugar to a carboxylic acid functionality. The subsequently formed carboxylic acid (an aldonic acid) is exhaustively methylated. We note that the methyl groups attached to the hydroxyl oxygens will not be disrupted by hydrolysis, since these are part of ether linkages. However, the acetal linkage between the two sugars will be broken and the carboxylic acid functionality will be regenerated. For example, in the hydrolysis of the methyl ester of methylated maltobionic acid:

Hydrolysis of the methylated melibiose yields a tetra-O-methyl-gluconic acid (A) and a tetra-O-methylgalactose (B). Since the carboxylic acid is derived from glucose, it is glucose that has the hemiacetal linkage. Nitric acid is a potent oxidizing agent, and it will convert any unprotected (unmethylated) hydroxyl group to a carboxylic acid functionality. Thus, since nitric acid transforms the gluconic acid into the glucaric acid, it is the hydroxyl group on carbon number six that is involved in the acetal bond with galactose. At this point we can depict the glucopyranosyl fragment of the carbohydrate as:

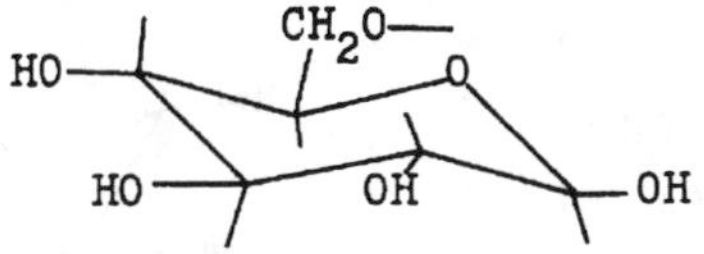

As seen previously, the acid hydrolysis of a methyl pyranoside will convert only the acetal linkage to a hydroxyl group. Since the hydroxyl group on carbon number five is not methylated, it must be involved in the intramolecular hemiacetal linkage and thus the galactose component of melibiose is a pyranose, bonding to glucose at carbon number one. Enzymes may promote the hydrolysis of carbohydrates. Since melibiose is hydrolyzed by an enzyme specific to α anomeric linkages (axial hydroxyl group at the anomeric carbon, which is the carbon involved in the hemiacetal formation), the acetal bond must be an α linkage. We designate melibiose in its Fischer projection form, Haworth projection and cyclohexane chair form in the following:

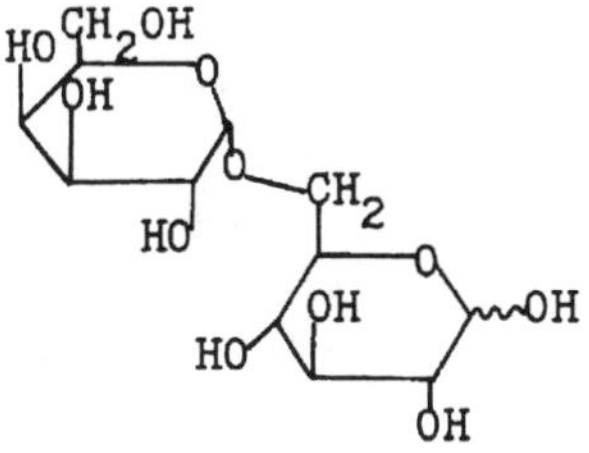

Haworth structure of melibiose

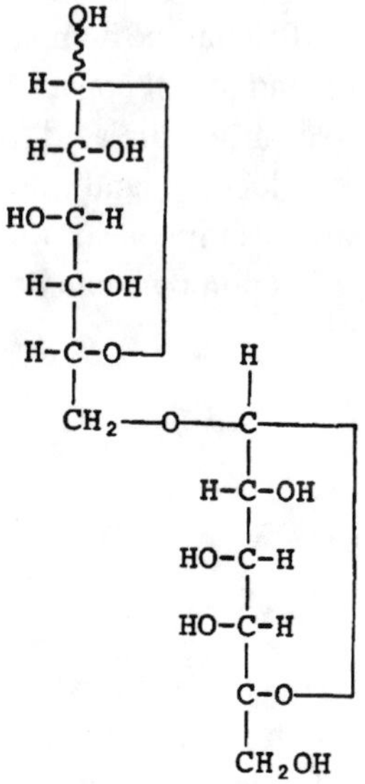

Fischer projection
of melibiose

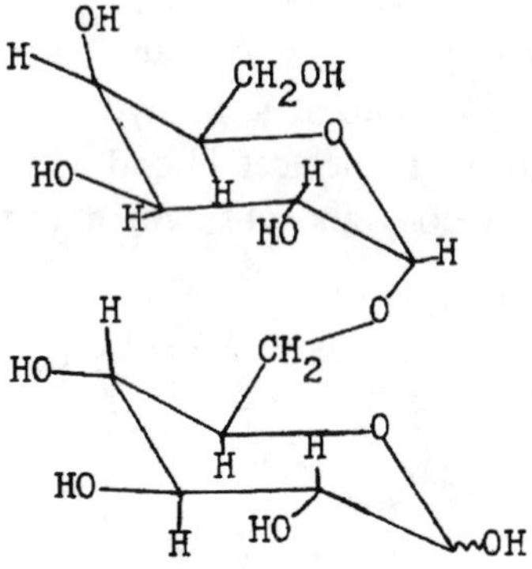

Cyclohexane conformation
of melibiose

Note the various ways in which the α anomeric linkage is designated in the different types of structures. Realizing the structure of melibiose, we can deduce compounds A and B and their related products.

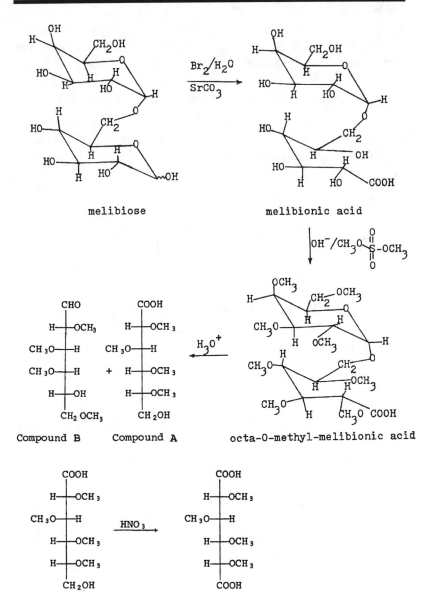

melibiose   melibionic acid

Compound B   Compound A   octa-O-methyl-melibionic acid

Compound A   tetra-O-methyl-glucaric acid

Q Paper napkins, tissues, and similar items are composed largely of cellulose. What is the structure of cellulose? At one time there was a television commercial that showed how "concentrated stomach acid" (hydrochloric acid) could rapidly destroy such paper articles. What chemical reaction was occurring during this dissolution?

A Cellulose is a large polymer found in nature that consists of β-D-glucosyl monomeric units. Its structure can be represented as:

Cellulose, like all glycosides, can be hydrolyzed in acid to its component sugars. Assuming complete hydrolysis, we depict this hydrolysis which is identical with the chemical reactions that destroy the paper articles as follows:

$$\frac{HCl}{H_2O} \longrightarrow \quad \text{n glucose units (both } \alpha \text{ and } \beta \text{)}$$

We should note that this is a chemical reaction that would occur in a beaker as well as a human stomach. The human stomach does not have β-glucosidase unlike the termite that can readily digest wood, a cellulose by-product.

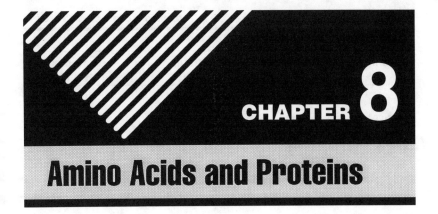

CHAPTER 8

Amino Acids and Proteins

8.1   Amino Acids

Amino acids constitute a particularly important class of difunctional compounds because they are the "building blocks" of proteins. The simplest amino acid does not contain an asymmetric carbon and hence is optically inactive. All the other amino acids contain an asymmetric carbon, can exist as enantiomers, and are optically active.

Problem Solving Example:

Q   Which of the amino acids in the table shown on the next two pages are "acidic" amino acids and which are "basic" amino acids? Which of the structures shown would have the most basic nitrogen? The least basic amino nitrogen? The most acidic and least acidic carboxyl group? Give the reasons for your choices.

Name	Abbreviation	Formula	Isoelectric point, pH units
glycine	Gly	$NH_2CH_2CO_2H$	6.0
alanine	Ala	$CH_3\underset{NH_2}{CH}CO_2H$	6.o
valine*	Val	$(CH_3)_2CH\underset{NH_2}{CH}CO_2H$	6.0
leucine*	Leu	$(CH_3)_2CHCH_2\underset{NH_2}{CH}CO_2H$	6.0
isoleucine*	Ileu	$CH_3CH_2\underset{CH_3}{CH}-\underset{NH_2}{CH}CO_2H$	6.0
phenylalanine*	Phe	$C_6H_5CH_2\underset{NH_2}{CH}CO_2H$	5.5
serine	Ser	$HOCH_2\underset{NH_2}{CH}CO_2H$	5.7
threonine*	Thr	$CH_3\underset{OH}{CH}-\underset{NH_2}{CH}CO_2H$	5.6
lysine*	Lys	$NH_2(CH_2)_4\underset{NH_2}{CH}CO_2H$	9.6
δ-hydroxylysine	Lys-OH	$NH_2CH_2\underset{OH}{CH}(CH_2)_2\underset{NH_2}{CH}CO_2H$	9.15
arginine	Arg	$\underset{NH_2}{\overset{NH}{C}}-NH(CH_2)_3\underset{NH_2}{CH}CO_2H$	11.2
apartic acid	Asp	$HO_2CCH_2\underset{NH_2}{CH}CO_2H$	2.8

[b]   refers to the lisomer. The D, L mixtures are usually less soluble.
*   must be included in diet for maintenance of proper nitrogen equilibrium in normal adult humans.

Name	Abbreviation	Formula	Isoelectric point, pH units
aparagine	Asp-NH$_2$	$NH_2COCH_2\underset{NH_2}{C}HCO_2H$	5.4
glutamic acid	Glu	$HO_2C(CH_2)_2\underset{NH_2}{C}HCO_2H$	3.2
glutamine	Glu-NH$_2$	$NH_2CO(CH_2)_2\underset{NH_2}{C}HCO_2H$	5.7
cysteine	CySh	$HSCH_2\underset{NH_2}{C}HCO_2H$	5.1
cystine	Cys	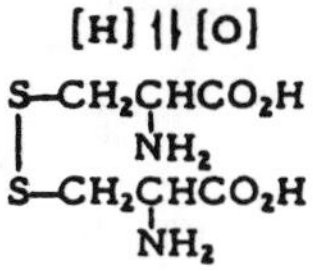 $[H] \updownarrow [O]$ $S-CH_2\underset{NH_2}{C}HCO_2H$ $\mid$ $S-CH_2\underset{NH_2}{C}HCO_2H$	5.0
methionine*	Met	$CH_3S(CH_2)_2\underset{NH_2}{C}HCO_2H$	5.7
tyrosine	Tyr	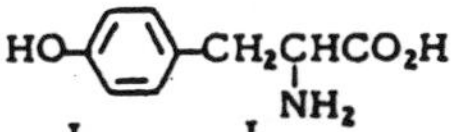 $HO-\!\!\bigcirc\!\!-CH_2\underset{NH_2}{C}HCO_2H$	5.7
thyroxine	Thy	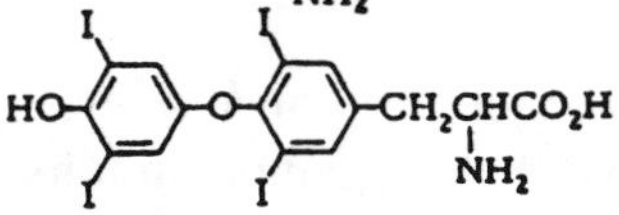	
proline	Pro	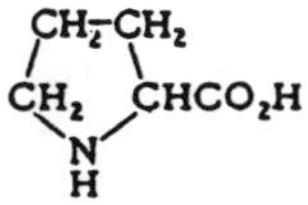	6.3
hydroxyproline	Hypro	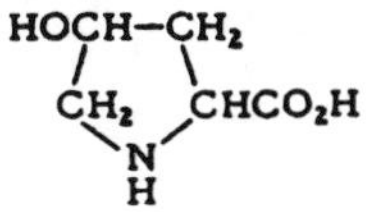	5.7
tryptophan*	Try	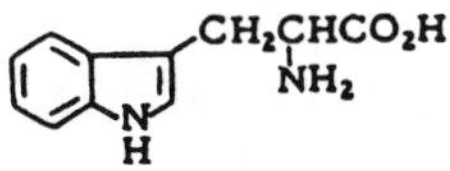	5.9
histidine	His	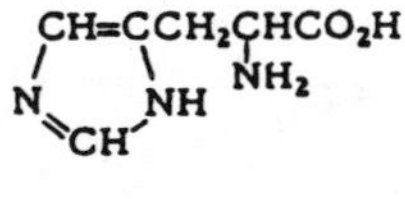	7.5

A   In order to determine the acidity and basicity of various amino acids, we must draw out their structural formulas and analyze them. The structure of a neutral amino acid is best represented by the zwitterion (dipolar ion) form:

$$R-\underset{\overset{|}{+NH_3}}{CH}-C\overset{\displaystyle O}{\underset{\displaystyle O^-}{\diagup\!\!\!\diagup}}$$

Therefore, in determining the acidity of the various amino acids, the zwitterion form will be used.

(a) glycine: $H-\underset{\overset{|}{+NH_3}}{CH}-C-O^-$ with $\overset{O}{/\!/}$

The hydrogen (R = H) has little effect on either ionic groups; therefore, glycine is approximately neutral.

(b) alanine: $CH_3-\underset{\overset{|}{+NH_3}}{CH}-C-O^-$ with $\overset{O}{/\!/}$

The methyl group, like all alkyl groups, is electron-releasing; therefore, it will stabilize the ammonium ion and destabilize the carboxylate anion:

(1) $G \leftarrow C\overset{\displaystyle O}{\underset{\displaystyle O^-}{\diagup\!\!\!\diagup}}$

G (substituent) withdraws electrons: stabilizes anion, strengthens acid.

(2) $G \rightarrow C\overset{\displaystyle O}{\underset{\displaystyle O^-}{\diagup\!\!\!\diagup}}$

G releases electrons: destabilizes anion, weakens acid.

(3) $G \leftarrow C - \overset{+}{N}H_3$

G withdraws electrons: destabilizes cation, weakens base.

(4) $G \rightarrow C - \overset{+}{N}H_3$

G releases electrons: stabilizes cation, strengthens base.

However, the inductive effect of alkyl groups is very slight and alanine is approximately neutral. Note that if the inductive effect of an electron-withdrawing substituent is very strong, it will stabilize the anion greatly and destabilize the cation greatly. This will result in the release of a proton by the carbonyl group much more readily than the acceptance of a proton by the amino group, causing the molecule, as a whole, to be acidic.

(c) valine:

$$\begin{array}{c} CH_3 \\ \diagdown \\ \quad\quad CH-CH-CH-C \overset{O}{\diagup\diagup} \\ \diagup \quad\quad | \quad\quad \diagdown \\ CH_3 \quad \overset{+}{N}H_3 \quad O^- \end{array}$$

The inductive effect of the alkyl group is again very small; therefore, valine is also approximately neutral.

(d) leucine:

$$\begin{array}{c} CH_3 \\ \diagdown \\ \quad\quad CH-CH_2-CH-C \overset{O}{\diagup\diagup}-O^- \\ \diagup \quad\quad\quad | \\ CH_3 \quad\quad \overset{+}{N}H_3 \end{array}$$

Same as alanine and valine.

(e) isoleucine: $CH_3-CH_2-CH - CH-C \overset{O}{\overset{\diagup\diagup}{}}-O^-$
$$\quad\quad\quad\quad\quad\quad\quad\quad | \quad\quad | \\ \quad\quad\quad\quad\quad\quad\quad CH_3 \quad \overset{+}{N}H_3$$

Same as alanine and valine.

(f) phenylalanine:

$$\text{C}_6\text{H}_5-\text{CH}_2-\underset{\overset{|}{\underset{\overset{+}{\text{NH}_3}}{}}}{\text{CH}}-\overset{\overset{\text{O}}{\|}}{\text{C}}-\text{O}^-$$

The benzyl group is slightly electron-withdrawing, but this effect is small and phenylalanine is approximately neutral.

(g) serine:

$$\text{HO}-\text{CH}_2-\underset{\overset{|}{\underset{\overset{+}{\text{NH}_3}}{}}}{\text{CH}}-\overset{\overset{\text{O}}{\|}}{\text{C}}-\text{O}^-$$

The alcohol group is electron-withdrawing, but the effect on the ionic groups is small; therefore, it is approximately neutral.

(h) threonine:

$$\text{CH}_3-\underset{\overset{|}{\text{OH}}}{\text{CH}}-\underset{\overset{+}{\text{NH}_3}}{\text{CH}}-\overset{\overset{\text{O}}{\|}}{\text{C}}-\text{O}^-$$

Same as serine.

(i) lysine:

$$\underset{\underset{\text{basic}}{\nwarrow}}{\text{H}_2\text{N}}\overset{\overset{}{}}{-}\text{CH}_2-\text{CH}_2-\text{CH}_2-\text{CH}_2-\underset{\overset{+}{\text{NH}_3}}{\text{CH}}-\overset{\overset{\text{O}}{\|}}{\text{C}}-\text{O}^-$$

The amino group on the side chain is basic. Therefore, lysine, as a whole, is basic.

(j) δ–hydroxylysine:

$$\underset{\underset{\text{basic}}{\nwarrow}}{\text{H}_2\text{N}}-\text{CH}_2-\underset{\overset{|}{\text{OH}}}{\text{CH}}-\text{CH}_2-\text{CH}_2-\underset{\overset{+}{\text{NH}_3}}{\text{CH}}-\overset{\overset{\text{O}}{\|}}{\text{C}}-\text{O}^-$$

Same as lysine—basic.

(k) arginine:

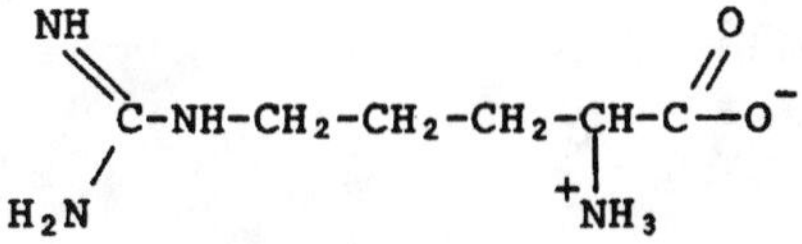

The imine group is basic; upon protonation, resonance stabilization of the cation renders arginine strongly basic.

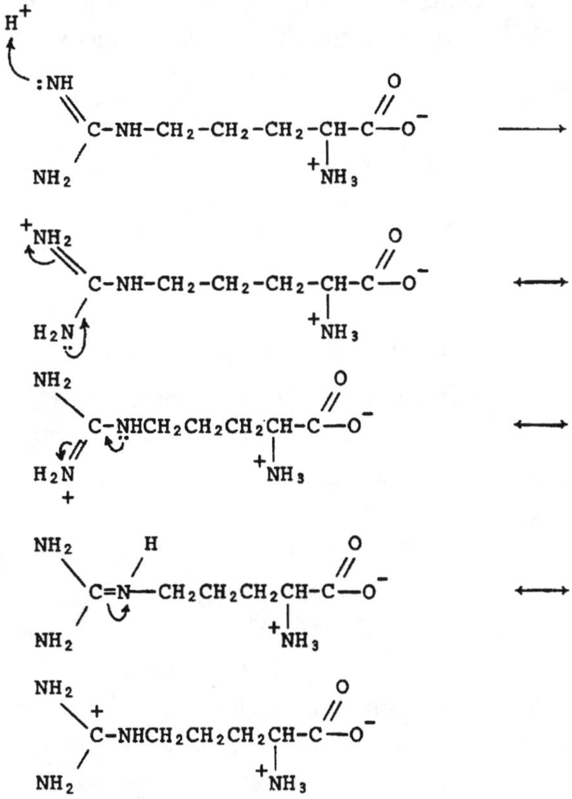

Delocalization of the positive charge causes the nitrogen on arginine to be the most basic, and hence its carboxylic group the least acidic of the amino acids in the table.

(1) aspartic acid:

$$\underset{HO}{\overset{O}{\underset{\diagup}{\overset{\diagdown\diagdown}{C}}}}-CH_2\underset{\overset{+}{NH_3}}{CH}-\overset{O}{\overset{/\!/}{C}}-O^-$$

The carboxyl group $\left(\overset{O}{\overset{/\!/}{C}}-OH\right)$ on the side chain is acidic; therefore, it should dissociate in solution, making aspartic acid an acid. Of greater importance is the strong electron-withdrawing inductive effect of the $\beta-\overset{O}{\overset{/\!/}{C}}-OH$ group:

$$\underset{HO}{\overset{O}{\underset{\diagup\beta}{\overset{\diagdown\diagdown}{C}}}}-CH_2\underset{\alpha}{CH}\underset{\overset{+}{NH_3}}{}-\overset{O}{\overset{/\!/}{C}}-O^-$$

electron-withdrawing: stabilizes anion, destabilizes cation.

This renders the carboxylic group of aspartic acid the most acidic, and hence its amino group the least basic of the amino acids in the table.

(m) asparagine:

$$NH_2-\overset{O}{\overset{||}{C}}-CH_2-\underset{\overset{+}{NH_3}}{CH}-\overset{O}{\overset{/\!/}{C}}-O^-$$

The amide group on the side chain is slightly electron-withdrawing, but this effect is small; therefore, Asparagine should be approximately neutral. Note that an amide group is not basic due to loss of resonance upon protonation:

$$\overset{\text{O}}{\underset{\text{NH}_2-\text{C}-\text{R}}{\|}} \longleftrightarrow \overset{\text{O}^-}{\underset{^+\text{NH}_2=\text{C}-\text{R}}{|}}$$

resonance-stabilized amide

$^+\text{NH}_3-\overset{\overset{\text{O}}{\|}}{\text{C}}-\text{R}$: electrons on nitrogen can no longer participate in resonance.

(n) glutamic acid:

$$\text{HO}-\overset{\overset{\text{O}}{\|}}{\text{C}}-\text{CH}_2\text{CH}_2-\overset{\overset{}{\underset{\underset{^+\text{NH}_3}{|}}{\text{CH}}}}{}-\overset{\overset{\text{O}}{\|}}{\text{C}}-\text{O}^-$$

Same as aspartic acid, except the inductive effect of the $\overset{\overset{\text{O}}{\|}}{\text{C}}-\text{OH}$ is now decreased due to the longer distance of the side chain. This will cause glutamic acid to be less acidic than aspartic acid.

(o) glutamine:

$$\text{NH}_2-\overset{\overset{\text{O}}{\|}}{\text{C}}-\text{CH}_2\text{CH}_2\overset{\overset{}{\underset{\underset{^+\text{NH}_3}{|}}{\text{CH}}}}{}-\overset{\overset{\text{O}}{/\!/}}{\text{C}}-\text{O}^-$$

Same as asparagine.

(p) cysteine:

$$\text{HSCH}_2\overset{\overset{}{\underset{\underset{^+\text{NH}_3}{|}}{\text{CH}}}}{}-\overset{\overset{\text{O}}{/\!/}}{\text{C}}-\text{O}^-$$

The thiol group (–SH) on the side chain has little effect on the acidity of this amino acid; therefore, it should be approximately neutral.

(q) cystine:

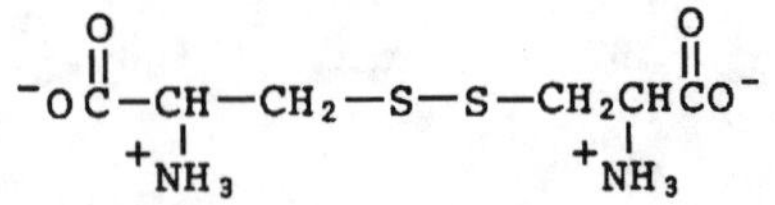

$$^-O\overset{\overset{\displaystyle O}{\|}}{C}-\underset{\overset{|}{^+NH_3}}{CH}-CH_2-S-S-CH_2\underset{\overset{|}{^+NH_3}}{CH}\overset{\overset{\displaystyle O}{\|}}{C}O^-$$

The disulfide group has little effect on the acidity of the amino acid, therefore cystine is approximately neutral.

(r) methionine:

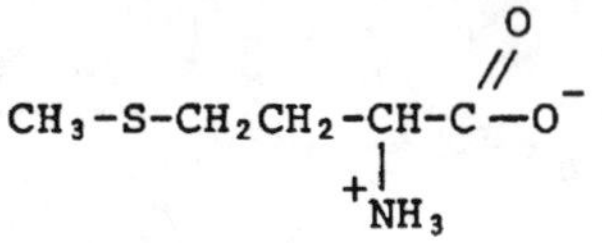

$$CH_3-S-CH_2CH_2-\underset{\overset{|}{^+NH_3}}{CH}-\overset{\overset{\displaystyle O}{/\!/}}{C}-O^-$$

The sulfide group has little effect on the acidity of the amino acid, therefore methionine is approximately neutral.

(s) tyrosine:

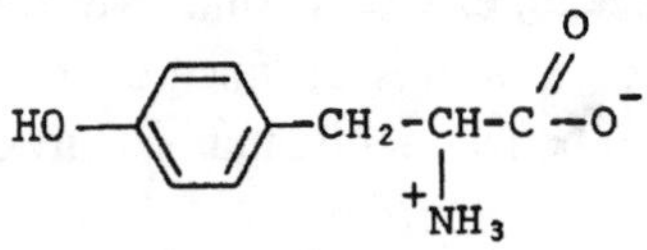

$$HO-\underset{}{\bigotimes}-CH_2-\underset{\overset{|}{^+NH_3}}{CH}-\overset{\overset{\displaystyle O}{/\!/}}{C}-O^-$$

The side chain is strongly electron-withdrawing, thereby stabilizing the anion and destabilizing the cation, causing tyrosine to be acidic.

(t) thyroxine:

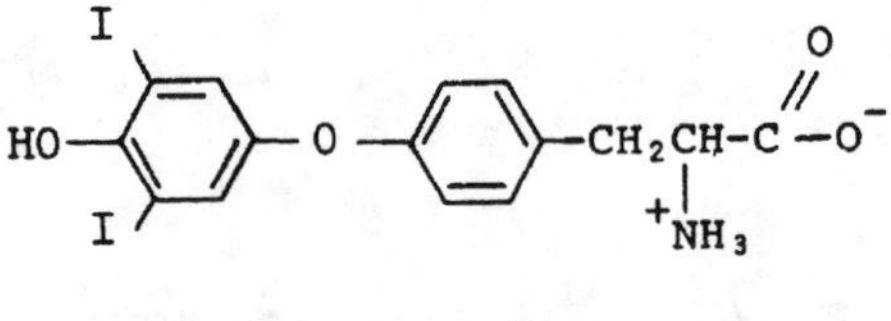

$$HO-\underset{I}{\overset{I}{\bigotimes}}-O-\bigotimes-CH_2\underset{\overset{|}{^+NH_3}}{CH}-\overset{\overset{\displaystyle O}{/\!/}}{C}-O^-$$

Same as tyrosine, acidic.

(u) proline:

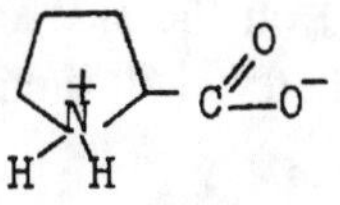

$$\underset{H \quad H}{\overset{+}{N}}\bigg\rangle-\overset{\overset{\displaystyle O}{\|}}{C}-O^-$$

The alkyl ring has little effect on the acidity; therefore, proline should be approximately neutral.

(v) hydroxyproline

$$HO \overset{}{\underset{\overset{+}{N}H_2}{\bigsqcup}} - C \overset{O}{\underset{O^-}{\diagdown}}$$

Same as proline.

(w) tryptophan:

$$\text{(indole)} CH_2 CH - C \overset{O}{\underset{O^-}{\diagup}} \\ \overset{+}{N}H_3$$

The side chain has little effect on the acidity; therefore, the amino acid is approximately neutral. Note that protonation of the side chain nitrogen would disrupt the aromatic ring and therefore does not occur:

$$\text{(indole, R, N-H)} \quad \overset{}{\nrightarrow} \quad \text{(indole, R, } \overset{+}{N}H_2)$$

Aromatic—10 π electrons (Hückel number) in two fused aromatic rings (the lone-pair electrons on nitrogen participate in the π cloud).

The five-member ring is not aromatic, since the lone-pair electrons on nitrogen can no longer participate in a cyclic π cloud.

(x) histidine:

$$\text{(imidazole)} - CH_2 - CH - C \overset{O}{\underset{}{\diagup}} - O^- \\ \overset{+}{N}H_3$$

The side chain has little effect on the acidity; therefore, the amino acid is approximately neutral. Note that (N–2) is not basic for the same reason as the nitrogen in tryptophan. The azole nitrogen (N–4) is only

slightly basic (pK$_a$ of imidazole is 7) for two reasons: (1) the

lone-pair electrons on the azole nitrogen occupy sp^2 orbitals; therefore, they are held more tightly than amino lone-pair electrons in sp^3 orbitals, rendering them much less basic; and (2) in solution, tautomers are in rapid equilibrium, rendering the two nitrogens equivalent:

tautomerism

This further lowers the basicity of the imidazole ring.

In summary:

Acidic amino acids: aspartic acid, glutamic acid, tyrosine, thyroxine.

Basic amino acids: lysine, δ-hydroxylysine, arginine, asparagine.

Most basic nitrogen: arginine.

Least basic amino nitrogen: aspartic acid.

Most acidic carboxyl group: aspartic acid.

Least acidic carboxyl group: arginine.

8.2 Properties of Amino Acids

$$\underset{\underset{NH_2}{|}}{R-CH-COOH} \qquad \underset{\underset{+NH_3}{|}}{R-CH-COO^-}$$

The unionized amino acid is converted to a dipolar amino acid by an internal hydrogen ion transfer. Amino acids occur in the dipole ion form in aqueous solution or in the solid state.

Amino acids are generally high-melting solids because of the strong intermolecular attractions that can exist between the dipolar molecules of an amino acid. Amino acids decompose, rather than melt, at fairly high temperatures. They are more soluble in water than in nonpolar organic solvents.

Amino acids in the dipolar ion form are amphoteric, which means they react with both acids and bases.

$$\underset{\underset{NH_3^+}{|}}{R-CH-COO^-} \xrightarrow{H^+} \underset{\underset{NH_3^+}{|}}{R-CH-COOH} \text{ (Acid Solution)}$$

$$\xrightarrow{OH^-} \underset{\underset{NH_2}{|}}{R-CH-COO^-} \text{ (Basic Solution)}$$

The isoelectric point of an amino acid is the hydrogen ion concentration of the solution in which the particular amino acid does not migrate under the influence of an electric field.

Neutral amino acids have isoelectric points at pH 5.5 to 6.3. Basic amino acids have isoelectric points at a high pH (around 10). Acidic amino acids have isoelectric points at a low pH (around 3).

All naturally occurring amino acids belong to the L-series configuration.

$$H_2N-\overset{\displaystyle CO_2H}{\underset{\displaystyle R}{|}}-H \qquad HO-\overset{\displaystyle CHO}{\underset{\displaystyle CH_2OH}{|}}-H$$

L-Amino acid   L-Glyceraldehyde

Problem Solving Examples:

Q In quite alkaline solution, an amino acid contains two basic groups, $-NH_2$ and $-COO^-$. Which is the more basic? To which group will a proton preferentially go as acid is added to the solution? What will the product be?

A An amino acid is any compound containing both an acid group and an amino group; however, it is generally used to mean an α-amino carboxylic acid of the general formula:

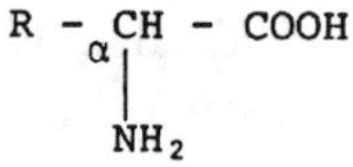

$$R - \underset{\alpha}{CH} - COOH$$
$$| $$
$$NH_2$$

Their physical and chemical properties include:

(a) They are nonvolatile crystalline solids that do not melt but rather decompose at fairly high temperatures.

(b) They are insoluble in nonpolar solvents and are appreciably soluble in water.

These properties suggest that they are salt-like molecules, and therefore, exist as zwitterions (dipolar ions) of the form:

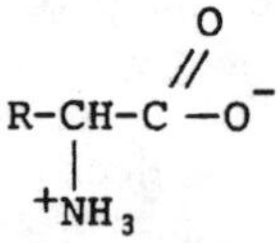

$$R-CH-\overset{O}{\overset{\|}{C}}-O^-$$
$$|$$
$$^+NH_3$$

We can quantitatively conclude from this that the $-NH_2$ group is more basic than the $-\overset{O}{\overset{\|}{C}}-O^-$ group.

(1)

$$R-CH-\overset{O}{\overset{\|}{C}}-OH \quad \underset{\longleftarrow}{\longrightarrow} \quad R-CH-\overset{O}{\overset{\|}{C}}-O^-$$
$$| \qquad\qquad\qquad\qquad\qquad\qquad |$$
$$NH_2 \quad \text{stronger acid} \qquad\qquad ^+NH_3 \quad \underline{\text{weaker base}}$$

$$\underline{\text{stronger base}} \qquad\qquad\qquad \text{weaker}$$
$$\text{acid}$$

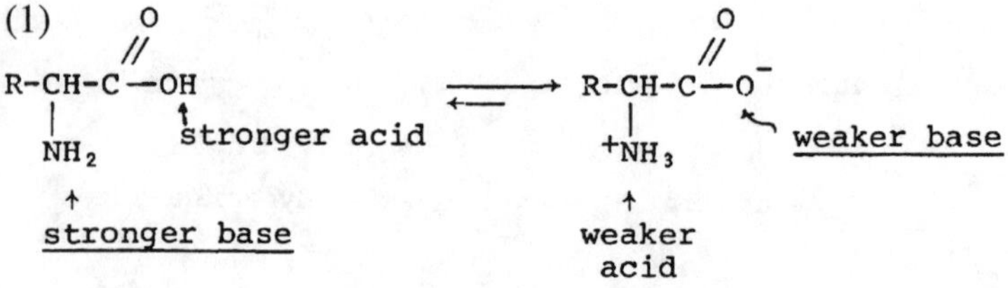

Experimental data is necessary to obtain their respective K_b's. Using glycine as an example, it has

$$K_a = 1.6 \times 10^{-10} \quad \text{and} \quad K_b = 2.5 \times 10^{-12}$$

Looking at the acid-base equations:

(2)
$$\overset{+}{H-NH_2}-\overset{R}{\underset{}{CH}}-\overset{O}{\underset{}{C}}-O^- \quad \rightleftarrows \quad H_3O^+ + H_2NCHRC-O^-$$

acid   conjugate base

(3)
$$\overset{+}{H_3NCHRC}-\overset{O}{\underset{}{C}}-O^- \quad \rightleftarrows \quad \overset{+}{H_3NCHRC}-OH \quad + \quad OH^-$$

base   conjugate acid

We see that the measured K_a actually refers to the acidity of the ammonium ion $(-\overset{+}{N}H_3)$, which is 1.6×10^{10}, while the K_b refers to the basicity of the carboxylate ion $\left(\overset{O}{\underset{}{-C}}-O^- \right)$, which is 2.5×10^{-12}. In order to get the K_b of the $-NH_2$ group, we must use the expression: $K_a \times K_b = 10^{-14}$; that is, in aqueous solution, the acidity and basicity of an acid and its conjugate base are related by the expression

$$K_a \times K_b = K_w = 10^{-14}.$$

From equation (2), we see that the conjugate base of $-\overset{+}{N}H_3$ is NH_2, therefore using the K_a value of 1.6×10^{-10} for $-NH_3$,

$$K_b = \frac{10^{-14}}{1.6 \times 10^{-10}} = 6.3 \times 10^{-5}$$

we can conclude that NH is more basic than

$$\overset{\overset{\displaystyle O}{\displaystyle \|}}{-C}-O^-:$$

$$K_b(NH_2) = 6.3 \times 10^{-5} \gg K_b \left(\overset{\overset{\displaystyle O}{\displaystyle \|}}{C}-O^- \right) = 2.5 \times 10^{-12}$$

As acid is added to the solution, the proton will preferentially go to the stronger base, $-NH_2$. The product will be the zwitterion:

$$\underset{\underset{\displaystyle H^+}{\displaystyle \longleftarrow}}{H_2\overset{..}{N}-\overset{\overset{\displaystyle R}{\displaystyle |}}{C}H-\overset{\overset{\displaystyle O}{\displaystyle \|}}{C}-O^-} \longrightarrow \overset{+}{H_3}\overset{\overset{\displaystyle R}{\displaystyle |}}{N}-\overset{}{C}H-\overset{\overset{\displaystyle O}{\displaystyle \|}}{C}-O^-$$

(a) What contributing structure(s) would account for the double-bond character of the carbon-nitrogen bond in amides?
(b) What does this resonance mean in terms of orbitals?

(a) The lone-pair electrons on nitrogen can participate in the π-orbital of the carbonyl double-bond, giving the C-N bond partial double-bond character.

$$-\overset{\overset{\displaystyle \overset{..}{O}\cdot}{\displaystyle \|}}{C}\underset{\underset{\displaystyle H}{\displaystyle |}}{\overset{..}{N}-} \longleftrightarrow -\overset{\overset{\displaystyle \overset{-}{O}}{\displaystyle /}}{C}\underset{\underset{\displaystyle H}{\displaystyle |}}{\overset{\displaystyle \|}{\overset{+}{N}-}} \equiv -\overset{\overset{\displaystyle \overset{\delta-}{O}}{\displaystyle \|}}{C}\underset{\underset{\displaystyle |}{\displaystyle }}{\overset{\displaystyle \delta+}{N}-}$$

(b) Participation in the π-orbital of the carbonyl double bond would require the lone-pair electrons of nitrogen to be in p orbitals, thereby permitting overlap:

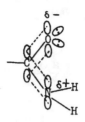

Note: Both the carbon and nitrogen are sp^2 hybridized.

8.3   Preparation of Amino Acids

A)   Ammonolysis of α-halo acids

$$CH_3CH_2COOH + Br_2 \xrightarrow{P} CH_3CHCOOH + HBr$$

propionic acid   Br

α -bromopropionic acid

$$\xrightarrow{NH_3(excess)} CH_3CHCOOH + NH_4Br$$

NH$_2$

alanine

B)   Strecker synthesis

$$RCHO + NH_3 + HCN \rightarrow RCHCN + HOH \xrightarrow{H_3O^+} RCHCOOH$$

NH$_2$ NH$_2$

α -aminonitrile   α-amino acid

C)   Acetylaminomalonate synthesis

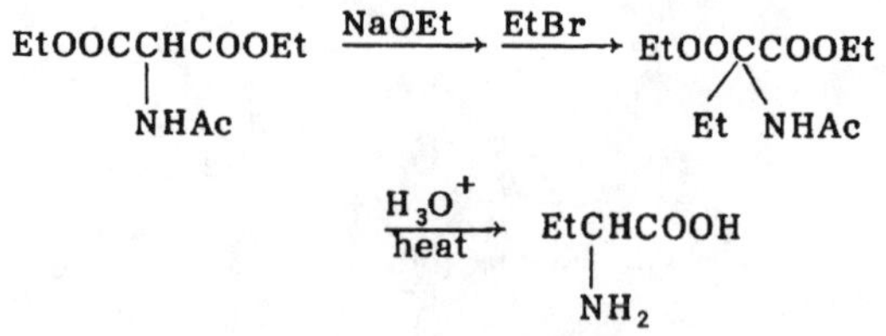

EtOOCCHCOOEt $\xrightarrow{\text{NaOEt}}$ $\xrightarrow{\text{EtBr}}$ EtOOCCOOEt
|
NHAc
Et NHAc

$\xrightarrow[\text{heat}]{\text{H}_3\text{O}^+}$ EtCHCOOH
|
NH$_2$

The starting material can be made from malonic ester.

CH$_2$(COOEt)$_2$ $\xrightarrow{\text{HONO}}$ ON–CH(COOEt)$_2$ $\xrightarrow{\text{H}_2,\text{Ni}}$

H$_2$NCH(COOEt)$_2$ $\xrightarrow{\text{Ac}_2\text{O}}$

AcNHCH(COOEt)$_2$

D) Reductive amination of keto acids

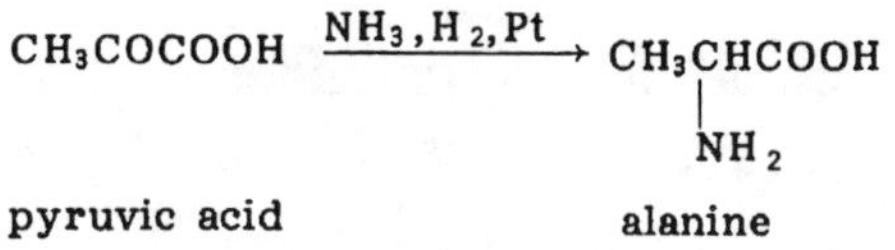

CH$_3$COCOOH $\xrightarrow{\text{NH}_3,\text{H}_2,\text{Pt}}$ CH$_3$CHCOOH
|
NH$_2$

pyruvic acid   alanine

E) The phthalimidomalonic ester method

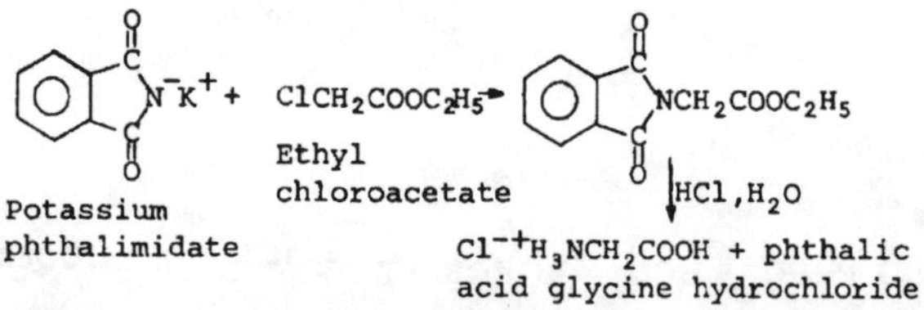

Potassium phthalimidate N^-K^+ + ClCH$_2$COOC$_2$H$_5$ → NCH$_2$COOC$_2$H$_5$

Ethyl chloroacetate

$\downarrow$HCl,H$_2$O

Cl^{-+}H$_3$NCH$_2$COOH + phthalic acid glycine hydrochloride

Problem Solving Example:

Show how the following amino acids can be prepared from the indicated starting materials.

(a) Leucine from isobutyl alcohol

(b) Lysine from 1,4-dibromobutane

(c) Proline from adipic acid

(d) Glutamic acid from α-ketoglutaric acid

A In order to synthesize an amino acid from the indicated starting material, we must analyze the product and see how it is derived from the starting material.

(a)

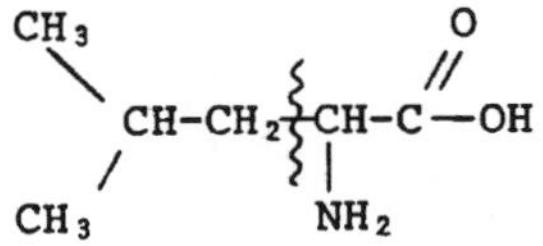

$$CH_3-CH(CH_3)-CH_2-CH(NH_2)-C(=O)-OH$$

isobutyl

leucine

Since the side chain is an isobutyl group, we must attach an amino acid root to the isobutyl alcohol. This can be accomplished via the phthalimidomalonic ester synthesis:

(1)

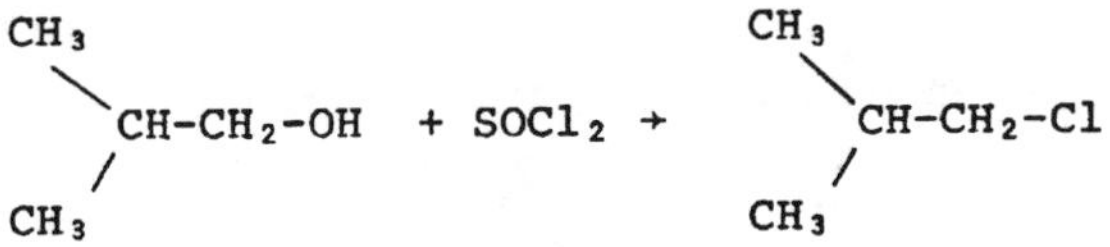

$$CH_3-CH(CH_3)-CH_2-OH \ + \ SOCl_2 \ \rightarrow \ CH_3-CH(CH_3)-CH_2-Cl$$

Substitution reaction to form the corresponding alkyl halide.

(2)

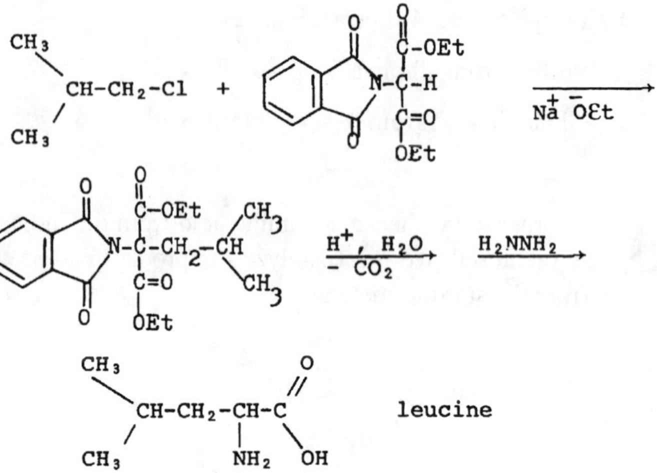

leucine

This is a modified malonic ester synthesis made possible by (1) the high acidity of the α-hydrogens of malonic ester and (2) the extreme ease with which malonic acid and substituted malonic acids undergo decarboxylation.

(b) 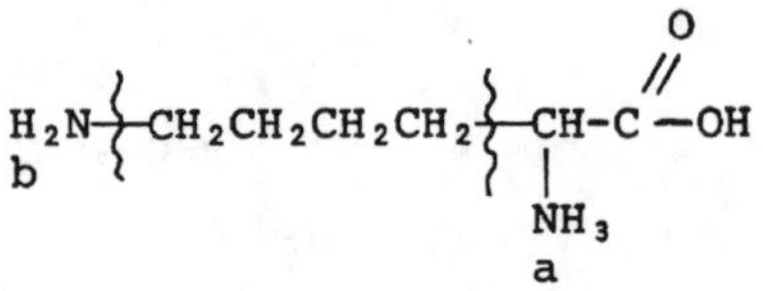 $H_2N \}_b CH_2CH_2CH_2CH_2 \}$ CH-C—OH   lysine

with NH_3 (a) and O and O.

Substituent a can be attached to the n-butyl group by phthalimidomalonic ester synthesis, while substituent b can be attached by a substitution reaction with ammonia with the carboxylic acid group protected. Therefore:

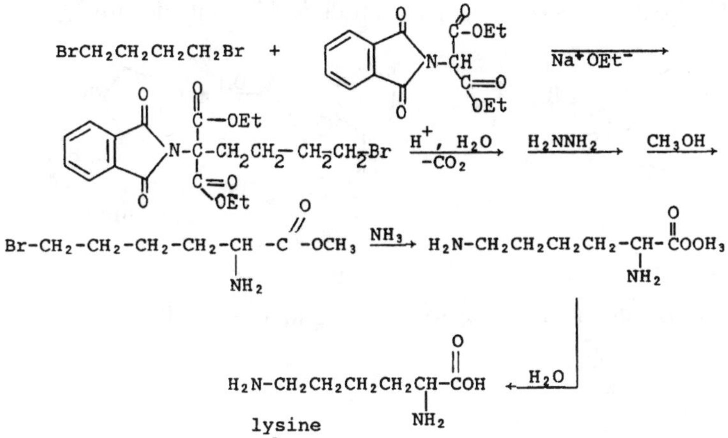

$$BrCH_2CH_2CH_2CH_2Br \quad + \quad \text{(phthalimide derivative)} \quad \xrightarrow{Na^+OEt^-}$$

$$\text{(phthalimide intermediate)} \quad \xrightarrow[-CO_2]{H^+,\ H_2O} \quad \xrightarrow{H_2NNH_2} \quad \xrightarrow{CH_3OH}$$

$$Br-CH_2-CH_2-CH_2-CH_2-\underset{NH_2}{CH}-\overset{O}{\overset{\|}{C}}-OCH_3 \quad \xrightarrow{NH_3} \quad H_2N-CH_2CH_2CH_2CH_2-\underset{NH_2}{CH}-\overset{O}{\overset{\|}{C}}OOH_3$$

$$\underset{\text{lysine}}{H_2N-CH_2CH_2CH_2CH_2\underset{NH_2}{CH}-\overset{O}{\overset{\|}{C}}OH} \quad \xleftarrow{H_2O}$$

(c)

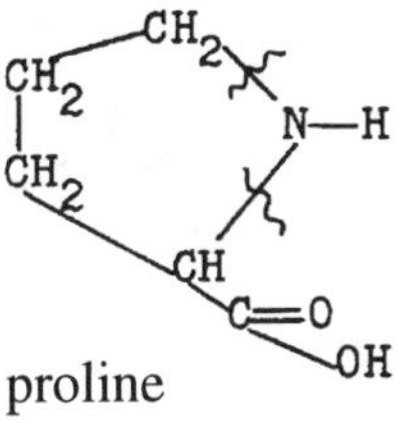

proline

Proline can be synthesized from adipic acid by substituting an amine group for a carboxyl group and ring closure at the α-carbon to the carboxyl group. Part (1) can be accomplished by the Schmidt reaction, while part (2) can be accomplished by α-halogenation followed by ring closure.

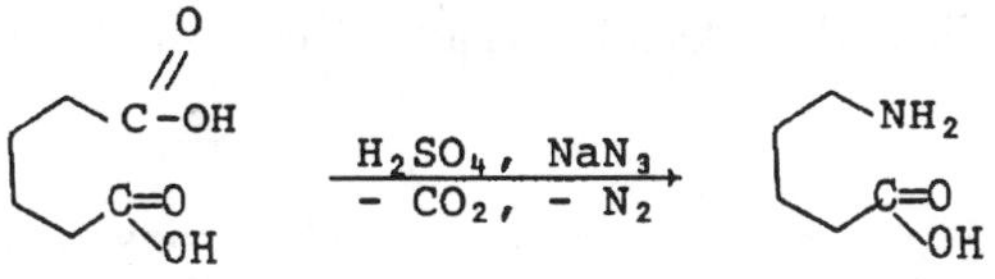

adipic acid

$$\xrightarrow[-CO_2,\ -N_2]{H_2SO_4,\ NaN_3}$$

This is the Schmidt reaction of the general type:

$$R-\overset{O}{\overset{\|}{C}}-OH \quad \xrightarrow[H_2SO_4]{NaN_3} \quad RNH_2 + N_2 + CO_2$$

The next step is α-halogenation followed by ring closure:

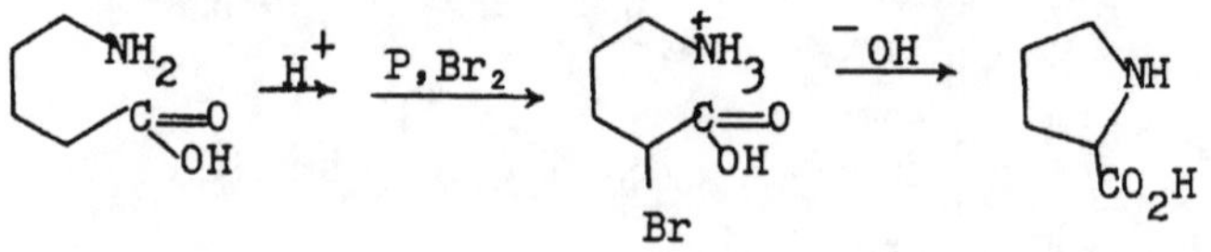

proline

(d) 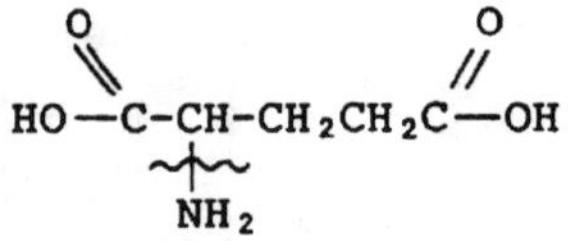 $HO-\overset{O}{\overset{\|}{C}}-\underset{\underset{NH_2}{|}}{CH}-CH_2CH_2\overset{O}{\overset{\|}{C}}-OH$ glutamic acid

One method of synthesizing glutamic acid from α-ketoglutaric acid is via the Strecker synthesis of the general type:

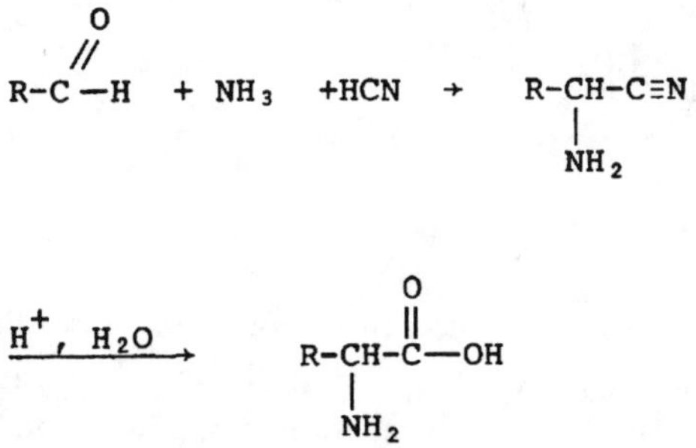

$$R-\overset{O}{\overset{\|}{C}}-H \ + \ NH_3 \ + HCN \ \rightarrow \ R-\underset{\underset{NH_2}{|}}{CH}-C\equiv N$$

$$\xrightarrow{H^+, \ H_2O} \ R-\underset{\underset{NH_2}{|}}{CH}-\overset{O}{\overset{\|}{C}}-OH$$

This is followed by decarboxylation. Therefore:

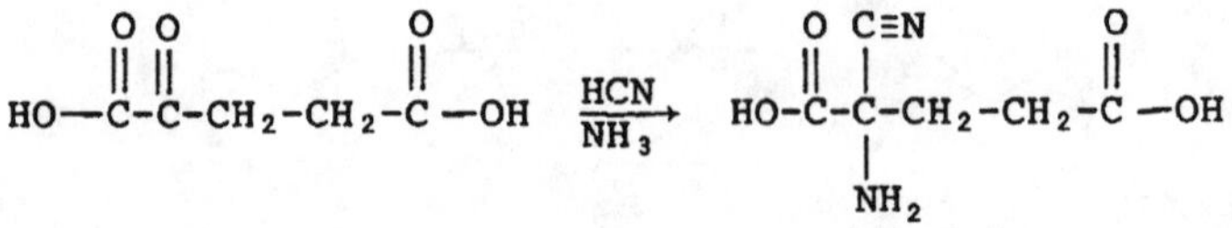

$$HO-\overset{O}{\overset{\|}{C}}-\overset{O}{\overset{\|}{C}}-CH_2-CH_2-\overset{O}{\overset{\|}{C}}-OH \ \xrightarrow{\underset{NH_3}{HCN}} \ HO-\overset{O}{\overset{\|}{C}}-\underset{\underset{NH_2}{|}}{\overset{\overset{C\equiv N}{|}}{C}}-CH_2-CH_2-\overset{O}{\overset{\|}{C}}-OH$$

α-ketoglutaric acid

$$\xrightarrow{\text{H}^+, \text{ H}_2\text{O}} \quad \underset{\text{NH}_2}{\text{HO}-\text{C}-\text{CH}-\text{CH}_2\text{CH}_2-\text{C}-\text{OH}} \quad \xrightarrow[-\text{ CO}_2]{\Delta} \quad \underset{\text{HO} \quad \text{NH}_2}{\text{C}-\text{CH}-\text{CH}_2\text{CH}_2\text{C}-\text{OH}}$$

glutamic acid

8.4 Reactions of Amino Acids

A) Acid-base equilibria

$$\underset{\text{NH}_2}{\text{CH}_3\text{CHCOO}^-} \quad \xrightarrow[\text{OH}^-]{\text{H}^+} \quad \underset{\text{NH}_3^+}{\text{CH}_3\text{CHCOO}^-} \quad \xrightarrow[\text{OH}^-]{\text{H}^+} \quad \underset{\text{NH}_3^+}{\text{CH}_3\text{CHCOOH}}$$

B) Acylation

$$\text{H}_3\text{N}-\text{CH}_2-\overset{\text{O}}{\overset{\|}{\text{C}}}\text{O}^- \quad \xrightarrow{^-\text{OH}} \quad \text{H}_2\text{N}-\text{CH}_2-\overset{\text{O}}{\overset{\|}{\text{C}}}\text{O}^-$$

$$\xrightarrow{\text{R}\overset{\text{O}}{\overset{\|}{\text{C}}}\text{Cl}} \quad \text{R}-\overset{\text{O}}{\overset{\|}{\text{C}}}-\text{NH}-\text{CH}_2-\overset{\text{O}}{\overset{\|}{\text{C}}}\text{OH}$$

C) Esterification

a) $$\underset{\text{R}}{\text{H}_3\overset{+}{\text{N}}-\text{CH}-\overset{\text{O}}{\overset{\|}{\text{C}}}\text{O}^-} \quad \xrightarrow[\underset{\text{CH}_3\overset{\|}{\text{C}}\text{Cl}}{\overset{\text{O}}{}}]{\text{NaOH}} \quad \text{CH}_3\overset{\text{O}}{\overset{\|}{\text{C}}}\text{NH}\underset{\text{R}}{\text{CH}}-\overset{\text{O}}{\overset{\|}{\text{C}}}\text{OH}$$

$$\xrightarrow{\text{SOCl}_2} \quad \xrightarrow{\text{R}'\text{OH}} \quad \text{CH}_3\overset{\text{O}}{\overset{\|}{\text{C}}}\text{NH}\underset{\text{R}}{\text{CH}}-\overset{\text{O}}{\overset{\|}{\text{C}}}\text{OR}'$$

b) $$\text{H}_3\overset{+}{\text{N}}\text{CH}_2\overset{\text{O}}{\overset{\|}{\text{C}}}\text{O}^- \quad \xrightarrow{\text{HCl}} \quad \text{H}_3\overset{+}{\text{N}}\text{CH}_2\overset{\text{O}}{\overset{\|}{\text{C}}}\text{OH}$$

$$\xrightarrow[\text{HCl}]{\text{CH}_3\text{OH}} \quad \text{H}_3\overset{+}{\text{N}}\text{CH}_2\overset{\text{O}}{\overset{\|}{\text{C}}}\text{OCH}_3$$

c)
$$R-CH(NH_2)-COOH + CH_3CH_2OH \xrightarrow{H^+} R-CH(NH_2)-COOCH_2CH_3 + H_2O$$

D) Reaction with nitrous acid (Van Slyke method)

$$R-CH(NH_2)-COOH \xrightarrow[H_2O]{HNO_2} R-CH(OH)-COOH + N_2 \uparrow$$

E) Reaction with ninhydrin (Quantitative test)

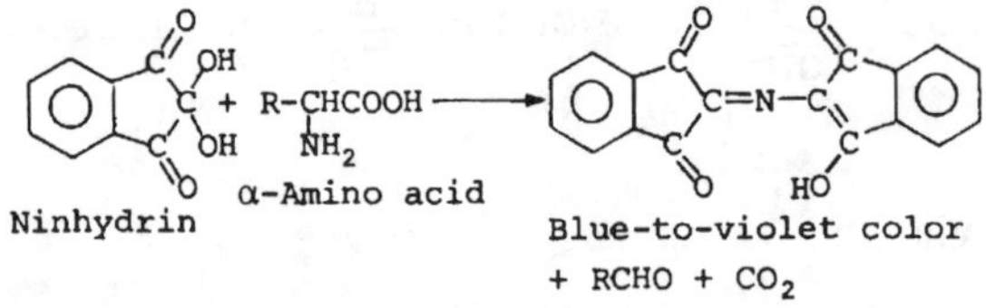

Ninhydrin + α-Amino acid → Blue-to-violet color + RCHO + CO_2

F) Cyclic amides (or lactams)

a) α-amino acids form diketopiperazines.

Example

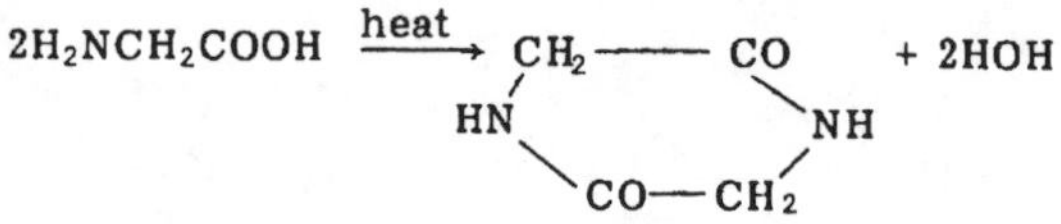

$$2H_2NCH_2COOH \xrightarrow{heat} \text{(diketopiperazine)} + 2HOH$$

b) β-amino acids react with ammonia to form unsaturated acids.

Example

$$H_2NCH_2CH_2COOH \rightarrow CH_2=CHCOOH + NH_3$$

c) γ-amino acids form lactams.

Example

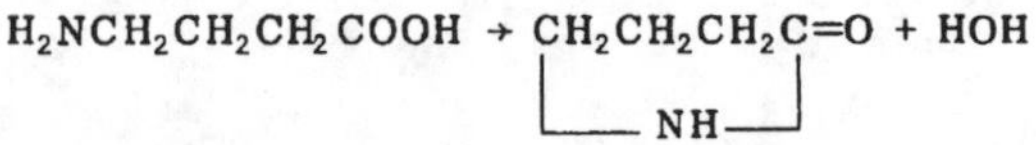

$$H_2NCH_2CH_2CH_2COOH \rightarrow \text{(lactam)} + HOH$$

Problem Solving Examples:

Q Suggest a way to separate a mixture of amino acids into three fractions: monoamino monocarboxylic acids, monoamino dicarboxylic acids (the acidic amino acids), and diamino monocarboxylic acids (the basic amino acids).

A The separation of a mixture of amino acids can be accomplished in quite a few ways. Perhaps the method most often used is ion-exchange chromatography. This method uses a column of sulfonated polystyrene resin, a cation exchanger (other cation-exchange resins and anion-exchange resins may also be used), into which the amino acid mixture is placed at pH 3. Under these conditions, the individual amino acids are positively charged and will displace some of the sodium ions from the $-SO_3^-Na^+$ form of the column. Hence, the amino acids are bound to the column material by electrostatic forces. Since the most basic amino acids have the most positive charge; they will be bound most tightly, while the least basic amino acids are bound least tightly. The column is then developed by gradually increasing the pH and ionic strength of the buffers with which the column is washed. This causes a gradual neutralization of the positive charges on the amino acids, and thereby weakens the salt linkages of the amino acids to the column material. Therefore, as the column is developed, the acidic amino acids (e.g., aspartic acid—a monoamino dicarboxylic acid) are the first ones to be removed, followed by the neutral amino acids (e.g. glycine-a monoamino monocarboxylic acid), followed, finally, by the basic amino acids (e.g., glysine—a diamino monocarboxylic acid). As the amino acids are eluted from the column, they are collected individually in separate flasks. Once separated, the individual amino acids are quantitated by reaction with ninhydrin and measurement of the resultant color:

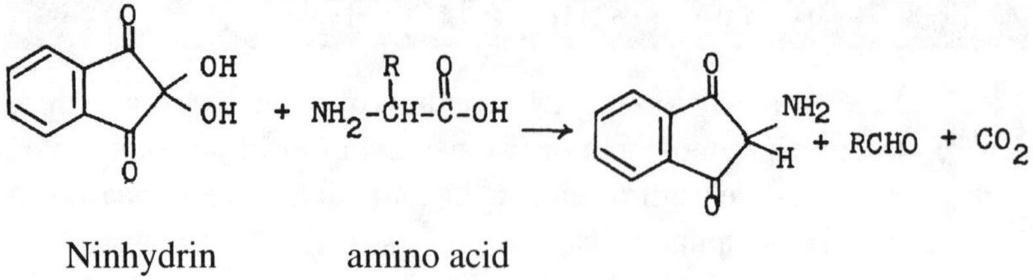

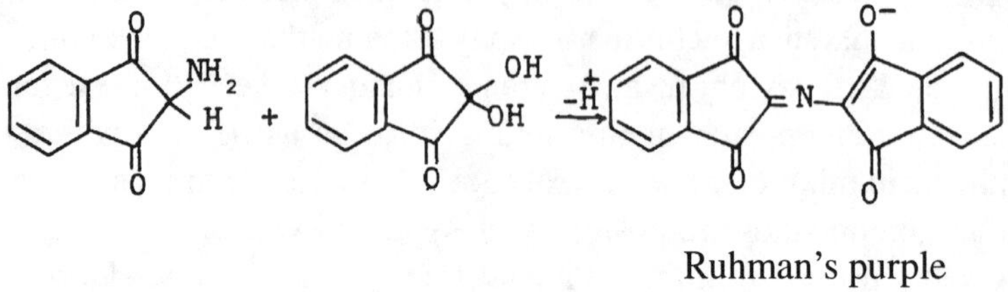

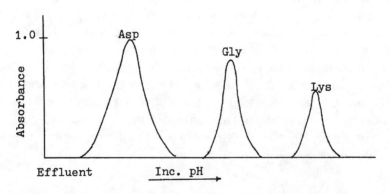

Ninhydrin   amino acid

Ruhman's purple

This absorption is nearly a linear function of the amount of amino acids originally present; therefore, a quantitative colorimetric assay is possible. Automatic amino acid analyzers, some of which use this principle, are commercially available (see figure below).

Amino Acid Chromatogram by an Automatic Amino Acid Analyzer

Other amino acid separation techniques include:

(1) Paper chromatography: Amino acids are separated by their differences in partition coefficients between water and an organic solvent. The amino acid mixture is applied to an appropriate paper support, and then irrigated with a developing liquid composed of aqueous and organic components. The organic phase migrates up the paper by capillary action while the aqueous phase is stationary. Therefore, amino acids that are more soluble in the organic phase move more rapidly than those that are less soluble. In two-dimensional chromatography, the chromatogram is developed first in one direction with one solvent system, then rotated 90°, and developed a second time with a second solvent system. The individual amino acids are located by the use of ninhydrin (see figure below).

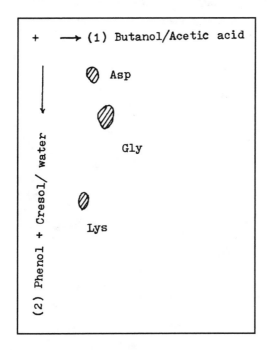

Two-Dimensional Chromatogram of an Amino Acid Mixture

(2) Electrophoresis: an Amino acid mixture on a suitable support is placed in an electric field. Due to the charge differences between the amino acids at appropriate values of pH, different amino acids will migrate differently in the electric field. For example, at neutral pH, acidic amino acids (e.g., aspartic acid—monoamino dicarboxylic acid) will migrate toward the anode, neutral amino acids (e.g., glycine—a monoamino monocarboxylic acid) will migrate to an area between the anode and the cathode, while the basic amino acids (e.g., lysine—a diamino monocarboxylic acid) will migrate toward the cathode.

 Propose a mechanism for the reaction between ninhydrin and an α-amino acid. Show all intermediates.

 Ninhydrin reagent produces a blue color upon reaction with α amino acids. The reaction proceeds as follows:

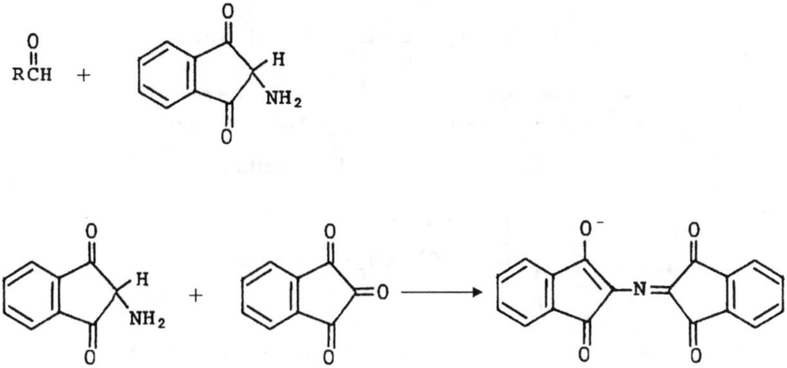

blue color

8.5 Peptides

Peptides are amides formed by interaction between amino groups and carboxyl groups of amino acids. The amide group $-\overset{\overset{\displaystyle O}{\|}}{C}-NH$ is referred to as the peptide linkage.

A combination of two amino acids would give a dipeptide, three amino acids would give a tripeptide, and so on. A polypeptide is produced when a large number of amino acids are joined by peptide linkages.

The N-terminal amino acid residue that contains the free amino group is by convention written at the left end. The C-terminal amino acid residue that contains the free carboxyl group is usually written at the right end.

Example

$$NH_2—\underset{\underset{\text{N-terminal residue}}{\underbrace{\hspace{3cm}}}}{\overset{\overset{R}{|}}{CH}—\underset{\underset{O}{\|}}{C}}—NH—\overset{\overset{R'}{|}}{CH}—\underset{\underset{O}{\|}}{C}—\underset{\underset{\text{C-terminal residue}}{\underbrace{\hspace{3cm}}}}{NH—\overset{\overset{R''}{|}}{CH}—COOH}$$

A typical tripeptide

$$^+H_3N\overset{\overset{O}{\|}}{\underset{\underset{R}{|}}{CHC}}—(NH\underset{\underset{R}{|}}{CH}CO)_n-NH\underset{\underset{R}{|}}{CH}COO^-$$

A polypeptide (n=1,2,3...)

When naming peptides, the amino acids present are listed starting with the N-terminal amino acid and going along the chain to the C-terminal amino acid. The amino acid suffix "-ine," is replaced by the suffix "-yl" for all the amino acids, except the C-terminal residue. For example,

$$NH_2—\underset{\underset{\text{Alanine}}{\underbrace{\hspace{2cm}}}}{\overset{\overset{CH_3}{|}}{CH}—\underset{\underset{O}{\|}}{C}}—\underset{\underset{\text{Glycine}}{\underbrace{\hspace{2cm}}}}{NH—CH_2—COOH} \qquad \text{Alanylglycine}$$

$$NH_2CH_2—\underset{\underset{\text{Glycine}}{\underbrace{\hspace{1.5cm}}}}{\underset{\underset{O}{\|}}{C}}—NH—\underset{\underset{\text{Serine}}{\underbrace{\hspace{1.5cm}}}}{\overset{\overset{CH_2OH}{|}}{CH}—\underset{\underset{O}{\|}}{C}}—\underset{\underset{\text{Cysteine}}{\underbrace{\hspace{1.5cm}}}}{NH—\overset{\overset{CH_2SH}{|}}{CH}—COOH}$$

Glycylserylcysteine

The structural formula of a peptide can be represented by using the standard abbreviations of the constituent amino acids.

Example

Isoleucyllysylmethionyltyrosine (Ileu-Lys-Meth-Tyr)

Glutathione (glutamylcysteinylglycine) (Glu-CySH-Gly)

Glycylvalylcysteinylproline (Gly-Val-CySH-Pro)

Problem Solving Examples:

The complete structure of gramicidin S, a polypeptide with antibiotic properties, has been worked out as follows:

(a) Analysis of the hydrolysis products gave an empirical formula of Leu,Orn,Phe,Pro,Val. (Ornithine, Orn, is a rare amino acid of formula $^+H_3NCH_2CH_2CH_2CH(NH_2)CO_2^-$.) It is interesting that the phenylalanine has the unusual D-configuration.

Measurement of the molecular weight gave an approximate value of 1,300. On this basis, what is the molecular formula of gramicidin S?

(b) Analysis for the C-terminal residue was negative; analysis for the N-terminal residue using DNFB yielded only $DNPNHCH_2CH_2CH_2CH(N^+H_3)COO^-$. What structural feature must the peptide chain possess?

(c) Partial hydrolysis of gramicidin S gave the following di- and tripeptides:

Leu.Phe   Phe.Pro   Phe.pro.Val   Val .Orn.Leu

Orn.Leu   Val   Om/Pro.Val.Orn

What is the structure of gramicidin S?

(a) The molecular formula of any compound is an integral multiple of the compound's empirical formula. Hence, the molecular weight of a compound is an integral multiple of the molecular weight corresponding to the empirical formula. In the problem we are given the empirical formula and the approximate molecular weight of gramicidin S. By calculating the molecular weight of the empirical formula, the molecular formula of gramicidin S can be determined. This is done as follows:

Empirical formula of gramicidin S:

Leu Orn Phe Pro Val

Molecular weight (MW) of empirical formula =

MW Leu + MW Orn + MW Phe + MW Pro + MW Val

MW Leucine $\left(\begin{array}{c} \text{COOH} \\ | \\ \text{NH}_2\text{CH-C(CH}_3)_3 \end{array}\right)$ = 131

MW Ornithine $\left(\begin{array}{c} \text{COOH} \\ | \\ \text{NH}_2\text{(CH}_2)_3\text{CHNH}_2 \end{array}\right)$ = 132

MW Phenylalanine $\left(\begin{array}{c} \text{COOH} \\ | \\ \text{NH}_2\text{CH-CH}_2\text{C}_6\text{H}_5 \end{array}\right)$ = 165

MW Proline $\left(\begin{array}{c} \text{NH} \\ \text{COOH} \end{array}\right)$ = 115

MW Valine $\left(\begin{array}{c} \text{COOH} \\ | \\ \text{NH}_2\text{CH-CH(CH}_3)_2 \end{array}\right)$ = 117

MW of gramicidin S empirical formula = 660

MW of gramicidin S ~ 1,300

$$\frac{\text{MW of gramicidin S}}{\text{MW of gramicidin S empirical formula}} \sim \frac{2}{1}$$

We find that the MW of gramicidin S is approximately twice that of the empirical formula. Hence, the molecular formula of gramicidin S must be twice the empirical formula:

2 X empirical formula = molecular formula

Therefore, gramicidin S has a molecular formula of:

$\text{Leu}_2\text{Orn}_2\text{Phe}_2\text{Pro}_2\text{Val}_2$.

(b) The negative result of the analysis for the C-terminal residue indicates that there is no free carboxylic acid group in gramicidin S.

DNFB (2,4-dinitrofluorobenzene or Sanger's reagent) reacted only with the ornithine portions of gramicidin S. The product was the dinitrophenyl derivative of ornithine at the δ (not the α) amino group. This means that the α amino group of ornithine does not exist as an amino group in gramicidin S. Hence, there are no C-terminal amino acid units or N-terminal amino acid units in gramicidin S. This suggests that gramicidin S is not a linear decapeptide, but is a cyclic one.

(c) The structure of gramicidin S may be deduced by analyzing the products of its partial hydrolysis. The peptide fragments must be arranged in a sequence that has identical amino acid units, overlapping with each other. To have a continuous array of amino acid units we must start and finish the sequence with the same amino acid. The peptide fragments that resulted from the partial hydrolysis of gramicidin S can be arranged in an overlapping manner as follows:

Leu • Phe

    Phe • Pro

    Phe • Pro • Val

              • Val • Orn • Leu

                   Orn • Leu

          • Val • Orn

       Pro • Val • Orn

The resulting sequence is:

Leu • Phe • Pro • Val • Orn • Leu

There is still one unit each of Phe, Pro, Val, and Orn to be accounted for. The only sequence that is in accord with the partial hydrolysis products is Phe • Pro • Val • Orn. This is concluded from the following arrangement of peptide fragments:

Phe	•	Pro				
Phe	•	Pro	•	Val		
				Val	•	Orn
		Pro	•	Val	•	Orn

The two major peptide fragments must be joined in a manner that is consistent with all the data, that is, the Leu • Phe • Pro • Val • Orn • Leu sequence must be continuous throughout the molecule. Hence, the cyclic decapeptide, gramicidin S, has the structure of

```
      Leu - Phe - Pro - Val
     /                      \
   Orn                       Orn
     \                      /
      Val - Pro - Phe - Leu
```

Q A pentapeptide on complete hydrolysis yields three moles of glycine, one mole of alanine, and one mole of phenylalanine. Among the products of partial hydrolysis are found H•Ala•Gly•OH and H•Gly•Ala•OH. What structures are possible for this substance on the basis of its giving no nitrogen in the Van Slyke determination?

A The structure of a peptide may be deduced by analyzing the complete and partial hydrolysis products. The products of complete hydrolysis of the peptide will give us the relative number of each amino acid in the peptide. Typically, complete hydrolysis is accomplished by exposing the amino acid to .03 N HCl at 11°C over a long period of time (24–28 hrs.). Partial hydrolysis, on the other hand, may he brought about by .03N HCl over shorter periods of time, or acid-specific enzyme peptide cleavers. Complete hydrolysis of the pentapeptide (a peptide consisting of five amino acid units) yielded three moles of glycine (NH_2CH_2COOH), one mole of alanine (NH_2CHCH_3COOH) and one mole of phenylalanine $\left[\begin{array}{c} NH_2CHCOOH \\ | \\ CH_2C_6H_5 \end{array}\right]$. From this data we can conclude that the pentapeptide is composed of three glycine units, one alanine unit, and one phenylalanine unit.

The structure of the pentapeptide can be deduced by analyzing the peptide fragments that result from the partial hydrolysis of the pentapeptide. Two of the peptide fragments produced in the partial hydrolysis of the pentapeptide are H•Ala•Gly•OH and H•Gly•Ala•OH, where the N or amino terminal is symbolized by H• and the C or carboxyl terminal is symbolized by •OH. Hence, the two peptide fragments may be represented as:

$$NH_2-CH-\overset{\overset{\displaystyle O}{\|}}{C}-NH-CH_2-\overset{\overset{\displaystyle O}{\|}}{C}OH$$
$$\underset{\displaystyle CH_3}{|}$$

H•Ala•Gly•OH

$$NH_2-CH_2-\overset{\overset{\displaystyle O}{\|}}{C}-NH-CH-\overset{\overset{\displaystyle O}{\|}}{C}OH$$
$$\underset{\displaystyle CH_3}{|}$$

H•Gly•Ala•OH

Since there were two types of alanine-glycine peptide fragments, one with alanine as the N-terminal amino acid and the other with alanine as the C-terminal amino acid, alanine must be bonded to two glycine portions in the pentapeptide. This means that we have accounted for a glycine-alanine-glycine portion, which can be written as:

$$-NH-CH_2-\overset{\overset{\displaystyle O}{\|}}{C}-NH-CH-\overset{\overset{\displaystyle O}{\|}}{C}-NH-CH_2-\overset{\overset{\displaystyle O}{\|}}{C}-$$
$$\underset{\displaystyle CH_3}{|}$$

•Gly•Ala•Gly•

The pentapeptide is composed of three glycine, one alanine, and one phenylalanine, which leaves us with one glycine unit and one phenylalanine unit to account for.

We are given additional information in this problem; the pentapeptide gave no nitrogen in the Van Slyke determination. The Van Slyke determination involves the treatment of an amino acid or a peptide with nitrous acid. Amino acids and peptides will react with nitrous acid as primary, secondary, and/or tertiary amines. Primary aliphatic amines react with nitrous acid to produce nitrogen gas via the diazonium salt.

Secondary amines react to produce the nitroso compound and do not give off nitrogen gas. Tertiary amines do not react with nitrous acid. All linear properties have an N-terminal amino acid unit that makes it a primary amine. Hence, all linear peptides will react with nitrous acid to release nitrogen gas. Since the pentapeptide does not react with nitrous acid to produce nitrogen, it cannot be a linear peptide, or more specifically, it is a peptide that lacks a free amino group. We have already accounted for a glycine-alanine-glycine portion. The pentapeptide must be a cyclic peptide in order to be consistent with the data. The cyclic peptide will have only secondary amino portions and will therefore react with nitrous acid to produce the corresponding penta-nitroso compound and release no nitrogen. There are two possible cyclic pentapeptide compounds that differ with respect to the positions of the remaining glycine and phenylalanine units. The two structural possibilities for the pentapeptide are:

```
        Ala                      Ala
       /   \                    /   \
    Gly     Gly             Gly       Gly
       \   /                   \     /
     Gly —Phe                Phe — Gly

        A                        B
```

Note that structures A and B are stereoisomers and differ only with respect to orientation.

8.6   Structure Determination of Peptides

The structural formula of peptides is determined by a combination of terminal residue analysis and partial hydrolysis.

Terminal residue analysis identifies the amino acid residues at the ends of the peptide chain. These procedures are used because the N- and C-terminal residues (which are at their respective ends of the peptide structure) differ from each other as well as from all the other residues.

A very successful method used to identify the N-terminal residue is pictured below:

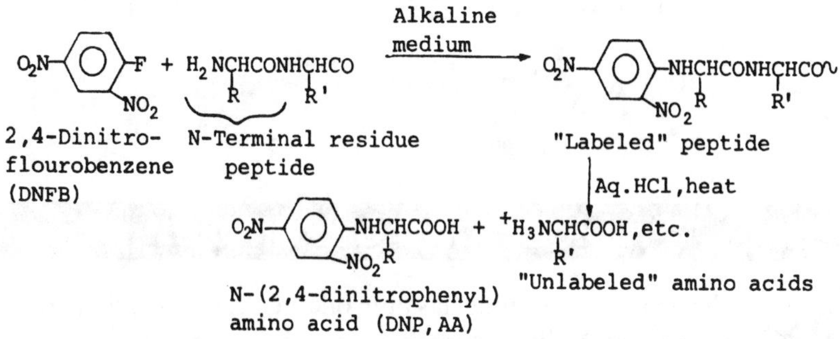

O_2N—⟨○⟩—F + H_2NCHCONHCHCO — 2,4-Dinitro-flourobenzene (DNFB) / N-Terminal residue peptide →(Alkaline medium)→ "Labeled" peptide

N-(2,4-dinitrophenyl) amino acid (DNP, AA) + "Unlabeled" amino acids

By this method, the rest of the peptide chain is left intact; then the analysis is repeated and the new terminal group of the shortened peptide is identified.

The C-terminal residue is determined enzymatically rather than chemically. The C-terminal residue of the peptide is selectively cleaved from the chain by using the enzyme carboxypeptidase as follows:

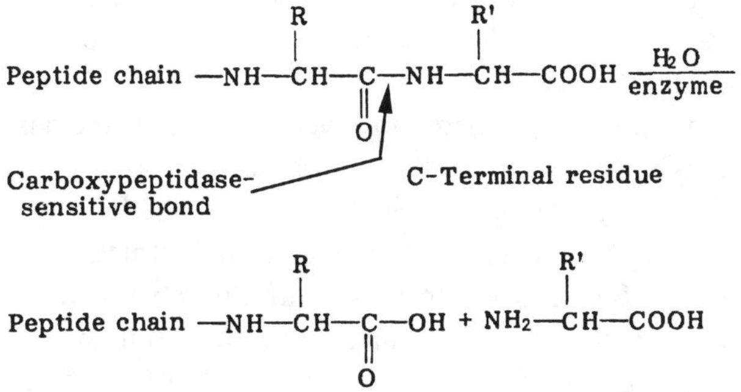

The structure of a peptide is determined by complete hydrolysis of the chain.

Hydrolysis of the DNP-peptide gives the DNP derivative of the N-terminal amino acid.

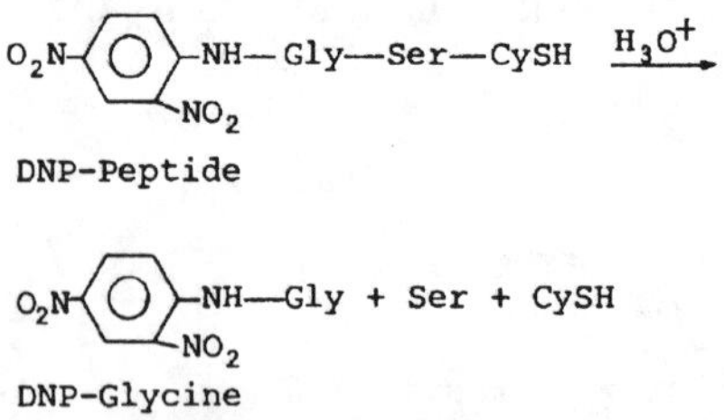

$$O_2N-\langle\bigcirc\rangle-NH-Gly-Ser-CySH \xrightarrow{\;H_3O^+\;}$$

with a NO_2 group

DNP-Peptide

$$O_2N-\langle\bigcirc\rangle-NH-Gly + Ser + CySH$$

with a NO_2 group

DNP-Glycine

Problem Solving Example:

Q The tripeptide eisenine has only one free carboxyl group, does not react with dinitrofluorobenzene, and on complete hydrolysis yields two moles of L-glutamic acid, one mole of L-alanine, and one mole of ammonia. With anhydrous hydrazine, it forms L-alanine but no glutamic acid. Write a structure for eisenine that is in accord with the information given.

A The information given directly in the problem states that the tripeptide eisenine consists of two moles of glutamic acid and one mole of alanine. The fact that upon reaction with anhydrous hydrazine, the liberation of alanine results is indicative of alanine's position at the C-terminus. Therefore, the amino acid sequence must be Glu-Glu-Ala. (Structure A)

A problem arises here, however, in that eisenine has only one free carboxyl group. The fact that ammonia is released upon hydrolysis would indicate that the carbonyl is tied up in the form of an amide. Another factor in the determination of eisenine's structure is its inability to react with Sanger's reagent (2,4-dinitrofluorobenzene). This would be indicative of the amino terminus being tied up in an amide form. Hence, a possible form for eisenine that would fit the data would be structure B.

$$
\begin{array}{c}
\quad\quad\ \overset{\displaystyle O}{\underset{\displaystyle\parallel}{\ }}\ H\ \quad\quad\ \overset{\displaystyle O}{\underset{\displaystyle\parallel}{\ }}\ H\ \ CH_3\ \ \overset{\displaystyle O}{\underset{\displaystyle\parallel}{\ }} \\
NH_2-CH-C-N-CH-C-N-CH-\!\!-C-OH \\
\quad\ |\quad\quad\quad\quad\ | \\
\quad\ CH_2\quad\quad\quad CH_2 \\
\quad\ |\quad\quad\quad\quad\ | \\
\quad\ CH_2\quad\quad\quad CH_2 \\
\quad\ |\quad\quad\quad\quad\ | \\
\quad\ C=O\quad\quad\ COOH \\
\quad\ | \\
\quad\ OH
\end{array}
$$

Structure A: Glu -Glu -Ala

$$
\begin{array}{c}
\quad\quad H \\
\quad\quad N \\
\quad/\ \ \backslash\quad\ O\quad\quad\quad O\ \ CH_3\ \ O \\
O=C\quad\ CH-C-N-CH-C-N-C\ -\ C-NH_2 \\
\quad\ |\quad\ |\quad\quad\quad\ \ | \\
\ CH_2-CH_2\quad\quad CH_2 \\
\quad\quad\quad\quad\quad\ | \\
\quad\quad\quad\quad\quad CH_2 \\
\quad\quad\quad\quad\quad\ | \\
\quad\quad\quad\quad\quad C=O \\
\quad\quad\quad\quad\quad\ | \\
\quad\quad\quad\quad\quad NH_2
\end{array}
$$

Structure B: Eisenine

8.7 Preparation of Peptides

A) Chloroacyl chloride method.

Example

$$ClCH_2COCl + CH_3\underset{\underset{\displaystyle NH_2}{|}}{CH}COOCH_3 \rightarrow ClCH_2CONHCHCOOCH_3 \\ \quad\quad\quad\quad\quad\quad\quad\quad\quad\quad\quad\quad\quad\quad\quad\underset{\displaystyle CH_3}{|}$$

$$\xrightarrow{NH_3} H_2NCH_2CONHCHCOOCH_3 \\ \quad\quad\quad\quad\quad\quad\quad\quad\quad\underset{\displaystyle CH_3}{|}$$

In the synthesis of peptides, a protecting group is necessary. This "protection" is required to insure that the desired reaction is not inadvertently carried out on the incorrect functional group of the amino acid.

B) Carbobenzoxy synthesis. The protection of an amino acid by the carbobenzoxy group.

$$C_6H_5CH_2OH + ClCOCl \rightarrow C_6H_5CH_2OCOCl + HCl$$

Phosgene   Carbobenzoxy chloride
(Carbonyl   (Benzyl chlorocarbonate)
chloride)

The carbobenzoxy synthesis has been used to make glycylalanine as shown below.

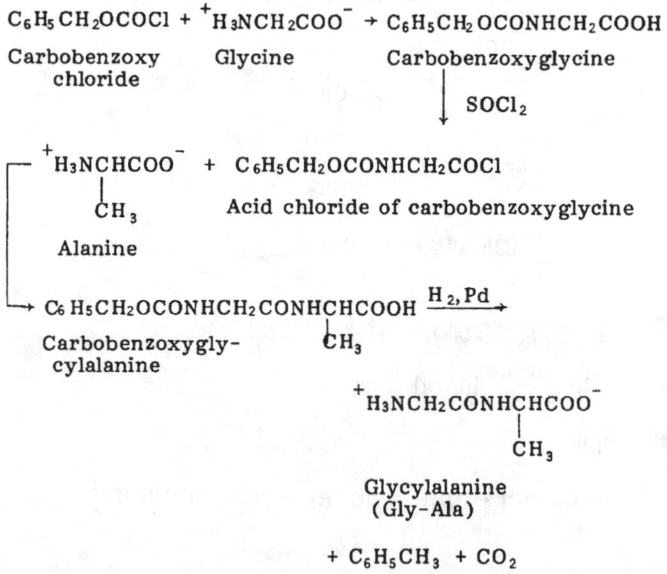

$$C_6H_5CH_2OCOCl + {}^+H_3NCH_2COO^- \rightarrow C_6H_5CH_2OCONHCH_2COOH$$

Carbobenzoxy   Glycine   Carbobenzoxyglycine
chloride

$$\downarrow SOCl_2$$

$$\left[\begin{array}{l} {}^+H_3NCHCOO^- \quad + \quad C_6H_5CH_2OCONHCH_2COCl \\ \quad\;\; | \\ \quad\; CH_3 \qquad\qquad \text{Acid chloride of carbobenzoxyglycine} \\ \text{Alanine} \end{array} \right.$$

$$\longrightarrow C_6H_5CH_2OCONHCH_2CONHCHCOOH \xrightarrow{H_2, Pd}$$

Carbobenzoxygly- CH_3
cylalanine

$${}^+H_3NCH_2CONHCHCOO^-$$
$$\qquad\qquad\qquad\qquad |$$
$$\qquad\qquad\qquad\qquad CH_3$$

Glycylalanine
(Gly-Ala)

$$+ \; C_6H_5CH_3 \; + \; CO_2$$

C) The protection of an amino acid by the phthaloyl group.

$$\text{Phthalic anhydride} + NH_2-\overset{R}{\underset{}{CH}}-COOH \longrightarrow \text{Phthaloyl derivative}$$

Phthalic anhydride   Phthaloyl derivative

$$\text{Phthaloyl-protected dipeptide} + NH_2-NH_2 \longrightarrow$$

Phthaloyl-protected   Hydrazine
dipeptide

$$NH_2-\overset{R}{\underset{}{CH}}-\overset{O}{\underset{}{C}}-NH-\overset{R'}{\underset{}{CH}}-COOH + \text{phthalhydrazide}$$

Dipeptide
                    phthalhydrazide

D) Solid-phase peptide synthesis (or the Merrifield method). The protection of an amino acid by a t-butoxycarbonyl group.

t-BuOCONHCHCOO⁻ + ClCH₂C₆H₄-P ⟶ t-BuOCONHCHCOOCH₂C₆H₄-P
        | |
        R   R
t-Butoxycarbonyl group   (P=polymer)

$$\xrightarrow{H^+} H_2N\underset{R}{CHCOOCH_2C_6H_4}-P \underset{\text{dicyclohexylcarbodiimide (DCC)}}{\overset{N=C=N}{\diagup\diagdown}} + HOOC\overset{R'}{\underset{}{CHNHOCO}}-t-Bu$$

t-BuOCONHCHCONHCHCOOCH₂C₆H₄-P $\xrightarrow{H^+}$
        | |
        R'   R
                 R"
$\underbrace{\xrightarrow{H^+} DCC + HOOC\overset{R"}{CHNHOCO}-t-Bu}_{\text{To be repeated many times}}$ $\underset{\text{Final step}}{\overset{HBr}{\longrightarrow}}$

H₂N-CHCO∿∿ NHCHCONHCHCONHCHCOOH + BrCH₂C₆H₄-P
      | | | |
      Rᴸ   R"   R'   R
Last residue
to be added

Problem Solving Example:

Show how each of the following substances can be synthesized starting with the individual amino acids.

(a) Glycylalanylcysteine

(b) $HO_2C(CH_2)_2CH(NH_2)CONHCH_2CO_2H$

(c) Glutamine from glutamic acid

A Peptide synthesis consists of three steps: (1) those amino and carboxyl groups (as well as other reactive groups) that are not involved in the reaction must be chemically blocked in order to prevent side reactions; (2) the free amino and carboxyl functionalities must be linked to form a peptide bond; and (3) the blocking groups must be removed. In performing the chemical block, we must use a blocking agent that 1) reacts quantitatively and without racemization of the amino acid; 2) is stable, in the blocked amino acid form, to conditions of the condensation process; and 3) is easily removed under conditions in which the peptide bond is stable and in which racemization of the peptide does not occur. Good blocking agents for amino groups include benzyl-chlorocarbonate, triphenyl methylchloride, t-butylchlorocarbonate and trifluoroacetic anhydride. Carboxyl groups are generally blocked through formation of the corresponding methyl, ethyl, or benzyl esters; esterifications and subsequent hydrolysis are performed by classical means (e.g., Fischer esterifications). With this in mind, let us synthesize the tripeptide, glycylalanylcysteine.

The first step in this synthesis is to block the amino group of glycine. This can be achieved by using t-butoxycarboxazide (BOCN$_3$),

$$(CH_3)_3C-O-\overset{\overset{\displaystyle O}{\displaystyle \|}}{C}-N_3$$, which converts an amino group into its t-butoxylcarbonyl (BOC) derivative:

$$CH_3-\underset{\underset{CH_3}{|}}{\overset{\overset{CH_3}{|}}{C}}-O-\overset{\overset{O}{||}}{C}-N_3 \quad + \quad H_2NCH_2\overset{\overset{O}{||}}{C}-OH \qquad \xrightarrow[\text{(2) acid}]{\text{(1) base}}$$

(BOCN$_3$)   glycine

$$CH_3-\underset{\underset{CH_3}{|}}{\overset{\overset{CH_3}{|}}{C}}-O-\overset{\overset{O}{||}}{C}-NH-CH_2-\overset{\overset{O}{||}}{C}-OH \qquad \equiv \qquad BOC-NH-CH_2-\overset{\overset{O}{||}}{C}-OH$$

N-protected glycine

Mixing an aliphatic amine with a carboxylic acid at room temperature results in the formation of a salt:

$$R-\overset{\overset{O}{/\!/}}{C}-\underset{\curvearrowleft}{O}\text{-}H \curvearrowleft :NH_2R \qquad \rightarrow \qquad R\overset{\overset{O}{/\!/}}{C}-O^- \; {}^+NH_3-R$$

acid   base   salt

Conversion of this salt to an amide requires temperatures too high for the survival of the peptide; therefore, we must convert the carboxyl group to a more reactive acyl derivative, that is, we must "activate" the carboxyl function. However, we do not have to activate the carboxyl group prior to amidation; instead, we treat the N-protected amino acid with the ester of the second amino acid (remember, we must use the ester form in order to block the carboxyl function) in the presence of dicyclohexylcarbodiimide (DCC), a potent nonacidic dehydrating agent:

$$BOC-NH-CH_2-\overset{\overset{O}{||}}{C}-OH \quad + \quad H_2N-\overset{\overset{CH_3}{|}}{C}H-\overset{\overset{O}{||}}{C}-OEt$$

N-protected glycine   C-protected alanine

$$\xrightarrow[CH_2Cl_2]{DCC} \quad BOC-NH-CH_2-\overset{\overset{O}{||}}{C}-NH-\overset{\overset{CH_3}{|}}{C}H-\overset{\overset{O}{||}}{C}-OEt$$

glycylalanine derivative

The mechanism for this remarkable reaction is:

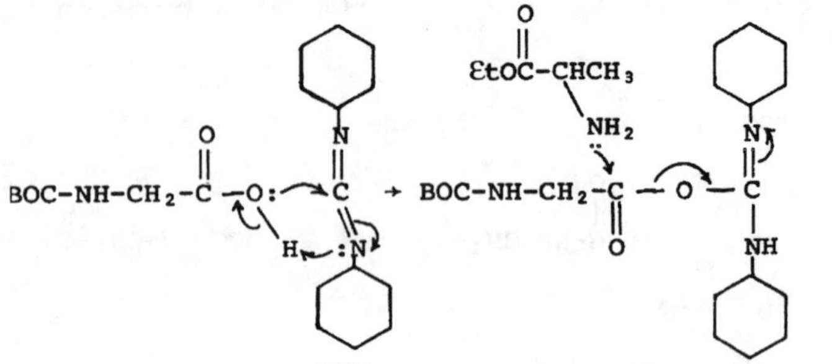

DCC

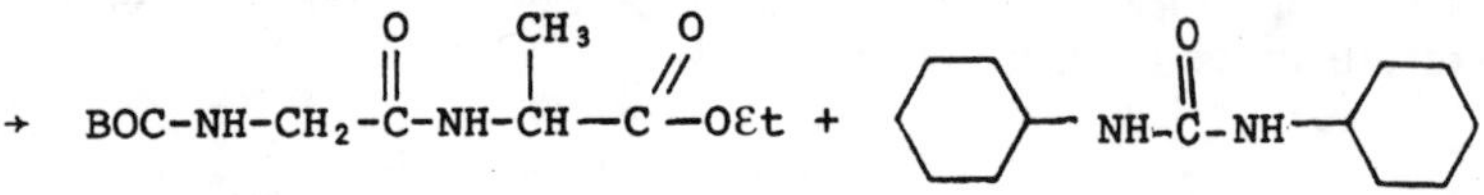

glycylalanine derivative   N,N'-dicyclohexyl urea

The next step is removal of the ester carboxyl-protecting group:

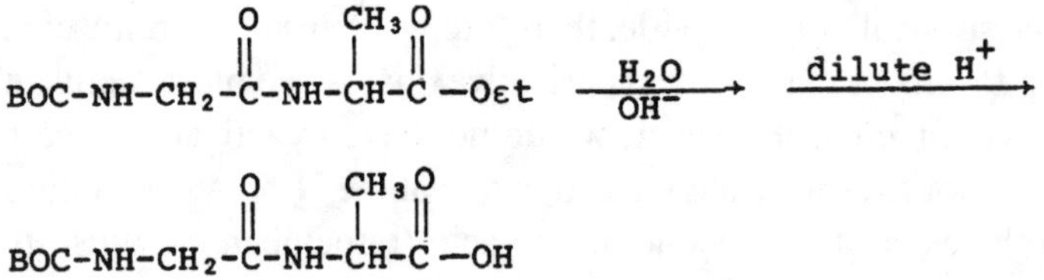

$$BOC-NH-CH_2-\overset{\overset{O}{\|}}{C}-NH-\overset{\overset{CH_3}{|}}{CH}-\overset{\overset{O}{\|}}{C}-OEt \xrightarrow[OH^-]{H_2O} \xrightarrow{\text{dilute } H^+}$$

$$BOC-NH-CH_2-\overset{\overset{O}{\|}}{C}-NH-\overset{\overset{CH_3}{|}}{CH}-\overset{\overset{O}{\|}}{C}-OH$$

glycylalanine derivative

Before we can add the cysteine residue to the dipeptide, we must protect the carboxyl group and the thiol group. The blocking of the thiol function is achieved by reacting cysteine hydrochloride with triphenylmethyl chloride:

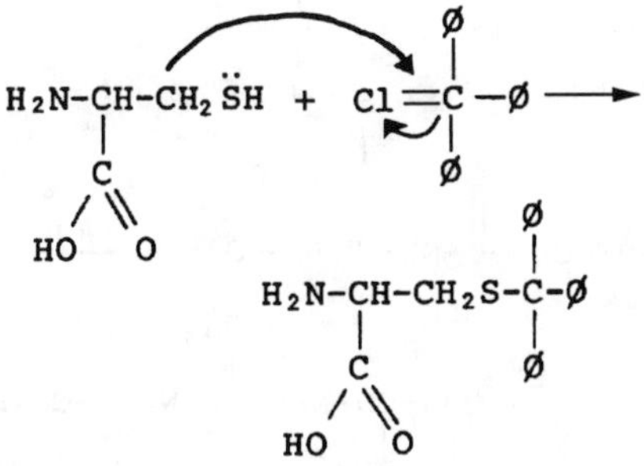

$$H_2N-CH-CH_2 \overset{..}{S}H \; + \; Cl=\overset{\phi}{\underset{\phi}{C}}-\phi \longrightarrow$$

with the cysteine (HO, O, C) groups, leading to:

$$H_2N-CH-CH_2S-\overset{\phi}{\underset{\phi}{C}}-\phi$$

Note that the amino function is not blocked in this reaction because nitrogen is not as nucleophilic as sulfur. The carboxyl group is protected by esterification. The formation of a peptide bond between the cysteine derivative and the dipeptide can be achieved by using the method that we used before:

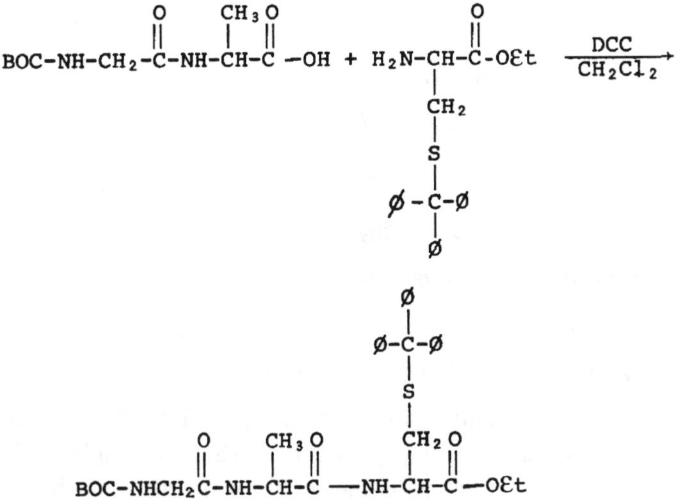

$$BOC-NH-CH_2-\overset{O}{\overset{||}{C}}-NH-\overset{CH_3}{\underset{|}{CH}}-\overset{O}{\overset{||}{C}}-OH \; + \; H_2N-\overset{}{\underset{}{CH}}-\overset{O}{\overset{||}{C}}-OEt \quad \xrightarrow[CH_2Cl_2]{DCC}$$

$$\phi-\overset{\phi}{\underset{\phi}{C}}-\phi$$
$$|$$
$$S$$
$$|$$
$$CH_2$$

$$BOC-NHCH_2\overset{O}{\overset{||}{C}}-NH-\overset{CH_3}{\underset{|}{CH}}-\overset{O}{\overset{||}{C}}-NH-\overset{CH_2}{\underset{|}{CH}}-\overset{O}{\overset{||}{C}}-OEt$$

glycylalanylcysteine derivative

At this point we must remove all the blocking agents:

$$\text{BOC-NHCH}_2\overset{\overset{\displaystyle O}{\|}}{C}\text{-NH-}\overset{\overset{\displaystyle CH_3}{|}}{CH}\text{-}\overset{\overset{\displaystyle O}{\|}}{C}\text{-NH-}\overset{\overset{\displaystyle CH_2}{|}}{CH}\text{-}\overset{\overset{\displaystyle O}{/\!/}}{C}\text{-O}\varepsilon\text{t} \xrightarrow{\text{AgNO}_3}$$

where the CH_2 bears $-S-\overset{\overset{\displaystyle \varnothing}{|}}{C}\text{-}\varnothing$ with $\varnothing$

$$\varnothing\text{-}\overset{\overset{\displaystyle \varnothing}{|}}{\underset{\underset{\displaystyle \varnothing}{|}}{C}}\text{+NO}_3^- \;+\; Ag^{+-}S\text{-CH}_2\text{-}\overset{\overset{\displaystyle }{|}}{CH}\text{-NHC-CH-NH-C-CH}_2\text{NH-BOC} \xrightarrow{\text{HCl}}$$

$$\text{HS-CH}_2\text{-CH-NHC-CH-NH-C-CH}_2\text{NH-BOC} \xrightarrow[\text{OH}^-]{\text{H}_2\text{O}} \xrightarrow{\text{dil H}^+}$$

$$\text{BOC-NHCH}_2\text{-C-NH-CH-C-NH-CH-C-OH} \xrightarrow{\dfrac{\text{HCl}}{\text{HOAc}}}$$

with SH on the CH_2

$$\text{NH}_2\text{CH}_2\text{C-NHCH-C-NH-CH-C-OH}$$

glycylalanylcysteine

Recently, this classical approach has been augmented by a solid-phase-peptide synthesis (SPPS) technique. The basic idea is to synthesize the peptide in a stepwise fashion, while the growing chain is attached to a solid support. The main advantage of this method is that purification of each product is avoided. Note that in the classical approach, for a sequence containing 100 steps, each of which has a 90% yield, we will get an overall yield of $0.90^{100} \times 100\%$ or .003%. Since purification of each product in the synthesis is avoided in the solid-phase technique, the overall yield is much higher. The overall procedure for this technique consists of the following steps: 1) attachment of a suitable functional group to a solid particle (e.g., polystyrene resin); 2) a suitably blocked amino acid is added to the functional group of the particle; 3) blocking agent removed; and 4) a second blocked amino acid is condensed with the first (see below).

$$X - polymer$$

$$\downarrow$$

$$protected — amino\ acid_1 — polymer$$

$$\downarrow$$

$$amino\ acid_1 — polymer$$

$$\downarrow$$

$$protected\ amino\ acid_2 - amino\ acid_1 - polymer$$

$$\downarrow$$

$$peptide$$

Solid-Phase Peptide Synthesis

Using this method, let us again synthesize glycylalanylcysteine.

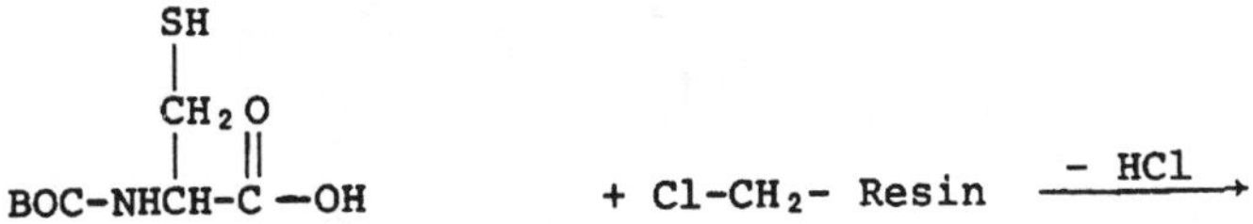

$$
\begin{array}{c}
SH \\
| \\
CH_2\ O \\
|\quad || \\
BOC-NHCH-C-OH
\end{array}
\qquad + Cl-CH_2-\ Resin \xrightarrow{\ -\ HCl\ }
$$

protected cysteine

$$
\begin{array}{c}
CH_2SH\ \ O \\
|\qquad\quad || \\
BOC-NH-CH\ ——\ C-O-CH_2-\ Resin
\end{array}
$$

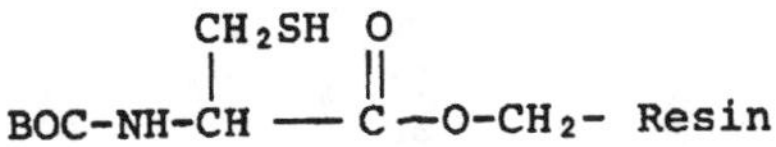

The amino group is now deblocked and reacted with the suitable form of alanine, etc.

$$\text{BOC-NH-CH} \overset{\overset{\displaystyle CH_2SH}{|}}{\underset{}{}} \overset{\overset{\displaystyle O}{\|}}{C} -OCH_2- \text{Resin} \xrightarrow{\text{HCl/AcOH}}$$

$$\text{H}_2\text{NCH-C} \overset{\overset{\displaystyle SH}{|}}{\underset{\overset{\displaystyle CH_2}{|}}{}} \overset{\overset{\displaystyle O}{\|}}{} -O-CH_2- \text{Resin}$$

$$\xrightarrow[]{\overset{\overset{\displaystyle CH_3}{|}}{\text{BOC-NHCH-C}} \overset{\overset{\displaystyle O}{\|}}{} -OH/DCC/DMF}$$

$$\text{BOC-NHCH} \overset{\overset{\displaystyle CH_3}{|}}{\underset{}{}} \overset{\overset{\displaystyle O}{\|}}{C} -NH-CH \overset{\overset{\displaystyle SH}{|}}{\underset{\overset{\displaystyle CH_2}{|}}{}} \overset{\overset{\displaystyle O}{\|}}{C} -O-CH_2- \text{Resin}$$

$$\xrightarrow{\text{HCl/AcOH}} \text{H}_2\text{NCH} \overset{\overset{\displaystyle CH_3}{|}}{\underset{}{}} \overset{\overset{\displaystyle O}{\|}}{C} -NH-CH \overset{\overset{\displaystyle SH}{|}}{\underset{\overset{\displaystyle CH_2}{|}}{}} \overset{\overset{\displaystyle O}{\|}}{C} -O-CH_2- \text{Resin}$$

$$\xrightarrow[]{\overset{\overset{\displaystyle O}{\|}}{\text{BOC-NHCH}_2\text{-C}} -OH/DCC/DMF} \text{BOC-NH-CH}_2-\overset{\overset{\displaystyle O}{\|}}{C}-NH-\underset{\underset{\displaystyle CH_3}{|}}{CH}-\overset{\overset{\displaystyle O}{\|}}{C}-NH-\underset{\underset{\overset{\displaystyle CH_2}{|}}{\overset{|}{SH}}}{CH}-\overset{\overset{\displaystyle O}{\|}}{C}-OCH_2\underset{\underset{N}{\underset{I}{\underset{S}{\underset{E}{R}}}}}{|}$$

$$\xrightarrow{\text{HBr/TFA}} \text{NH}_2-\text{CH}_2-\overset{\overset{\displaystyle O}{\|}}{C}-NH-\underset{\underset{\displaystyle CH_3}{|}}{CH}-\overset{\overset{\displaystyle O}{\|}}{C}-NH-\underset{\underset{\overset{\displaystyle CH_2}{|}}{\overset{|}{SH}}}{CH}-\overset{\overset{\displaystyle O}{\|}}{C}-OH \quad + \quad CO_2 \quad +$$

$$(CH_3)_2C = CH_2 + BrCH_2- \text{Resin}$$

The abbreviations used in this synthesis are: AcOH = acetic acid; DCC = dicyclohexylcarbodiimide; DMF = dimethylformamide; BOC = t-butoxycarbonyl; and TFA = trifluoroacetic acid. Note that in the last step of the synthesis, the mixture of hydrogen bromide and trifluoroacetic acid cleaves the polypeptide from the resin, as well as removing the protecting group from the N-terminal amino acid.

(b) Before we can synthesize this dipeptide of glutamic acid and glycine, we must block the side chain carboxyl group and the amino group. Acid-catalyzed esterification of glutamic acid will yield only the side-chain ester if the esterification is not driven to completion. The side-chain carboxyl group is more reactive in acid-catalyzed esterification because it is farther from the positively charged α-amino group; the positive charge on this function hinders the protonation of the a-carboxyl oxygen:

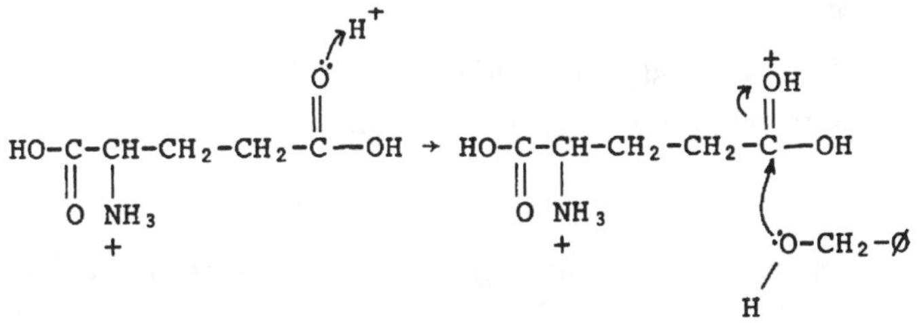

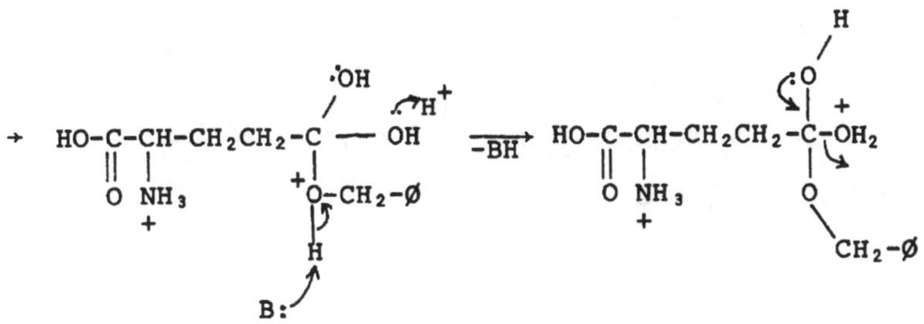

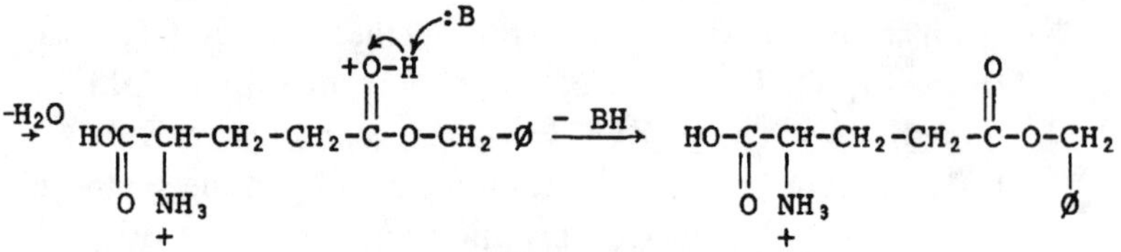

$$-H_2O \rightarrow \quad HOC-CH-CH_2-CH_2-C-O-CH_2-\phi \xrightarrow{-BH} \quad HO-C-CH-CH_2-CH_2-C-O-CH_2$$

Now we can proceed with the rest of this dipeptide synthesis:

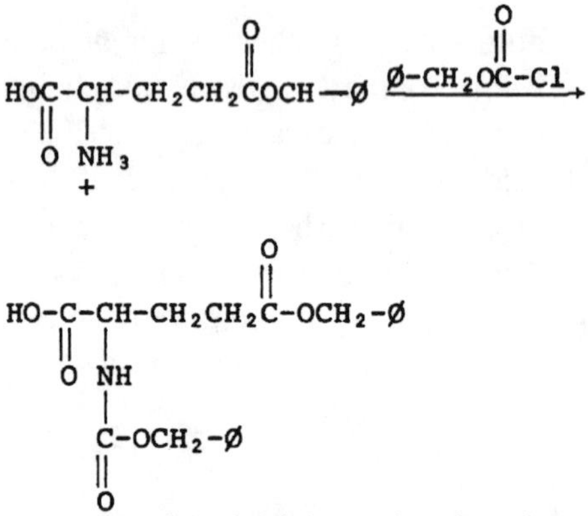

$$HOC-CH-CH_2CH_2COCH-\phi \xrightarrow{\phi-CH_2OC-Cl}$$

$$HO-C-CH-CH_2CH_2C-OCH_2-\phi$$

This last step blocked the amino function; the next step is the formation of the peptide bond.

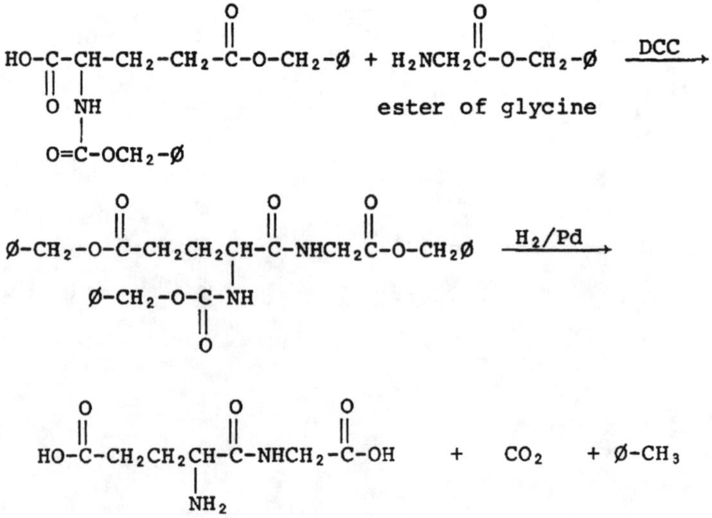

$$HO-C-CH-CH_2-CH_2-C-O-CH_2-\phi + H_2NCH_2C-O-CH_2-\phi \xrightarrow{DCC}$$

ester of glycine

$$\phi-CH_2-O-C-CH_2CH_2CH-C-NHCH_2C-O-CH_2\phi \xrightarrow{H_2/Pd}$$

$$HO-C-CH_2CH_2CH-C-NHCH_2-C-OH \quad + \quad CO_2 \quad + \quad \phi-CH_3$$

(c) Preferential oxidation of the y-carboxyl group can be accomplished by using the γ-ester of glutamic acid:

$$\underset{\underset{O}{\|}}{HO-C}-CH_2CH_2\underset{\underset{NH_2}{|}}{CH}-\underset{\underset{O}{\|}}{C}-OH \xrightarrow[HCl]{\emptyset-CH_2OH}$$

$$\emptyset-CH_2-O-\underset{\underset{O}{\|}}{C}-CH_2CH_2\underset{\underset{NH_3^+}{|}}{CH}-\underset{\underset{O}{\|}}{C}-OH \xrightarrow{\emptyset-CH_2-O-\overset{O}{\overset{\|}{C}}-Cl}$$

$$\emptyset-CH_2-O-\underset{\underset{O}{\|}}{C}-CH_2CH_2-\underset{\underset{\underset{\emptyset-CH_2-O-C=O}{|}}{N-H}}{CH}-\underset{\underset{O}{\|}}{C}-OH \xrightarrow{NH_3}$$

$$\emptyset-CH_2-O-\overset{O}{\overset{\|}{C}}-CH_2CH_2-\underset{\underset{\underset{\emptyset-CH_2-O-C=O}{|}}{NH}}{CH}-\overset{O}{\overset{\|}{C}}-O^{-\,+}NH_4 \longrightarrow$$

$:\!NH_3$

non-reactive toward acyl substitution (O^{-2} will not leave)

$$NH_2-\overset{O}{\overset{\|}{C}}-CH_2CH_2\underset{\underset{\underset{\emptyset-CH_2-O-C=O}{|}}{NH}}{CH}-\overset{O}{\overset{\|}{C}}-O^- \xrightarrow{H^+} \xrightarrow{H_2/Pd}$$

$$H_2N-\overset{O}{\overset{\|}{C}}-CH_2CH_2\underset{\underset{NH_2}{|}}{CH}-\overset{O}{\overset{\|}{C}}-OH$$

glutamine

8.8   Proteins

Proteins are polypeptides of great size and weight. They differ from other polypeptides by having higher molecular weights (over 10,000) and more complex structures. Some of the chemical behavior of proteins is similar to that of polypeptides. For example, the techniques of hydrolysis and end-group analysis can be applied to proteins.

Many proteins undergo denaturation (or precipitation), which is a nonchemical transformation that may alter the physical or biological properties of a protein. Few, if any, changes in the chemical structure result from this process. Protein denaturation can be caused by heat, ultraviolet light, organic solvents, extremes of pH, and by many chemical agents. Denaturation is either reversible or irreversible, depending upon the agent used. The coagulation of an egg white by heating is an example of irreversible denaturation.

Polypeptides do not undergo denaturation, probably due to their smaller and less complex nature.

Pure proteins are usually amorphous solids. Proteins are optically active and divided into two basic classes: fibrous proteins, which are insoluble in water and are resistant to denaturation, and globular proteins, which are soluble in water and in aqueous solutions of acids, bases, or salts. Globular proteins are more sensitive to denaturation than are fibrous proteins.

When proteins are dissolved in water, they form colloidal solutions due to their large molecular size.

Molecules of fibrous proteins are long and thread-like and tend to lie side by side to form fibers; in some cases they are held together by hydrogen bonds. The intermolecular forces are, therefore, very strong.

Fibrous proteins function as the main structural materials of animal tissues because of their insolubility and fiber-forming tendency.

Globular proteins serve functions that require mobility and solubility. These functions are related to the maintenance and regulation of the life processes. Such proteins can be enzymes.

Proteins, like amino acids, have isoelectric points. These are the specific pH's at which the acid functional groups neutralize the amino functional groups. As with amino acids, solubility is at minimum at the isoelectric point. Proteins, like amino acids, are amphoteric; this is due to the fact that they contain both free carboxyl (acidic) and free amino (basic) groups.

Problem Solving Example:

 Of what importance are proteins to biological systems?

Proteins serve two important biological functions. On the one hand, they serve as structural material. The structural proteins tend to be fibrous in nature. That is, the long polypeptide chains are lined up more or less parallel to each other and are bonded to each other by hydrogen bonds. Depending on the actual three-dimensional structure of the individual protein molecule and its interaction with other similar molecules, a variety of structural forms may result. Examples are the protective tissues such as hair, skin, nails, and claws (α- and β-keratins), connective tissues such as tendon (collagen), or the contractile material of muscle (myosin). Fibrous proteins are usually insoluble in water.

The other important feature of proteins is their role as biological regulators. They are responsible for regulating the speed of biochemical reactions and the transport of various materials throughout the organism. The catalytic proteins (enzymes) and transport proteins tend to be globular in nature. The polypeptide chain is folded around itself in a manner that gives the entire molecule a rounded shape. Each globular protein has its own characteristic geometry, which is a result of interactions between different sites on the chain. The interactions within a protein may be of three types: disulfide bridging, hydrogen bonding, or van der Waals attraction.

The globular proteins are usually water soluble. Sometimes a globular protein consists of a single long polypeptide chain twisted about and folded back upon itself. In other cases, the molecule is composed of several subunits. Each subunit is a single polypeptide chain that has adopted its own unique three-dimensional geometry. Several of the subunits are then bounded together by secondary forces (hydrogen bonding and van der Waals attraction) to give the total globular unit.

Some globular proteins carry a nonprotein molecule (the prosthetic group) as a part of their structure. The prosthetic group may be covalently bonded to the polypeptide chain, or it may be held in place by other forces. These proteins are called conjugated proteins. Examples of conjugated proteins are nucleoproteins (prosthetic group-nucleic acids) and mucoproteins (prosthetic group-carbohydrates).

8.9 Structure of Proteins

Primary Structure of Proteins

The primary structure of a protein is the amino acid sequence of the protein, which is genetically determined for every protein.

The primary structure of a protein may contain more than one amino acid chain. These chains are bonded to each other at specific points by disulfide, -S-S-, linkages. The following diagram shows how two amino acid chains can be joined by these linkages.

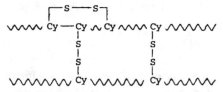

The amino acid sequence for any given protein is always the same, and any slight change in its sequence may have dramatic biological consequences. A change of 1 amino acid out of 150 completely alters the properties of the protein.

Example   Val-His-Leu-Thr-Pro-(Glu)-Glu-Lys...

Normal hemoglobin

Val-His-Leu-Thr-Pro-(Val)-Glu-Lys...

Sickle-cell hemoglobin

Sickle-cell anemia is the hereditary disease caused by the substitution of a valine for a glutamic acid in hemoglobin.

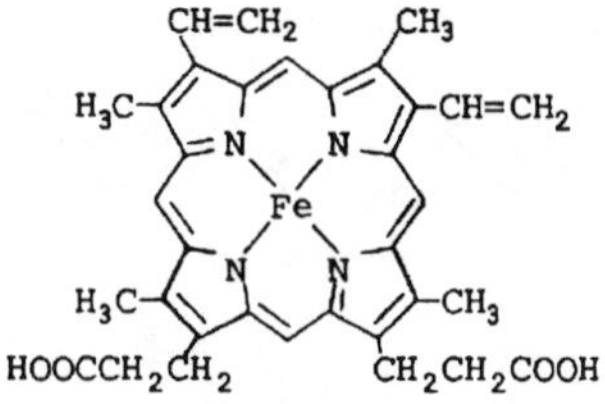

Heme

A prosthetic group is the nonpeptide part of a conjugated protein molecule. The prosthetic group is concerned with the biological action of the protein. An example of a prosthetic group is the heme in hemoglobin.

Many enzymes require co-factors to exert their catalytic effects. If the organic co-factors are covalently bonded to the enzyme, they are called coenzymes; these too are prosthetic groups.

The coenzyme nicotinamide adenine dinucleotide (NAD) is associated with a number of dehydrogenation enzymes. The characteristic biological function of these dehydrogenation enzymes is to convert the nicotinamide portion of NAD or NADP into the dihydro structure.

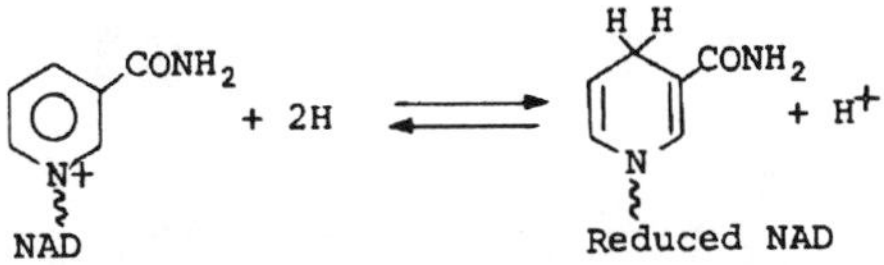

Many of the molecules that make up coenzymes are vitamins.

Secondary Structure of Proteins

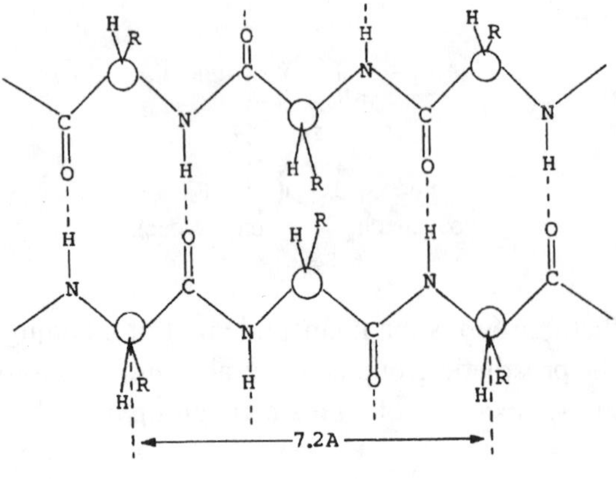

A Hypothetical Flat-Sheet Structure for a Protein

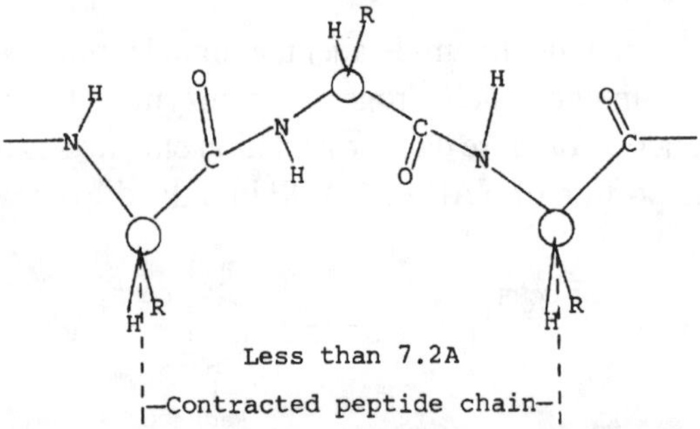

Less than 7.2A

—Contracted peptide chain—

This structure is called the beta-arrangement.
It is characterized by a pleated-sheet form.

Many amino acid chains exist in the form of a right-handed helical coil which is stabilized by hydrogen bonds. The coil is called the alpha-helix and contains 3.6 amino acid residues per turn of the helix.

α helix
(right-handed)

In addition to the x-ray diffraction patterns characteristic of the alpha- and beta-type proteins, there is a third kind called collagen, the protein of tendon and skin. Collagen is characterized by a high proportion of proline and hydroxy proline residues, and by frequent repetitions of the sequence Gly-Pro-Hypro. The pryrrolidine ring of proline and hydroxyproline can affect the secondary structure.

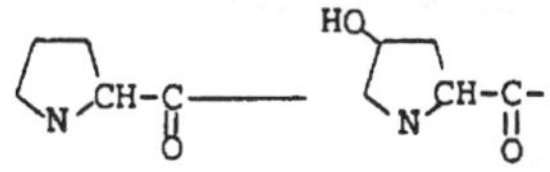

Proline residue   Hydroxyproline
                   residue

Tertiary Structure of Proteins

The tertiary structure is the manner in which the protein helical coils are arranged to give a gross protein structure. The protein is globular if the helical coil is intertwined into a sphere. Fibrous proteins are formed by the winding together of helical coils to form long strands.

Globular
protein

Fibrous protein

The mechanism of denaturation involves the disruption of the spherical shape (tertiary structure).

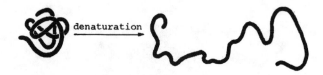

Quaternary Structure of Protein

The quaternary structure is the clumping together of the spherical units of globular proteins into specific shapes. The cause of the clustering is not as yet clearly understood; however, it may possibly be due to electrostatic attraction.

The following are examples of the quaternary structure of proteins:

1-

Myoglobin, which has no quarternary structure

2-

Hemoglobin

3-

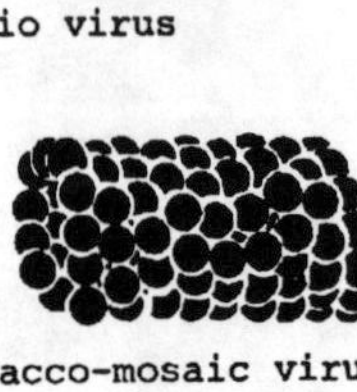

Polio virus

4-

Tobacco-mosaic virus

Quiz: Carbohydrates, Amino Acids, and Proteins

1. The following statements below concern the amino acids. Which one of the following statements is not true?

   (A) They are nonvolatile crystalline solids.

   (B) They are appreciably soluble in water, but insoluble in nonpolar solvents such as benzene.

   (C) Their aqueous solutions behave like solutions of high dipole moment.

   (D) Acidity and basicity constants are fairly high for $-COOH$ and $-NH_2$ groups.

   (E) The acidic group of a simple amino acid is $-NH_3^+$ and the basic group is $-COO^-$.

2. What type of bond links amino acid residues to form peptide chains?

   (A) disulfide   (D) saccharide

   (B) carbonyl   (E) denatured

   (C) amide

3. The carbonbenzoxy group is used as a protecting group for what functionality?

   (A) alcohol   (D) amine

   (B) carboxyl   (E) sulfhydryl

   (C) hydroxyl

4. A reagent used for the identification of the N–terminal residue of a peptide chain is

    (A) dimethyl formamide.

    (B) dicyclohexyl carbodiimide.

    (C) 2,4-Dinitrofluorobenzene.

    (D) t-Butoxycarbonyl.

    (E) trifluoroautic acid.

5. The C – terminal residue is usually determined enzymatically using which enzyme?

    (A) Hydroxylase

    (B) Catalase

    (C) Maltase

    (D) Carboxy peptidase

    (E) Amylase

6. The first step in the synthesis of the peptide Val-gly is to block the amino group of the valene amino acid. This can be accomplished with which of the following reagents?

    (A) t-butoxycarboxazide

    (B) Trifluoroacetic acid

    (C) Dicyclohexyl carbodiimide

    (D) Dimethyl formamide

    (E) N, N' – dicyclohexyl urea

7. Which one of the following amino acids is basic?

    (A) Lysine   (D) Glutamine

    (B) Alanine   (E) Methionine

    (C) Proline

8. The amino acid synthesized from the following reaction is

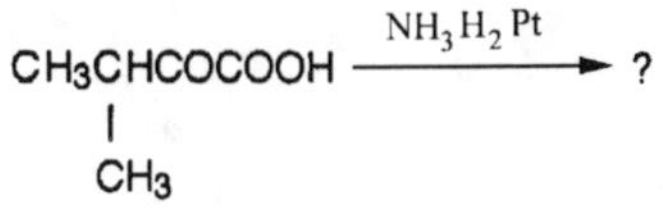

$$CH_3CHCOCOOH \xrightarrow{NH_3 \, H_2 \, Pt} ?$$
$$|$$
$$CH_3$$

    (A) serine.   (D) cysteine.

    (B) valine.   (E) isoleucine.

    (C) glycine.

9. All of the following are protection groups in the synthesis of peptides EXCEPT

    (A) carbobenzoxy.   (D) chloroacyl chloride.

    (B) phthaloyl.   (E) dinitrofluorobenzene.

    (C) t-butoxycarbonyl.

10. The amino acid sequence of a protein is referred to as its

    (A) primary structure.

    (B) secondary structure.

    (C) tertiary structure.

    (D) quaternary structure.

    (E) None of the above.

ANSWER KEY

1. (D)

2. (C)

3. (D)

4. (C)

5. (D)

6. (A)

7. (A)

8. (B)

9. (E)

10. (A)

CHAPTER 9

Spectroscopy

9.1 Mass Spectrum

Each kind of ion has a particular ratio of mass to charge, or m/e value. The m/e value is simply the mass of the ion.

	Molecular ion				
	$(C_4H_9)^+$	$(C_3H_5)^+$	$(C_2H_5)^+$	$(C_2H_3)^+$	and others
m/e:	57	41	29	27	
Relative intensity:	100	41.5	38.5	15.7	
	Base peak				

The peak of the largest intensity is called the base peak; its intensity is taken as 100. The intensities of the other peaks are expressed relative to the base peak. A mass spectrum is a plot; the abscissa is the m/e ratio; and the ordinate is the relative number of ions of relative intensity (height of each peak).

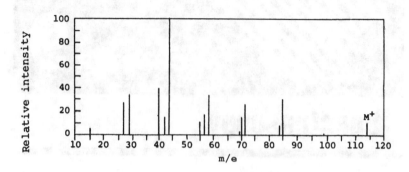

The Mass Spectrum of N-Octane

The mass spectrum is used to determine the molecular weight, molecular formula, and structure of a compound.

When a molecule loses an electron, a molecular ion is produced. The molecular ion peak is called the parent peak, and its m/e value is the molecular weight of the compound. The existence of isotopes produce small peaks with m/e values of M + 1, M + 2, etc.

$$M + e^- \rightarrow M + 2e^-$$

Molecular ion
(Parent ion)

m/e = mol. wt.

Problem Solving Example:

An unknown compound contains only carbon and hydrogen. Its mass spectrum is shown. Propose a structure for the compound.

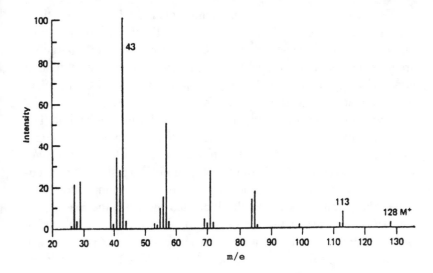

Mass spectroscopy is a technique used to determine the molecular weight, formula, and structure of a compound. It differs from infrared, raman, ultraviolet, and nuclear magnetic resonance spectroscopy in that it is a destructive spectroscopy; the sample is fragmented by the technique and cannot be recovered in its original form. Mass spectroscopy involves the bombardment of a sample with an electron beam of a particular energy. If the sample is subjected to a low-energy electron beam (about 10 eV—electron volts), the molecule will lose an electron to produce the molecular ion, $M^{\bullet+}$.

$$M: \xrightarrow{\text{10 eV}} e^- + M^{\bullet+}$$
$$\text{sample} \qquad \text{electron} \quad \text{molecular ion}$$

If the sample is subjected to a high-energy electron beam (about 70 eV) the molecule will undergo fragmentation to yield various ionic species.

$$A - B - C: \xrightarrow{\text{70 eV}} A - B - C^{\bullet+} + e^-$$
$$\text{sample} \qquad\qquad \text{molecular ion}$$

$$A - B - C^{\bullet+} \xrightarrow[\text{70 eV}]{\text{fragmentation}} A^+ + B - C^{\bullet}$$

The mass spectrum of a compound plots the relative intensity of the ionic species versus the mass-to-charge ratio of each species. Note that only cationic species (those species that bear positive charge(s)) are recorded in the mass spectrum. The mass spectrum of a compound is taken by sequentially bombarding the sample with a low-energy electron beam and a high-energy electron beam. Bombardment by a low-energy electron beam principally produces the molecular ion. The molecular ion has a charge of + 1 and has the same molecular weight as the sample compound and therefore its mass-to-charge ratio (m/e) will be the molecular weight of the compound. As a result, the molecular ion peak indicates the molecular weight of the compound and is called the parent peak (generally indicated by M^+ in the mass spectrum). One may reason that the peak with the greatest m/e is the parent peak, but this is not necessarily true. The existence of isotopes will produce small peaks with m/e values of M + 1, M + 2, etc. The relative abundance or intensity of these peaks reflects the relative abundance of the isotopes. For example, C^{13} and C^{12} are isotopes of carbon with relative abundances of 1% and 99%, respectively. In the mass spectrum of methane, CH_4, the parent peak will be at m/e = 16. There will, however, be a peak at m/e = 17, which was derived from the methane molecules containing C^{13}. The peak at 17 will be $1/_{99}$ the intensity of the peak at 16.

The parent peak is usually of moderate intensity and may sometimes be the peak of greatest intensity. There is another type of peak in the mass spectrum that is essential in determining the structure of the compound. This peak is the one that arises from the most abundant species that resulted from the fragmentation of the compound which occurred upon bombardment by the high-energy electron beam. This species is often the most stable one that resulted from the fragmentation of the compound. Its peak is the one of greatest intensity and is called the base peak. Sometimes the base peak and the parent peak may be the same, as in the case of methane, although this usually does

not happen. For example, the mass spectrum of isobutane

$$\left(\begin{array}{c} CH_3 \\ | \\ CH_3-C-CH_3 \\ | \\ H \end{array} \right)$$

has its parent peak at m/e = 58 and its base peak at m/e = 57. The species that gives rise to the base peak is

$$CH_3-\overset{\overset{\displaystyle CH_3}{|}}{\underset{+}{C}}-CH_3,$$

which was formed by the loss of a hydrogen atom from the molecular ion,

$$\left[\begin{array}{c} CH_3 \\ | \\ CH_3-C-CH_3 \\ | \\ H \end{array}\right]^{\bullet +}$$

Note that $CH_3-\overset{\overset{\displaystyle CH_3}{|}}{\underset{+}{C}}-CH_3$ is the most abundant fragment because of the relatively high stability of a tertiary carbocation, a trialkylated carbon bearing a positive charge. (The relative order of carbocation stability is:

$$\text{tertiary} > \text{secondary} > \text{primary} > CH_3^+.$$

With these basic principles of mass spectroscopy, we may now attempt to identify the compound whose mass spectrum is shown. The spectrum indicates that the parent peak (M^+) is the one with m/e = 128; hence, the molecular weight of the compound is 128. Since the compound is a hydrocarbon (contains only carbon and hydrogen), the molecular weight restricts the number of carbons to nine or ten. The possible molecular formulas are C_9H_{20} and $C_{10}H_8$. Note that the compound cannot contain eight carbons because the molecular formula would then be C_8H_{32}; the greatest number of hydrogens an eight carbon compound can accommodate is 18. Hence, the compound must contain more than eight carbons. A compound with a molecular formula of C_9H_{20} is a saturated hydrocarbon and is in the class of alkanes. A compound with a molecular formula of $C_{10}H_8$ has seven units of unsaturation. This is determined by the fact that a fully saturated ten-carbon compound has 22 hydrogens and for each unit of unsaturation gained, there is a loss of two hydrogens. Since the parent peak has a small relative intensity, the compound is more likely to be the saturated hydrocarbon (C_9H_{20}) than the unsaturated one ($C_{10}H_8$). This is because fragments with multiple bonds have higher stability and less tendency to rearrange than saturated fragments. The unsaturated compound will therefore have a great relative intensity for its parent peak whereas the saturated compound will not. We can conclude that the compound is a saturated hydrocarbon with the molecular formula C_9H_{20}.

Examining the mass spectrum further, we note that the peak with the greatest relative intensity (possessing a value of 100) has a m/e value of 43. This peak, the base peak, most probably represents the isopropyl cation

$$\left(\begin{array}{c} H \\ | \\ CH_3CCH_3 \\ + \end{array} \right)$$ and therefore the compound contains one or more iso-

propyl fragments. A n-propyl fragment would have the same molecular weight but would not have such a high relative intensity value. The absence of any significant peak for m/e = 29 tells us that the compound does not produce the ethyl cation ($C_2H_5^+$) and therefore is not of the type R–C_2H_5 because compounds of this type would produce the pertinent cation. Thus, the compound seems to have branched terminal groups, probably isopropyl portions. Assuming this, we have accounted for 86 grams of the molecular weight of 128 grams. Hence, we hypothesize the unknown compound to have one of the following three structures:

1   2   3

We can deduce that the compound has the third structure in that there is only a moderate peak at m/e = 113, which corresponds to the compound's loss of a methyl fragment. (The other structures have five or six methyl groups.) More importantly, we can relate the different moieties to their m/e values in the mass spectrum upon fragmentation of the molecule along the chain.

$$CH_3 - \underset{\underset{CH_3}{|}}{\overset{\overset{H}{|}}{C}} - CH_2\,CH_2\,CH_2 - \underset{\underset{CH_3}{|}}{\overset{\overset{H}{|}}{C}} - CH_3$$

9.2   Electromagnetic Spectrum

Energy is associated with the electromagnetic radiation at a particular frequency by the following expression:

$$\Delta E = h\nu,$$

where ΔE = gain in energy, ergs

h = Plank's constant, 6.5×10^{-27} ergs-sec

ν = frequency, Hz (cycles/sec)

Or by $\Delta E = \dfrac{hc}{\lambda},$

where c = speed of light, 3×10^{10} cm/sec

λ = wavelength, cm.

The higher the frequency (the shorter the wavelength), the greater the gain in energy.

The structure of a molecule determines the electronic, vibrational, and rotational levels in the molecule.

The electromagnetic spectrum of a compound shows how much electromagnetic radiation is absorbed at each frequency. It determines the compound's structure.

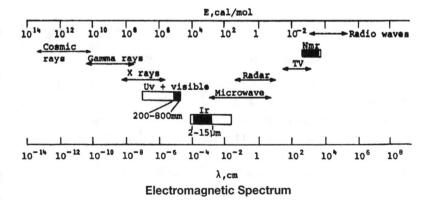

Electromagnetic Spectrum

9.3 Infrared Spectrum

The infrared spectrum of a compound is produced by the infrared absorption of its molecules, due to the degree of molecular vibration and rotation of its bonds when subjected to infrared irradiation.

The IR spectrometer records the percent transmittance ($\%T$) of incident light through the sample as a function of the wavelength of light, expressed in micrometers ($\mu m = 10^{-4}$ cm), or versus the frequency of the incident light, expressed in wave numbers. Transmittance is related to absorbance, A, by the equation $A = \log_{10} 1/T$. The wave number is the reciprocal of the wavelength in centimeters and is directly proportional to the energy of the light absorbed.

Functional Group	Infrared Absorption, μ
O—H	2.8-3.1
N—H	2.9-3.2
C—H	3.0-3.5
C ≡ C	4.5
C ≡ N	4.5
C=O	5.7-6.0
C=C (alkenes)	6.0-6.2
C⋯C (aromatic)	6.2-6.3 and 6.7-6.8
$\overset{+}{—N}\overset{\displaystyle O}{\underset{\displaystyle O^-}{\diagup\diagdown}}$	6.5 and 7.5

Infrared Absorptions of Diagnostic Value

The infrared spectrum reveals the molecular structure by indicating what groups are present in, or absent from, the molecule. This is based upon the fact that a group of atoms gives rise to a characteristic absorption band. The absorption band of a particular group of atoms can be shifted by various structural features such as angle strain, van der Waals strain, conjugation, electron withdrawal, and/or hydrogen bonding.

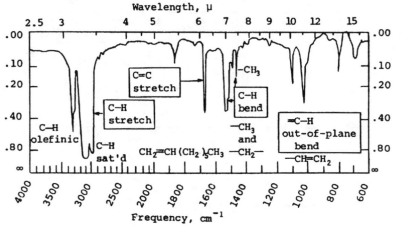

The Infrared Spectrum of 1-Octene

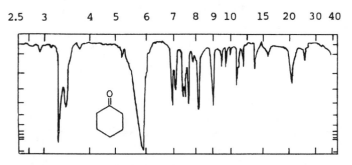

The Infrared Spectrum of Cyclohexanone

 Problem Solving Examples:

Q Give a structure or structures consistent with each of the following infrared spectra.

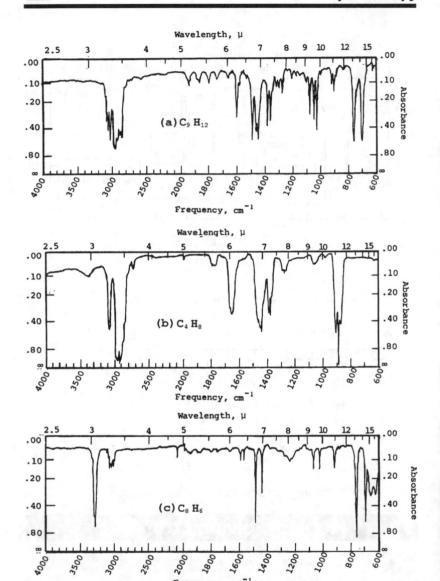

A Infrared (IR) spectra can reveal the molecular structure of an organic compound. It is based on the fact that a molecule is constantly vibrating. Chemical bonds stretch (and contract) and bend with respect to each other. The absorption of infrared light (that is, light lying beyond the red end of the visible spectrum—lower frequency, longer wavelength, less energy) causes changes in vibrations of a molecule.

An infrared spectrum may be referred to by its wavelength or, preferably, by its frequency which is expressed in wavenumbers, cm^{-1}, as can be seen in the spectra given in the problem. (Wavenumbers may be defined as the number of waves per centimeter.)

Two substances that possess identical infrared spectra are identical in thousands of different physical properties; they must almost certainly be the same compound. The infrared spectrum reveals the molecular structure by indicating what groups are present (or absent) in the molecule. This is based on the fact that a group of atoms gives rise to characteristic absorption bands. In other words, a particular group absorbs light of certain frequencies that are essentially the same from compound to compound. For example, the carbonyl group $(C = O)$ of ketones absorbs at $1,710 \ cm^{-1}$. Now, the absorption band of a particular group of atoms can be shifted by various structural features such as angle strain, van der Waals strain, conjugation, electron withdrawal, and hydrogen bonding. Hence, the interpretation of an infrared spectrum may not be easy. The table provided lists some characteristic infrared absorption frequencies. In solving for the molecular structure, try to correlate the frequencies in the table with those for bands indicated in the problem.

(a) A band is present in the $3,000$-$3,100 \ cm^{-1}$ region; this indicates an aromatic ring. This can be confirmed by the presence of a band in the $1,500$, $1,600$ regions which would indicate the stretch of the C—C bond in the ring. Such a band is present. Therefore, one knows that six of the nine carbons in C_9H_{12} belong to an aromatic portion: a benzene ring. The weak finger-like bands from $2,000 \ cm^{-1}$ to $1,700 \ cm^{-1}$ are characteristic of the absorption of monosubstituted aromatic rings. Hence, the remaining three carbons must be part of a single substituent. The aromatic bands account also for five hydrogen atoms, so that the substituent possesses seven hydrogen atoms. There exist two possibilities for an organic structure with three carbons and seven hydrogens: isopropyl ($-CH(CH_3)_2$) and propyl ($-CH_2CH_2CH_3$). That it is isopropyl can be determined from the isopropyl split at about 1,400 cm^{-1} So, the structure in (a) can be written as

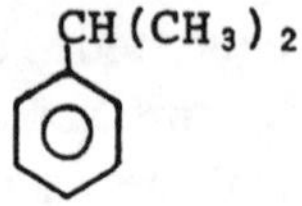

$$CH(CH_3)_2$$

The diagram below assigns groups of atoms to each bond for further clarity.

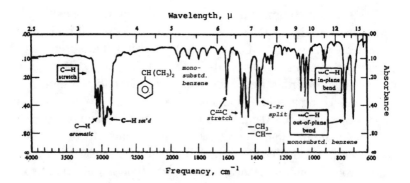

(b) That this is an alkene can be seen from the strong band present in the 1,640 to 1,680 cm^{-1} region, due to the C = C stretch. Also, the molecular formula given, C_4H_8, fits the general formula for an alkene, C_nH_8. There exists a band in the region of 3,020–3,080 cm^{-1} which indicates the C – H absorption for alkenes. The diagram below assigns structures to the bands which lead to the conclusion that the structure is $(CH_3)_2C = CH_2$, isobutylene.

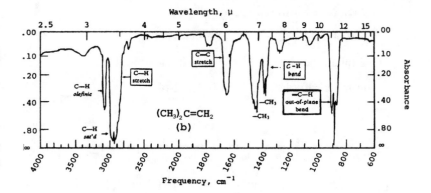

(c) That the structure is phenylacetylene may be determined by
examining two prominent bands. There exist the C – H stretch
of aromatics (confirmed by the C–C stretch at 1,500, 1,600
cm⁻¹) at 3,000 to 3,100 cm⁻¹ and the – C– C to stretch characteristic of alkynes at 2,100 to 2,260 cm⁻¹.

One concludes, therefore, that C_8H_6 represents 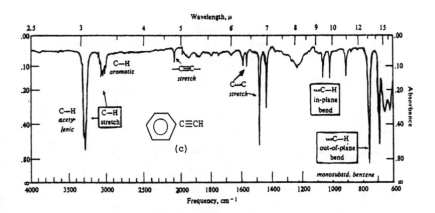 ⟨O⟩-C≡CH

phenylacetylene. Other bands are assigned structures in the diagram.

Which (if any) of the following compounds could give rise to each of the infrared spectra shown?

isobutyraldehyde   ethyl vinyl ether

2-butanone   cyclopropylcarbinol

tetrahydrofuran   3-buten-2-ol

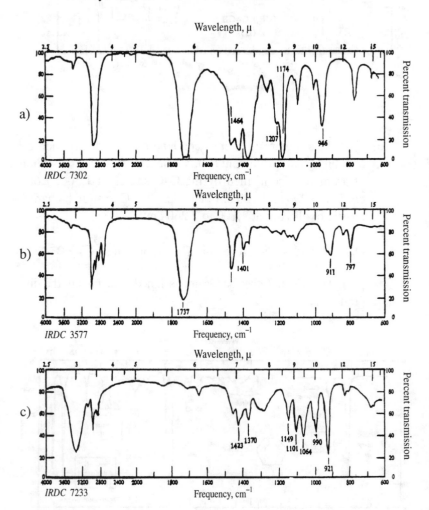

(a) This infrared spectrum shows a strong absorption near 1,720 cm^{-1}. This is the characteristic absorption of a C = O stretch for saturated aldehydes and saturated acyclic ketones. The weak absorption at 3,400 cm^{-1} is the C = O overtone and it further supports our conclusion that we are dealing with a carbonyl compound. The C = O overtone is noticeable in the spectrum because the C = O stretch is such an intense absorption. The only given carbonyl compounds are isobutyraldehyde (an aldehyde, RCHO) and 2-

$$\overset{O}{\overset{\|}{R C R'}}$$

butanone (a ketone, RCR'). The C–H stretch of the aldehyde proton characteristically shows absorptions at 2,820 cm^{-1} and 2,720 cm^{-1}. The IR spectrum shows only a strong band at 2,900 cm^{-1} in this vicinity. Hence, the compound is not the aldehyde and must be the ketone, 2-butanone. The band at 2,900 cm^{-1} is the characteristic absorption band of the C – H stretch of methyl and methylene groups. 2-butanone has the structure:

$$\overset{O}{\overset{\|}{CH_3-C-CH_2-CH_3}}$$

(b) This IR spectrum shows a strong absorption at 1,737 cm^{-1}. This is suggestive of the C = O stretch of a saturated aldehyde because the frequency at which aldehydes absorb infrared radiation is greater than that of ketones. The sharp bands at 2,820 cm^{-1} and 2,740 cm^{-1} are the absorptions of the C – H stretch of an aldehyde. Hence, the compound is an aldehyde and must be isobutyraldehyde.

$$\begin{array}{c} CH_3 \quad\quad O \\ \diagdown \quad\quad \| \\ CH-C-H \\ \diagup \\ CH_3 \end{array}$$

(c) This spectrum shows a strong broad band at 3,400 cm^{-1}, which is indicative of a bonded O–H stretch of an alcohol.

Our choice is therefore narrowed down to cyclopropylcar-binol ($\triangleright$–CH_2OH , a primary alcohol) and 3-buten-2-ol ($CH_2 = CH–CHOHCH_3$, a secondary alcohol). The C–O stretch of primary alcohols characteristically absorbs at 1,050 cm^{-1} and that of secondary alcohols absorbs at 1,100 cm^{-1}. Since the spectrum shows a complex absorption pattern in this range, we must look for other clues. There is a weak absorption at 1,640 cm^{-1}. This is the characteristic absorption of the C=C stretch of terminal alkenes ($RHC=CH_2$ or $R_2C=CH_2$). Hence, the compound must be 3-buten-2-ol because it possesses a double bond, whereas cyclopropyl-carbinol has none.

9.4 Ultraviolet Spectrum

Ultraviolet spectroscopy involves the excitation of an electron in its ground state level to a higher energy level. This is accomplished by irradiating a sample with ultraviolet light (electromagnetic radiation with wavelengths in the range of 200 nm to 400 nm).

A spectrum is described by the position of the maximum absorption (λ_{max}) and the intensity or the molar absorptivity of the incident light (ε_{max}, the extinction coefficient).

The molar absorptivity is related to absorbance by Beer's law.

$$\varepsilon = \frac{A}{Cl},$$

Where ε = molar absorptivity (molar extinction coefficient)

A = absorbance at λ_{max}

C = molar concentration of sample, $\dfrac{moles}{liter}$

l = path length of sample tube, cm.

$$A = \log(I_0/I),$$

Where I_0 = initial light intensity

I = final light intensity

Therefore,

$$\varepsilon = \frac{\log(I_0 I)}{Cl}$$

The percentage of light absorbed is related to the percentage of light transmitted by the following equation:

$$\%A = 100 - \%T,$$

Where $\%A$ = % absorbance

$\%T$ = % transmittance

Also,

$$T = \frac{I}{I_0}$$

Molecules contain bonding electrons that are directly involved in their bonding and which are in sigma (σ) or pi (π) molecular orbitals. σ^* and π^* are antibonding orbitals and are unstable where the electron density between the nuclei is very low. Many molecules that contain the O, S, N, Br, Cl, F, and I atoms contain nonbonding electrons, which are not directly involved in bonding and which are in unhybridized nonbonding orbitals, n.

The n $\rightarrow \pi^*$ and $\pi \rightarrow \pi^*$ transitions are the most observed and useful transitions in organic molecules. A chromophore is the molecular or functional group that gives rise to $\pi \rightarrow \pi^*$ and/or $n \rightarrow \pi^*$ transitions. C = C, C $\equiv$ C, C = O, N = O, C = S, and aromatic rings are typical chromophores.

In the $n \rightarrow \pi^*$ translation, the electron of an unshared pair goes to an unstable (antibonding) π^* orbital.

$$C = \ddot{O}: \rightarrow C \dot{=} \dot{O}: \quad n \rightarrow \pi^*$$

In the $\pi \rightarrow \pi^*$ transition, an electron goes from a stable (bonding) π orbital to an unstable π^* orbital.

$$\text{>C=\ddot{O}:} \quad \rightarrow \quad \text{C}\overset{\cdot}{\underset{}{\text{---}}}\ddot{\text{O}}: \qquad \pi \rightarrow \pi^*$$

Conjugation of double bonds lowers the energy required for the transition, and absorption moves to longer wavelengths.

Resonance stabilizes the excited state more than the ground state, and it reduces the difference between them.

The ultraviolet spectrum shows the relationships (mainly conjugation) between functional groups. It reveals the number and location of substituents attached to the carbons of conjugated systems.

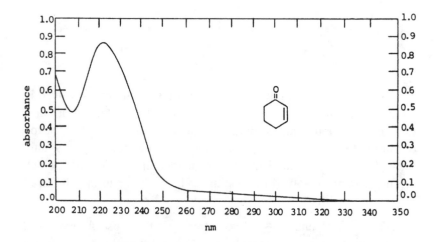

The Ultraviolet Spectrum of Cyclohexanone

Very highly conjugated molecules absorb light in the visible region, 400 to 800 nm. When absorption occurs in the visible region, the compound appears to be colored. The color of a compound corresponds to the wavelength of the light that remains after the absorbed light has been subtracted. The compound has a complementary color to the color of the light absorbed.

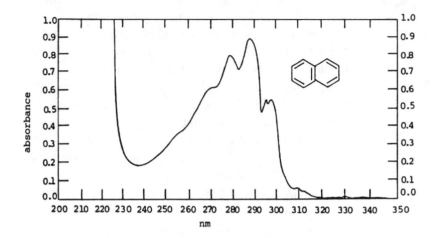

The Ultraviolet Spectrum of Naphthalene

Wavelength, nm	Color of the light	Complementary color
400	Violet	Yellow
450	Blue	Orange
500	Blue-green	Red
550	Yellow	Violet
600	Orange-red	Blue-green
700	Red	Green

Wavelengths of Visible Colors

Q Compounds A, B, and C have the formula C_5H_8, and on hydrogenation all yield n-pentane. Their ultraviolet spectra show the following values of λ_{max} : A, 176 nm; B, 211 nm; and C, 215 nm. (1-pentene has λ_{max} 178 nm.) (a) What is a likely structure for A? For B and C? (b) What kind of information might enable you to assign specific structures to B and C?

(a) The molecular formula of compounds A, B, and C is of the general form C_nH_{2n-2}. This formula indicates that the compounds are hydrocarbons with either one carbon-carbon triple bond (an alkyne), two carbon-carbon double bonds (a diene), two rings, or one ring and one carbon-carbon double bond; that is, the compounds have two units of unsaturation. Since hydrogenation of the compounds yields n-pentane, and hydrogenation of alkynes and dienes does not change the carbon skeleton, all the compounds must be straight-chained. Hence, the compounds are normal five-carbon alkynes or dienes.

Compound A has maximum absorption at a wavelength of 176 nm (λ_{max}). Since this is very close to the λ_{max} of 1-pentene, compound A must have a chromophore similar to that of 1-pentene. A chromophore is a functional group that gives rise to an absorption with characteristic λ_{max} and ε (molar absorptivity) values. Hence, compound A's chromophore, like 1-pentene, is an isolated double bond. The only structure of a normal five-carbon compound with two units of unsaturation and isolated double bonds is 1,4-pentadiene:

$$H_2C = CH - CH_2 - HC = CH_2$$

1,4-pentadiene

(Compound A)

Compounds B and C have λ_{max} of 211 nm and 215 nm, respectively. These values are greater than that of compound A and are of lower energies (longer wavelength ([λ], lower frequency, lower energy). One thing that could account for this is if B and C were conjugated dienes. Conjugated dienes are compounds that have two double bonds separated by one single bond ($R - CH = CH - CH = CH - R'$). The extra stability of conjugated dienes as compared to the analogous nonconjugated dienes is due to electron delocalization in the former. This extra stability lessens the energy difference between the ground state and excited state. Hence, the frequency necessary for the maximum absorption of the compound is less than expected and the wavelength (λ_{max}) will be greater. Therefore, compounds B and C must be geometric isomers of the conjugated compound, 1,3-pentadiene. These are the "E" and "Z" isomers:

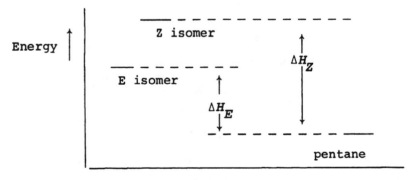

(b)   The E and Z isomers of the diene can be distinguished by their heats of hydrogenation. The Z isomer is less stable than the E isomer because of steric considerations, due to the proximity of the vinyl and methyl groups. This means that the Z isomer will have a higher heat of hydrogenation than the E isomer.

$$|\Delta H_Z| > |\Delta H_E|$$

Where

ΔH_z = heat of hydrogenation for the Z isomer,

ΔH_E = heat of hydrogenation for the E isomer.

9.5 Nuclear Magnetic Resonance (NMR) Spectrum

When an atom is placed in an external magnetic field, the magnetic field generated by the nucleus will be aligned with or against the external magnetic field. Alignment with the field is more stable. At some frequency of electromagnetic radiation, the nucleus will absorb energy and "flip" over so that it reverses its alignment with the external field. This is known as the nuclear magnetic resonance (NMR) phenomenon. It is concerned with the nuclear magnetic resonance of hydrogen atoms.

The energy required to flip the proton over depends upon the strength of the external magnetic field; the stronger the field, the greater the energy and the higher the frequency of radiation. This relationship is expressed by the following equations:

$$\upsilon = \frac{\lambda H_0}{2\pi}$$

Where υ = frequency, Hz

H_0 = strength of magnetic field, gauss

λ = a nuclear constant, the gyromagnetic ratio, 26,750 for the proton

$$\upsilon = \frac{2\mu H_0}{h}$$

Where μ = magnetic moment of the nucleus

h = Planck's constant

In nuclear magnetic resonance, spectroscopy molecules are placed in a strong magnetic field to create different energy states which are then detected by absorption of light of the appropriate energy.

Various aspects of the NMR spectrum are:

A) the number of signals, which tells us how many different "kinds" of protons are present in a molecule;

B) the positions of the signals, which tell us something about the electronic environment of each kind of proton;

C) the intensities of the signals, which tell us how many protons of each kind are present; and

D) the splitting of a signal into several peaks, which tells us about the environment of a proton with respect to other nearby protons.

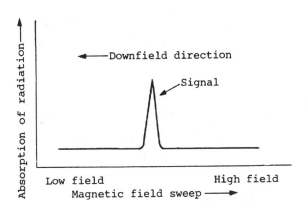

The NMR Spectrum

Number of Signals

A set of protons with the same environment are equivalent.

$$CH_3 - CH_2 - CH_2 - Cl \qquad\qquad CH_3 - CHCl - CH_3$$

   a   b   c   a   b   a

3 NMR signals   2 NMR signals

n-Propyl chloride   Isopropyl chloride

In order for protons to be chemically equivalent, they must be stereochemically equivalent.

2 NMR signals   3 NMR signals   4 NMR signals
Isobutylene   Vinyl chloride   Methylcyclopropane

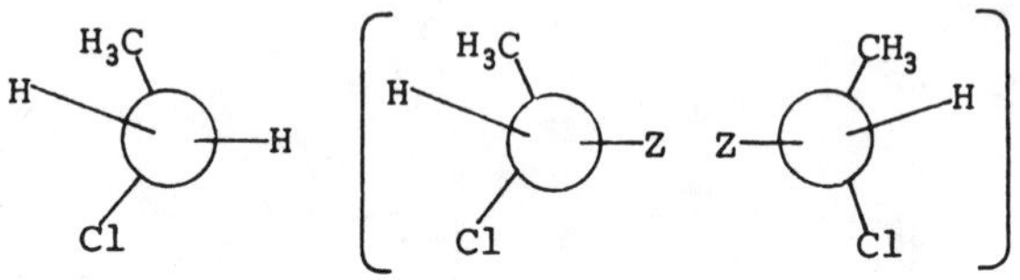

Ethyl chloride
Enantiotopic
protons

These two protons are mirror images of each other, in an achiral medium. They behave as if they were equivalent, and there will therefore be one NMR signal for the pair.

1,2-Dichloropropane
Diastereotopic
protons

These two protons are neither identical nor mirror images of each other. They are nonequivalent, and there will be a separate NMR signal for each one.

Positions of Signals

The positions of signals determine what kind of protons they represent: aromatic, aliphatic, primary, secondary, tertiary, benzylic, vinylic, acetylenic, or adjacent to halogen or other atoms or groups.

Circulation of electrons about the proton generates a field that opposes the applied field. The field felt by the proton is decreased, and the proton is shielded.

Circulation of electrons—specifically, π electrons—about nearby nuclei generates an induced field. If this induced field opposes the applied field, the proton is shielded. If the induced field strengthens the applied field, then the effective field of the proton is increased, and the proton is deshielded.

A shielded proton requires a higher applied field strength, and a deshielded proton requires a lower applied field strength. Shielding shifts absorption upfield, and deshielding shifts absorption downfield. Chemical shifts are shifts in the position of NMR absorptions that arise from shielding and deshielding by electrons.

The point from which chemical shifts are measured is the reference point.

The δ (delta) scale is the most commonly used scale and is calculated by the equation:

$$\delta = \frac{(\text{chemical shift in Hz}) \times 10^6}{\text{spectrometer frequency}}$$

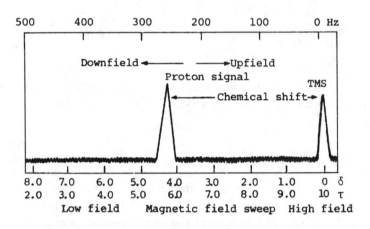

An Example of the NMR Spectrum

Positions of Absorption of Hydrogen Atoms in Nuclear Magnetic-Resonance Spectroscopy

Type of Hydrogen Atom	Absorption, in ppm from TMS
$(CH_3)_4Si$	0
$CH_3-\overset{\textstyle |}{\underset{\textstyle |}{C}}-$	0.6-1.5
$C-CH_2-C$	1.2-1.5
$C-\overset{\textstyle C}{\underset{\textstyle H}{C}}-C$	1.4-1.8
$CH_3-\overset{\textstyle }{\underset{\textstyle O}{\overset{||}{C}}}-$	1.9-2.5
$H-C\equiv C-$	2.5-3.0
$\overset{\textstyle H}{\underset{\textstyle H}{C=C}}\overset{}{\diagdown}$	4.5-6.6
⬡—H	6.5-8.0

Peak Area and Proton Counting

The area under an NMR signal is directly proportional to the number of protons that give rise to the signal.

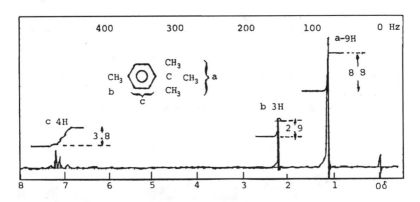

NMR Spectrum of p-tert-butylene. Proton Counting

The ratio of step heights, a:b:c, is

$$8.8:2.9:3.8 = 3.0:1.0:1.3 = 9.0:3.0:3.9.$$

Therefore, a = 9H's, b = 3H's, and c = 4H's.

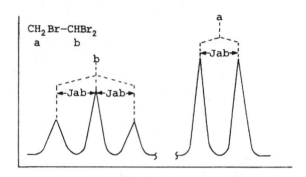

NMR Splitting of Signals. Spin-spin Splitting

An NMR signal is split into a doublet by one nearby proton, and into a triplet by two (equivalent) nearby protons.

The number of peaks (or multiplicity) observed for a given proton (or equivalent protons) is equal to $n + 1$, where n is the number of protons on adjacent atoms.

Spin-spin splitting is observed only between nonequivalent neighboring protons. Nonequivalent protons are protons with different chemical shifts. Neighboring protons are protons on adjacent carbons.

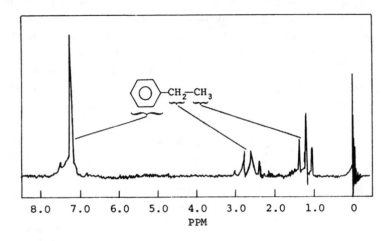

The NMR Spectrum of Ethyl Benzene

Coupling Constant

The distance between peaks in a multiplet is called spin coupling constant, J. It is a measure of the effectiveness of spin-spin coupling.

Peak separations that result from splitting remain constant, while peak separations that result from chemical shifts change.

The size of a coupling constant depends on the structural relationships between coupled protons. The size of J varies with the electronegativity of substituents.

Carbon-13 NMR

Carbon-13 NMR has several advantages over proton NMR in determining the structure of organic molecules (including biological compounds). Perhaps the most important advantage is that carbon-13 NMR provides a great deal more information about the structure of a molecule than does proton NMR, which only provides information on peripheral protons. This occurs since the backbone of organic molecules is composed of carbon.

Another advantage is that the chemical shift in most organic compounds is about 200 ppm (compare this with 10 to 20 ppm for proton NMR), so there is less chance that peaks will overlap. Due to the low abundance of carbon-13 in natural samples (1%), it is unlikely that two carbon-13 atoms will be next to each other on the same molecule; hence, homonuclear spin-spin coupling is not encountered. Since carbon-12 has zero quantum spin, heteronuclear spin-spin coupling is not a problem with this isotope. Also, good methods exist for decoupling the interactions between carbon-13 and protons.

Type of carbon	Chemical shift in ppm	Type of carbon	Chemical shift in ppm
CH_3–halogen	0 to 40	–C≡C–	75 to 95
$CH_3N=$	10 to 45	–C≡N	115 to 125
–CH_2–halogen	0 to 50	C=C alkene	110 to 145
=CH–halogen	35 to 75	C=C aromatic	110 to 160
–CH_2–N=	40 to 50	≡C–C≡	30 to 50
=CH–N=	60 to 80	=CH–C≡	30 to 45
CH_3–O	50 to 65	–CH_2–C≡	20 to 45
–CH_2–O	40 to 75	CH_3–C≡	5 to 30
=CH–O	75 to 85	C=O acid	170 to 180
		C=O aldehyde	195 to 205
		C=O ketone	195 to 215

Fourier Transform (FT) NMR

Carbon-13 NMR would never have been practical (due to the low abundance of carbon-13) had it not been for the advancement of FT NMR. In FT NMR, resolution of the elements in the spectrum is the result of very brief measurements that yield a time domain spectrum, rather than a frequency domain. This time domain spectrum can be obtained in less than a few seconds; hence, many (hundreds, or even thousands) can be taken and averaged in a relatively brief period of time. A frequency domain spectrum can then be obtained by a Fourier transform using a digital computer.

Problem Solving Example:

 The figure below shows the proton NMR spectrum of $(CH_3)_2C(OH)CH_2COCH_3$ with tetramethylsilane as standard. The stepped line is an electronic integration of the areas under the signal peaks. (a) List the chemical shift of each proton signal in ppm, and deduce, from the trace of the integrated areas, the identity of the protons that give rise to each line. (b) List the line positions in cps relative to tetramethylsilane expected at 100 Mcps. (c) Sketch out the spectrum and integral expected for $CH_3COC(CH_3)_2CHO$ at 60 Mcps.

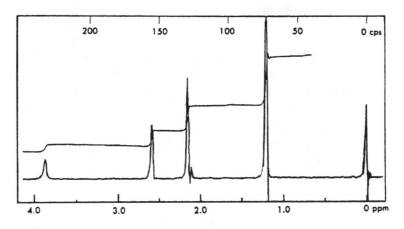

NMR spectrum of diacetone alcohol, $(CH_3)_2C(OH)CH_2COCH_3$, at 60 Mcps relative to TMS 0.00 ppm. The stepped line is the integrated spectrum.

A (a) To assign the identity of the protons to the spectral lines we must look at the different types of protons within the molecule. Once we have identified all the nonequivalent protons in the molecule, we must determine their relative chemical shifts. When we have deduced the relative order of chemical shifts, we can read off the chemical shift of each signal and assign these values to particular protons. Diacetone alcohol has the following structure:

$$CH_3-\underset{\underset{CH_3}{|}}{\overset{\overset{OH}{|}}{C}}-CH_2-\overset{\overset{O}{||}}{C}-CH_3$$

There are no two adjacent carbons in this molecule that bear nonequivalent hydrogens. This is supported by the fact that all the signals in the NMR spectrum are singlets; no spin-spin splitting took place. The methylene ($-CH_2-$) and hydroxyl ($-OH$) protons are unique in this molecule and hence, constitute their own type of proton.

The protons of the ketone methyl group ($-\overset{\overset{O}{||}}{C}-CH_3$) are nonequivalent to the protons of the other methyl groups by inspection. The protons of the two methyl groups bonded to the carbinol carbon (($CH_3)_2COH-$) are in identical electronic environments (made possible by the ability of the carbinol carbon to rotate about its single bonds) and are therefore equivalent. Hence, there are four sets of nonequivalent protons in diacetone alcohol; they are labeled below as a through d:

$$\underset{b}{}\;\; CH_3-\underset{\underset{b}{\underset{CH_3}{|}}}{\overset{\overset{a}{\overset{OH}{|}}}{C}}\text{---}\underset{c}{CH_2}-\overset{\overset{O}{||}}{C}-CH_3 \;\;\underset{}{d}$$

The NMR signal of proton a will be downfield from all the other signals; it will absorb energy at a lower field strength (higher chemical shift) than any other proton in the molecule. This is due to the powerful deshielding effect of oxygen; this proton's electron density is the least of all the protons in the molecule. The next most deshielded protons are those of the methylene (c) group. The electron-withdrawing effects of the

$$\overset{\text{O}}{\underset{\|}{}}$$

adjacent carbonyl group $(-\overset{\overset{\text{O}}{\|}}{\text{C}}-)$ and the nearby hydroxyl group (–OH) have a powerful combined deshielding effect. The ketone methyl group experiences deshielding from only the carbonyl group, and the b protons experience deshielding from only the nearby hydroxyl group. Since the carbonyl group has greater electron-withdrawing capabilities than the carbinol group $(-\overset{\overset{\text{OH}}{|}}{\underset{|}{\text{C}}}-)$, the ketone methyl group is more deshielded than the b protons. Hence, the relative order of deshielding and chemical shift values of the proton types a through d is:

$$a > c > d > b$$

By looking at the NMR spectrum of diacetone alcohol, we can see that the chemical shifts of the four signals are 3.85, 2.6, 2.15, and 1.2. By using the relative amounts of deshielding of each type of proton, we can assign the protons to a particular chemical shift as follows:

–OH proton, 3.85 ppm; –CH_2– proton, 2.6 ppm;

–$COCH_3$ proton, 2.15 ppm; $(CH_3)_2C$ – OH proton, 1.2 ppm.

(b) The signals in an NMR spectrum are measured relative to an internal standard such as tetramethylsilane $(Si(CH_3)_4)$ for matters of convenience. Traditionally, the chemical shift is measured in ppm (parts per million), but it can also be measured in cps (cycles per second) or H_z (Hertz). The line positions in cps relative to TMS (tetramethylsilane) expected at a frequency of 100 Mcps (1 mega cps = 1 million cps) can be calculated by using the line positions of the spectrum at 60 Mcps. This is accomplished by using a direct proportion:

$$\frac{60 \text{ Mcps}}{100 \text{ Mcps}} = \frac{\text{relative cps at } v = 60 \text{ Mcps}}{\text{expected cps at } v = 100 \text{ Mcps}}$$

The line positions in cps of the NMR spectrum shown at 60 Mcps are: –OH proton, 232 cps; –CH$_2$–proton, 157 cps; –COCH$_3$ proton, 128 ops; (CH$_3$)$_2$C–proton, 72 cps. We can calculate the expected line positions in cps relative to TMS at 100 Mcps as shown.

For the –OH proton:

$$\frac{60 \text{ Mcps}}{100 \text{ Mcps}} = \frac{232 \text{ cps (at } v = 60 \text{ Mcps)}}{\text{expected cps (at } v = 100 \text{ Mcps)}}$$

expected cps = 387

For the –CH$_2$–proton:

$$\frac{60 \text{ Mcps}}{100 \text{ Mcps}} = \frac{157 \text{ cps}}{\text{expected cps}}$$

expected cps = 262

For the –COCH$_3$ proton:

$$\frac{60 \text{ Mcps}}{100 \text{ Mcps}} = \frac{128 \text{ cps}}{\text{expected cps}}$$

expected cps = 213

For the (CH$_3$)$_2$C– proton:

$$\frac{60 \text{ Mcps}}{100 \text{ Mcps}} = \frac{72 \text{ cps}}{\text{expected cps}}$$

expected cps = 120

The expected line positions in cps relative to TMS at 100 Mcps are: – OH proton, 387 cps; –CH$_2$– proton, 262 cps; –COCH$_3$ proton, 213 cps; (CH$_3$)$_2$–C– proton, 120 cps.

(c) The compound CH$_3$COC(CH$_3$)$_2$CHO is 2,2-dimethyl-3-oxo-butanal and has the structure:

$$
\begin{array}{ccccc}
& O & CH_3 & O & \\
& \| & | & \| & \\
CH_3-C & - & C_2 & - & C_1-H \\
& & | & & \\
& & CH_3 & &
\end{array}
$$

The ketone methyl group is nonequivalent to the other methyl groups by inspection. These other methyl groups, bonded to carbon number 2, are in identical electronic environments (because of carbon 2's ability to rotate about its single bonds) and are hence, equivalent to each other. Therefore, there are three sets of nonequivalent protons in the compound. They are labeled as:

$$
\begin{array}{ccccccc}
& & & b & & & \\
& & O & CH_3 & O & & \\
c & & \| & | & \| & & a \\
& CH_3- & C & - C & - C & - H & \\
& & & | & & & \\
& & & CH_3 & & & \\
& & & b & & &
\end{array}
$$

At a frequency of 60 Mcps, an aldehydic proton (a) will characteristically absorb between 9.4 and 10.4 ppm. Hence, proton (a) will absorb in this range. The protons of a ketone methyl group (c) characteristically absorb between 2.1 and 2.4 ppm. The protons of the two equivalent methyl groups (b) are slightly deshielded by the carbonyl groups and will absorb between 1.0 and 1.4 ppm. Note that the signal of proton (c) will occur downfield from that of proton (b) due to greater deshielding effects; proton (c) is adjacent to the carbonyl group, whereas proton (b) is separated from the carbonyl groups by one carbon.

The integration curve of an NMR spectrum indicates the relative intensities or relative areas of the peaks. The signal for proton a will have a relative intensity of one. Likewise, the signals for protons (b) and (c) will have relative intensities of six and three, respectively. These relative intensities will be represented on the integration curve as the relative changes in height at each peak. The NMR spectrum of 2,2-dimethyl-3-oxo-butanal is sketched on the next page.

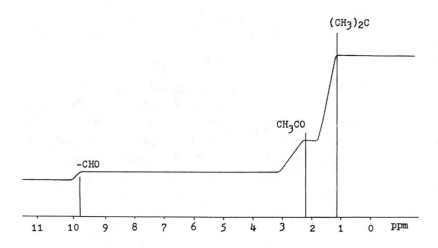

9.6 Electron Spin Resonance (ESR) Spectrum

The odd electron of a free radical generates a magnetic moment by spinning. Each electron of an electron pair generates a magnetic moment of equal and opposite magnitude (because the two electrons have equal and opposite spins). When a free radical is placed in a magnetic field, the magnetic moment generated by the odd electron can be aligned with or against the external magnetic field. When this system is exposed to electromagnetic radiation of the proper frequency, the odd electron absorbs radiation and reverses its spin; like the proton in NMR, the electron "flips" over in ESR. An absorption spectrum is obtained and is called an electron spin resonance (ESR) spectrum or an electron paramagnetic resonance (EPR) spectrum.

The signals of an ESR spectrum show splitting for the same reason that NMR signals split. The ESR signal will be split by n neighboring protons into $n + 1$ peaks.

ESR spectroscopy is used to detect the presence of free radicals, to measure their concentration, and to determine their structure. The protons that are responsible for splitting the signal indicate the distribution of the odd electron in the free radical.

The eight-line ESR signal of propane indicates that there are seven neighboring hydrogens about the odd electron.

$$CH_3CHCH_3$$

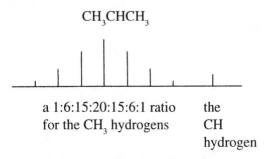

a 1:6:15:20:15:6:1 ratio   the
for the CH_3 hydrogens   CH
                                       hydrogen

Problem Solving Example:

Q Although all electrons spin, only molecules containing unpaired electrons—only free radical—give ESR spectra. Why is this? (Hint: Consider the possibility (a) that one electron of a pair has its spin reversed, or (b) that both electrons of a pair have their spins reversed.)

A Electron spin resonance (ESR) is a phenomenon involving electrons that occurs in a similar fashion as nuclear magnetic resonance. The odd electron of a free radical generates a magnetic moment by spinning; the free radical, therefore, has a net magnetic moment. Each electron of an electron pair also generates a magnetic moment but of equal and opposite magnitude (because the two electrons have equal and opposite spins); the electron pair, therefore, has no net magnetic moment. When a free radical is placed in a magnetic field, the magnetic moment generated by the odd electron may be aligned with or against the external magnetic field. When this system is exposed to electromagnetic radiation of the proper frequency, the odd electron absorbs the radiation and reverses its spin; like the proton in NMR, the electron "flips" over in ESR. An absorption spectrum is obtained, which is called an electron spin resonance (ESR) spectrum or an electron paramagnetic resonance (EPR) spectrum.

The signals of an ESR spectrum may show splitting for the same reason that NMR signals may be split. The ESR signal will be split by n neighboring protons into $n + 1$ peaks. The protons that are responsible for splitting the signal indicate the distribution of the odd electron in the free radical.

(a) Every electron that resides in an atomic or molecular orbital can be described by a set of four quantum numbers. Quantum numbers are used to depict the relative energies and distribution of electrons. The first quantum number is called the principal quantum number (n); it gives the order of increasing distance of the average electron density from the nucleus. The second quantum number is the orbital quantum number (.l); it gives the subshell in which the electron resides and the spatial geometry of the electron distribution. The third quantum number is the magnetic quantum number (m); it describes the circulation of the electric charge which generates a magnetic moment. The fourth quantum number is the spin quantum number (s); it describes the orientation of electron spin and can be either $+\frac{1}{2}$ or $-\frac{1}{2}$.

The Pauli exclusion principle states that no two electrons in the same atom are identical. No two electrons in the same atom have an identical set of quantum numbers. Each electron of an electron pair differs only by its spin orientation; they have identical principal, orbital, and magnetic quantum numbers. One electron has a spin quantum number of $+\frac{1}{2}$ and the other one of the pair has a spin quantum number of $-\frac{1}{2}$. If one electron of a pair had its spin reversed, it would have a set of quantum numbers identical to the other electron. This is a violation of the Pauli exclusion principle and will hence, not occur. As a result, no ESR spectrum is expected to arise in this manner.

(b) If both electrons of a pair have their spins reversed, there is no net change in the energy because the resulting electron pair is identical to the initial electron pair in all respects. Hence, there will be no absorption signals and no ESR spectrum.

Quiz: Spectroscopy

1. The diagram below represents the mass spectra of a hydrocarbon, where m/e is the compound's molecular weight. Which one of the following statements concerning the mass spectra is incorrect?

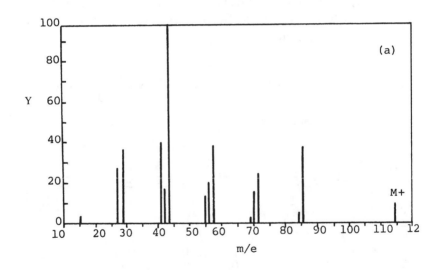

(A) The Y-axis represents the relative intensity of the compound(s).

(B) The mass spectra may be used to prove the identity of compounds.

(C)   Mass spectral studies help to establish the structure of new compounds.

(D) M^+ (the molecular ion), once identified, gives the most accurate molecular weight attainable for any compound.

(E) M^+ is always referred to as the base peak.

2.   Light of wavelength 3×10^{-3} cm falls in which band?

(A)   Microwave
(B)   Infrared
(C)   Visible
(D)   Ultraviolet
(E)   X-ray

3.   The product of the reaction below was subject to an infrared spectral analysis. Based on its IR spectra pictured below, one can deduce that the product is

Reaction: $CH_3CH_2Cl + Mg \xrightarrow{\text{ether}} \xrightarrow{CO_2} \xrightarrow[\text{H}_2\text{O}]{\text{H}_2\text{SO}_4} \text{product}$

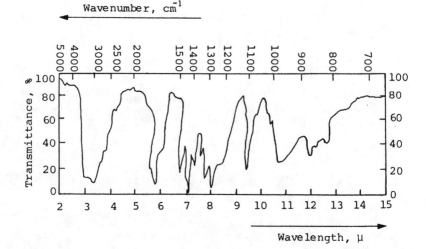

(A) acetaldehyde.   (D) acetone.

(B) propanol.   (E) propionic acid.

(C) ethanoic acid.

4. The structure shown below for NH_4^+ is referred to as

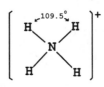

(A) trigonal.   (D) tetrahedral.

(B) trigonal planar.   (E) octahedral.

(C) planar.

5. CMR is less sensitive than proton NMR because

(A) carbon is heavier than hydrogen.

(B) ^{13}C has a low natural abundance.

(C) spectra are taken at lower frequencies.

(D) C – C splitting is not observed.

(E) the carbon nucleus contains neutrons.

6. The infrared spectrum (IR) is frequently used for the analysis of organic compounds. Which of the statement(s) below concerning IR spectroscopy is (are) correct?

I. This technique (IR analysis) by itself yields more information about the compound's structure than any other technique.

II. Absorption due to carbon-hydrogen stretching occurs at the lower frequency end of the IR spectrum.

III. It is the absorption of infrared light that causes the changing vibrations of a molecule.

    (A) I only.   (D) II and III only.

    (B) I and II only.   (E) None of the above.

    (C) II only.

7. Consider the following statements about the NMR spectrum, then indicate which statement(s) is (are) correct.

    I. The number of NMR signals tells how many different types of protons are present in a molecule.

    II. The area under an NMR signal is directly related to the number of protons causing the signal.

    III. The position of an NMR signal tells us something about the electronic environment of each kind of proton.

    (A) I only.   (D) II and III only.

    (B) I and II only.   (E) I, II, and III.

    (C) I and III only.

8. Which one of the following compounds below exhibits three signals in the nuclear magnetic resonance spectrum?

    (A) CH_3CH_2Cl (D) $CH_3CH_2CH_2Cl$

    (B) $CH_3CHClCH_3$ (E) $CH_3 \overset{\displaystyle Cl}{\underset{\displaystyle Cl}{C}} - CH_3$

    (C) $CH_3 - \overset{\displaystyle H}{\underset{\displaystyle Cl}{C}} - \overset{\displaystyle H}{\underset{\displaystyle Cl}{C}} - CH_3$

9. A compound with molecular formula $C_3H_6Cl_2$ shows these peaks in its NMR spectrum: δ, 1.3(triplet); δ, 2.0(multiplet); and δ, 4.5(triplet). The compound could be which one of the following?

   (A) $ClCH_2CH_2CH_2Cl$ (D) $CH_3CCl_2CH_3$

   (B) $ClCH_2CHClCH_3$ (E) None of the above.

   (C) $Cl_2CHCH_2CH_3$

10. The absorption spectrum obtained from the spinning of the odd electron of a free radical is called

   (A) electromagnetic.

   (B) infrared.

   (C) nuclear magnetic resonance.

   (D) Fourier transform NMR.

   (E) electron spin resonance.

ANSWER KEY

1.	(E)	6.	(E)
2.	(B)	7.	(E)
3.	(E)	8.	(D)
4.	(D)	9.	(C)
5.	(B)	10.	(E)

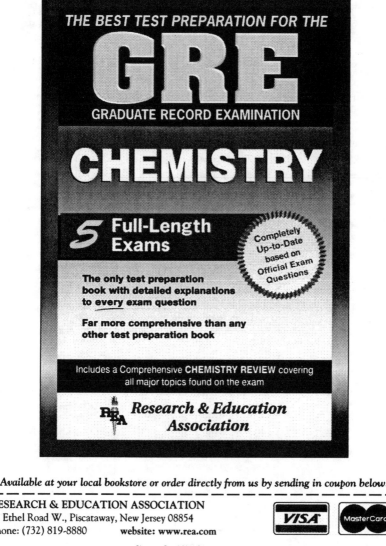

REA's Test Preps
The Best in Test Preparation
